W9-BRT-160
S V Beer

# FUNDAMENTALS OF BACTERIAL PLANT PATHOLOGY

# FUNDAMENTALS OF **BACTERIAL PLANT PATHOLOGY**

*Masao Goto*

Faculty of Agriculture
Shizuoka University
Shizuoka, Japan

**ACADEMIC PRESS, INC.**
*Harcourt Brace Jovanovich, Publishers*
San Diego New York Boston London
Sydney Tokyo Toronto

This book is printed on acid-free paper. ♾



Translation of Fundamentals of Bacterial Plant Pathology

Academic Press, Inc.
1250 Sixth Avenue, San Diego, California 92101-4311

*United Kingdom Edition published by*
Academic Press Limited
24–28 Oval Road, London NW1 7DX

Library of Congress Cataloging-in-Publication Data

Goto, Masao, date
Fundamentals of bacterial plant pathology / Masao Goto.
p. cm.
Includes index.
ISBN 0-12-293465-2
1. Bacterial diseases of plants. 2. Phytopathogenic bacteria.
I. Title.
SB734.G63 1992
632'.32--dc20 92-6312
CIP

PRINTED IN THE UNITED STATES OF AMERICA
92 93 94 95 96 97 EB 9 8 7 6 5 4 3 2 1

# *Contents*

CHAPTER EIGHT



CHAPTER NINE



CHAPTER TEN

# *Preface to the English Edition*

Recent remarkable advances in molecular plant–microbe interactions have greatly impacted bacterial plant pathology. The emergence of this new field substantially affects not only the analysis of pathogenesis in plant pathogenic prokaryotes, but also the clarification of their taxonomic relationships and the strategies of disease control.

I wrote the Japanese edition of the textbook *Fundamentals of Bacterial Plant Pathology* (Yokendo Co. Ltd., Tokyo, 1990) to provide comprehensive and up-to-date information on the rapidly developing field of bacterial plant pathology for university students and general plant pathologists. In this volume, I emphasized the basic features of plant pathogenic prokaryotes because they are essential to understanding new concepts emerging in plant bacteriology and to extending these concepts to the applied field of bacterial plant pathology.

This type of textbook has been absent in plant pathology. I wanted to provide an English edition of this volume for students and plant pathologists in other countries. Revisions have been made to update the contents and to facilitate easier understanding of the subjects for the beginner.

I am deeply indebted to Dr. Arthur Kelman for his consistent encouragement and support of my academic activities in the international community of plant pathology. Thanks are due to the many plant pathologists and microbiologists who have kindly contributed their valuable photographs and illustrations for this volume.

*Masao Goto*

# *Preface to the Japanese Edition*

In 1981, *New Bacterial Plant Pathology* was published (Soft Science Co. Ltd., Tokyo). Since then, remarkable advances have been made in the genetic analysis of plant pathogenic bacteria and molecular elucidation of host–parasite interactions. Consequently, considerable gaps occur between the descriptions in that volume and current understanding in plant bacteriology. This trend is particularly notable in the fields of taxonomy, genetics, and pathogenesis.

Under such circumstances, I want to provide a new introductory textbook for students of bacterial plant pathology based on recent advances. Dr. H. Ikegami, professor at Gifu University, kindly introduced my plan to the president, Mr. Kiyoshi Oikawa, and the director, Mr. Kouichi Ohtsu, of Yokendo Co. Ltd., making it possible to publish the present volume *Fundamentals of Bacterial Plant Pathology*.

This volume was written primarily as a comprehensive textbook including current progress on plant pathogenic bacteria and bacterial plant diseases. I describe the contents of this volume in language that is accessible to the reader, but students may meet with some difficult expressions. Wherever readers become aware of any difficult and/or inappropriate descriptions, I would appreciate their being pointed out so they may be revised in future editions. I hope that student interest in bacterial plant pathology will be stimulated by the ideas in this volume.

In preparation of the manuscript, I referred to a number of publications. I would like to acknowledge these excellent studies and the authors. I deeply thank those who provided their valuable photographs, illustrations, and suggestions for this volume. I am also indebted to Mr. Kazumi Oikawa of Yokendo Co. Ltd., who was instrumental in the publication of this volume.

*Masao Goto*

CHAPTER
ONE

# Introduction

## 1.1 The Scope of Bacterial Plant Pathology

Bacterial plant pathology is a branch of plant pathology dealing with plant diseases caused by prokaryotes such as eubacteria, actinomycetes, spiroplasmas, and mycoplasmalike organisms. In this science, taxonomy, morphology, physiology, genetics, serology, and predator–prey relationships make up the rather basic field directly related with the pathogen. This field may be called *plant bacteriology*. In contrast, host–parasite interactions that include pathogenesis, the resistance and infection process, epidemiology that includes the survival and spread of pathogens and the effect of environments on disease development, and disease management that includes diagnosis, loss assessment, and control are the somewhat applied field primarily associated with the disease. This field is an essential part of plant pathology that constitutes the science of plant protection. A better understanding of such broad and diverse fields of bacterial plant pathology may be facilitated by a basic knowledge in such various fields of science as microbiology, botany, genetics, chemistry, physics, biochemistry, molecular biology, plant physiology, agronomy, soil science, and meteorology.

In addition to the subjects of bacterial plant pathology mentioned above, such topics as the infection process of symbiotic root nodule bacteria, the dynamics and functions of epiphytic and endophytic bacteria, and the evaluation of plant pathogenic prokaryotes as a useful genetic resource are often referred to in modern bacterial plant pathology with respect to pathogenesis, biological control, and industrial use of bacterial components and metabolites. Such recent trends have been significantly expanding the fields of interests in bacterial plant pathology.

## 1.2 Importance of Bacterial Plant Diseases

The loss assessment of bacterial plant diseases is an immature field of bacterial plant pathology. The available statistical data on yield loss are consequently very limited. Table 1.1 shows gross dollar loss in 1976 due to plant pathogenic prokaryotes in the United States (Kennedy and Alcorn, 1980). In another assessment, global loss from plant pathogens in 1976 was estimated to be $49.6 billion. It is obvious that great financial losses are a result of reduced crop yields, and additional expenses for disease control are annually incurred.

The economic importance of bacterial plant diseases varies depending on the regions of the globe or countries in which they occur because the economic importance of crops may vary in each region or country. For example, several prokaryote pathogens that cause significant yield loss of rice in Asia are not included in Table 1.1. On the other hand, the table lists losses due to some prokaryotes that have not yet been reported in other regions of the globe.

The importance of bacterial diseases for individual farmers does not necessarily coincide with the national gross loss or total affected acreage. Diseases may destroy locally cultivated crops that are of economic importance only for particular regions or districts. Recent outbreaks of bacterial canker of kiwi (*Pseudomonas syringae* pv. *actinidiae*) in some regions of Japan are one such example.

Chemical control of bacterial plant diseases is very difficult in general because of the low effectiveness of the available bactericides and the absence of effective chemicals that can be applied in field conditions. Why success is so rare in the development of effective bactericides for practical use is a question that has not yet been fully answered.

## 1.3 Historical Review of Bacterial Plant Pathology

Antony van Leeuwenhoek (1632–1723) discovered the new microbial world with a simple handmade microscope in the middle of 17th century. It took about a century, however, before microbiology was established as a modern science through the remarkable contributions of Louis Pasteur (1822–1895) on the abandonment of spontaneous generation, of John Tyndall (1820–1893) on the development of discontinuous sterilization called *Tyndallization*, and of Robert Koch (1843–1910) on the develop-

**Table 1.1**
Loss Estimates for Plant Pathogenic Prokaryotes in 1976[a]

| Prokaryote | Disease name | Loss (millions of dollars) |
|---|---|---|
| *Pseudomonas solanacearum* | Bacterial wilt of tobacco and tomato | 9.4 |
| *P. syringae* pv. *glycinea* | Bacterial blight of soybean | 65 |
| *P. syringae* pv. *syringae* | Bacterial leaf blight of wheat | 18 |
| *Xanthomonas campestris* pv. *malvacearum* | Bacterial blight of cotton | 15 |
| *Agrobacterium tumefaciens* | Crown gall of fruit and nut | 23 |
| *Erwinia amylovora* | Fire blight of pear | 4.7 |
| *E. carotovora* subsp. *carotovora* and/or subsp. *atroseptica* | Soft rot and/or blackleg of potato | 14 |
| *Clavibacter michiganensis* subsp. *insidiosus* | Bacterial wilt of alfalfa | 17 |
| *C. michiganensis* subsp. *nebraskensis* | Goss's bacterial wilt and blight of corn | 3 |
| *C. xyli* subsp. *xyli* | Ratoon stunt of sugarcane | 10 |
| *Xylella fastidiosa* | Phony peach | 20 |
| | Pierce's disease of grape | 3 |
| *Spiroplasma citri* | Stubborn disease of citrus | 1 |
| Mycoplasmalike organism | Pear decline | 1.6 |
| | Lethal yellowing of coconut | 3 |

[a] Adapted from B. W. Kennedy and S. M. Alcorn (1980). Estimates of U.S. crop losses to prokaryote plant pathogens. *Plant Disease* **64,** 674-676.

ment of a pure culture method and Koch's postulates on the definition of pathogens. The first golden era of classical bacteriology in the 19th century was thus brought about with the discovery of bacteria that caused such human and animal diseases as anthrax, cholera, and tetanus.

The history of bacterial plant pathology can also be traced to the late 19th century. Table 1.2 lists some noticeable milestones in this science. T. J. Burrill (1880) proved by an inoculation test that fire blight of pear was caused by infection of a bacterium, which he named *Micrococcus amylovorus* in 1882. In 1883, J. H. Wakker reported a yellowing disease of hyacinths caused by *Bacterium hyacinthi.* A series of descriptions of plant pathogenic bacteria was subsequently reported by E. F. Smith. Such quick movement defining bacteria as plant pathogens raised questions about the pathogenicity of bacteria among plant pathologists who were working mainly on fungal diseases in this era. A heated controversy arose between E. F. Smith and A. Fischer from 1897 to 1901 regarding the validity of

**Table 1.2**
Historical Events in Bacterial Plant Pathology

| Year | Name | Event |
|---|---|---|
| 1882 | T. J. Burrill | Publication of the fire blight pathogen (*Micrococcus amylovorus*) |
| 1896 | J. Omori | Publication of the soft rot pathogen of wasabi (*Bacillus alliariae*) |
| 1901 | E. F. Smith | Validation of bacteria as plant pathogens in the controversy with A. Fischer |
| 1910 | C. O. Jensen | Demonstration of equivlence between crown gall of plants and cancer of animals |
| 1912 | S. Hori | Publication of soft rot pathogen of orchid (*Bacillus cypripedii*) (First publication of valid bacterial names from Japan) |
| 1925 | G. H. Coons and J. E. Kotila | Isolation of bacteriophages of *Bacillus carotovorus* |
| 1944 | W. D. Valleau *et al.* | Demonstration of saprophytic survival of plant pathogenic bacteria on rhizosphere of nonhost plants |
| 1944 | N. Okabe | Demonstration of biovars of *P. solanacearum* and avirulent colony mutants |
| 1954 | R. M. Klein | Elucidation of the basic process of transformation in crown gall |
| 1955 | A. C. Braun | Elucidation of wildfire toxin and its mode of action |
| 1962 | H. Stolp | Discovery of bdellovibrios |
| 1964 | Z. Klement *et al.* | Discovery of hypersensitive reaction |
| 1967 | Y. Doi *et al.* | Discovery of mycoplasmalike organisms |
| 1972 | A. K. Chatterjee and M. P. Starr | Demonstration of conjugative gene transfer in plant pathogenic bacteria |
| 1972 | P. B. New and A. Kerr | Success in biological control of crown gall with *A. radiobacter* strain K84 |
| 1974 | I. Zaenen *et al.* | Demonstration of Ti-plasmid in *A. tumefaciens* |
| 1974 | L. R. Maki *et al.* | Discovery of ice nucleation-active bacteria |
| 1975 | D. C. Graham and M. D. Harrison | Discovery of aerosol dispersal of potato soft rot bacteria |
| 1976 | L. C. Melton, *et al.* | Determination of structure of xanthan gum produced by *Xanthomonas campestris* |
| 1980 | D. W. Dye *et al.* | Introduction of the pathovar system in taxonomy of plant pathogenic bacteria |
| 1982 | L. Comai and T. Kosuge | Determination of virulence genes of *P. syringae* pv. *savastanoi* by cloning |
| 1984 | B. J. Staskawicz *et al.* | Demonstration of avirulence genes of *P. syringae* pv. *glycinea* |
| 1986 | P. B. Lindgren *et al.* | Demonstration of hypersensitive reaction and pathogenicity (*hrp*) genes |

bacteria as plant pathogens. Smith established the concept of bacterial plant pathogens during this controversy on the basis of clear presentations and convincing evidence. It can be said, therefore, that bacterial plant pathology was established in 1901 as a specific discipline of plant pathology.

In the initial stage of plant bacteriology, the major interests were focused on the descriptions of new bacterial pathogens, serological identification of bacteria, and descriptions of bacteriophages. From the 1940s to the 1960s, significant advances were made in the physiology of diseased plants and the ecology of plant pathogenic bacteria. The contribution of Braun and his colleagues (1955) on the structure and mode of action of wildfire toxin produced by *Pseudomonas syringae* pv. *tabaci* was a pioneer work of subsequent studies on the halo-producing toxins, although their work was partly revised later. The discovery of mycoplasma-like organisms by Doi *et al.* (1967) was also a remarkable contribution in these decades. Their work stimulated the search for real pathogens in so-called virus diseases with no convincing evidence of viral particles in plant tissues, resulting in the subsequent discovery of spiroplasmas and fastidious prokaryotes.

The notable contribution in the 1970s was the demonstration of conjugative transfer of bacterial genes in *Erwinia* (Chatterjee and Starr, 1972). Although it was a quarter century after the discovery of sexual reproduction in *Escherichia coli* (Lederberg and Tatum, 1946), the impact was so strong and attractive to plant bacteriologists that the main current of research since then has shifted to molecular genetics. The innovation in genetic engineering techniques from the late 1970s to the early 1980s greatly accelerated advances in this field, opening a new era in the history of bacterial plant pathology.

In bacterial plant diseases, crown gall is a unique disease in that major interests have been placed on the mechanisms of tumorigenesis from the beginning of study. Jensen (1910) had already indicated the autonomous growth of tumor tissues. This was consistently followed by such advances as clarification of the transformation process of tumor cells in the 1950s, the elucidation of hormone imbalance of tumor cells in the 1960s, the discovery of Ti (tumor-inducing) plasmid in the 1970s, and genetic analysis of T-DNA (transfer DNA) and virulence region (*vir*) of Ti-plasmid in the 1980s. The series of research on crown gall has thus been leading the investigation on pathogenesis of plant pathogenic bacteria.

Genetic analysis of recognition, or host–parasite interactions, at the initial stage of bacterial infection started in the 1980s. Great progress has been made in the past decade in the clarification of the hypersensitive reaction and pathogenicity (*hrp*) genes and avirulence (*avr*) genes which may genetically interpret host-specificity at the species and cultivar levels.

Research in Japan on bacterial plant diseases also started late in the 1880s. Omori studied soft rot of Japanese horseradish, or Wasabi (*Eutrema wasabi*), and named its pathogen *Bacillus alliariae.* His work was published in 1896 in the official gazette of the Japanese government. The bacterium was identical to Jones's *Bacillus carotovorus* from its described bacteriological properties. Although his study was 5 years ahead of Jones's publication, it was not adopted by the scientific community because of its invalid form of publication. By the 1920s, Uyeda, Hori, Miyake, Ishiyama, and others had described many plant pathogenic bacteria, for example, *P. conjac, P. oryzae, P. zingiberi, B. araliavorus, B. cypripedii, B. eriobotryae, B. harai, B. lili, B. oryzae,* and *B. theae.* Most of these, however, seem to have been misidentified with *E. herbicola,* and only *P. oryzae* (*X. campestris* pv. *oryzae*) and *B. cypripedii* (*E. cypripedii*) have been recognized as valid pathogens. This misidentification resulted from the inadequate method of isolation, which consisted of placing pieces of infected tissues on agar plates and picking as the pure culture bacterial growth that appeared around the pieces on agar slope.

From the 1920s to the 1940s, Nakata, Takimoto, Muko, and Okabe contributed to the establishment of bacterial plant pathology in Japan. Significant progress was made by Takimoto in the identification of many bacterial plant pathogens, by Nakata in the ecological behavior of *P. solanacearum,* and by Okabe in the biovar differentiation as well as colony-form mutations associated with pathogenesis in the same bacterium.

In the 1950s to 1960s, epidemiology was the main subject of interest in Japan, and bacterial leaf blight of rice, soft rot of Chinese cabbage, bacterial wilt of solanaceous plants, and citrus canker were intensively investigated. On the other hand, physiological and biochemical studies remained without noticeable progress until the 1970s, when Nishiyama and Sakai started to work on coronatine. Today, a number of scientists in Japan are working on the molecular biology of plant pathogenic bacteria and making notable contributions.

## 1.4 Bibliography

One of the objectives of this book is to provide an introductory knowledge of bacterial plant pathogens and the diseases that they cause. Therefore, only a limited number of references, primarily review articles, are listed at the end of each chapter. These reviews were selected on the basis of rapidly progressing areas and relatively new areas or concepts in bacterial plant pathology.

The books listed at the end of each chapter may be helpful to those readers wishing to acquire a general grasp of the subjects in a given chapter, and the multiauthor books may be helpful to further advance knowledge on certain specific topics in that chapter.

For day-to-day progress on specific topics, readers can also search the journals and periodicals such as *Annals of Phytopathological Society of Japan, Applied and Environmental Microbiology, International Journal of Systematic Bacteriology, Journal of Bacteriology, Journal of General Microbiology, Molecular Plant-Microbe Interactions, Physiological and Molecular Plant Pathology, Phytopathology, Plant Disease, Proceedings of the National Academy of Sciences of the United States, Annual Review of Genetics, Annual Review of Microbiology, Annual Review of Phytopathology*, and *Annual Review of Plant Physiology and Plant Molecular Biology*.

## Further Reading


Agrios, G. N. (1988). "Plant Pathology," 3rd ed. Academic Press, San Diego.

Billing, E. (1987). "Bacteria as Plant Pathogens." Van Nostrand Reinhold. Wokingham, (Berkshire) (UK).

Fahy, P. C., and Persley, G. J., eds. (1983). "Plant Bacterial Diseases: A Diagnostic Guide." Academic Press, Sydney.

Dowson, W. J. (1957). "Plant Diseases Due to Bacteria," 2nd ed. Cambridge University Press, Cambridge.

Goto, M. (1981). "New Bacterial Plant Pathology." (In Japanese). Soft Science Co. Ltd. Tokyo.

Lelliott, R. A., and Stead, D. E. (1987). "Methods for the Diagnosis of Bacterial Diseases of Plants." Blackwell Scientific Publications, Oxford.

Mount, M. S., and Lacy, G. H., eds. (1982). "Plant Pathogenic Prokaryotes," Volume 1 and Volume 2. Academic Press, New York.

Stapp, C. (1961). "Bacterial Plant Pathogens." Oxford University Press, Oxford.

Starr, M. P., ed. (1983). "Phytopathogenic Bacteria." Selections from "The Prokaryotes. A Handbook on Habitats, Isolation, and Identification of Bacteria" (M. P. Starr, H. Stolp, H. G. Trüper, A. Balows, and H. G. Schlegel, eds.). Springer-Verlag, New York.

Starr, M. P. (1984). Landmarks in the development of phytobacteriology. *Ann. Rev. Phytopathol.* **22,** 169–188.

Whitcomb, R. F., and Tully, J. G., eds. (1989). "The Mycoplasmas, Volume V, Spiroplasmas, Acholeplasmas, and Mycoplasmas of Plants and Arthropods." Academic Press, New York.

Van der Zwet, T., and Keil, H. L. (1979). "Fire Blight: A Bacterial Disease of Rosaceous Plants." U.S. Department of Agriculture, Washington, D.C.

CHAPTER
**TWO**

# *Morphology, Structure, and Composition*

The structure of the bacterial cell envelope is important for pathogenicity because host recognition is determined by the interactions between the host cell wall and extracellular polysaccharides or lipopolysaccharides of bacteria. In addition, the external as well as internal structures have major taxonomic value and provide essential information in the description of bacterial taxa.

## 2.1 Morphology

### 2.1.1 SHAPE AND SIZE

Eubacteria have three basic shapes, i.e., spherical, rod, and spiral, which are referred to as coccus (pl. cocci), bacillus (pl. bacilli), and spirillum (pl. spirilla), respectively. Most plant pathogenic bacteria are rod shaped and divide by binary fission. The size of bacterial cells varies, depending on various factors such as incubation temperature, culture medium, culture age, and the staining methods. Most plant pathogenic bacteria fall within the range of $1.0–5.0 \times 0.5–1.0$ $\mu$m. Bacterial cells gradually become smaller when cultures or lesions are aged. Sometimes elongated filamentous cells may be formed, however, because cell division or the separation of divided cells is likely to be inhibited in aged cultures.

*Streptomyces* is characterized by the formation of a highly branched

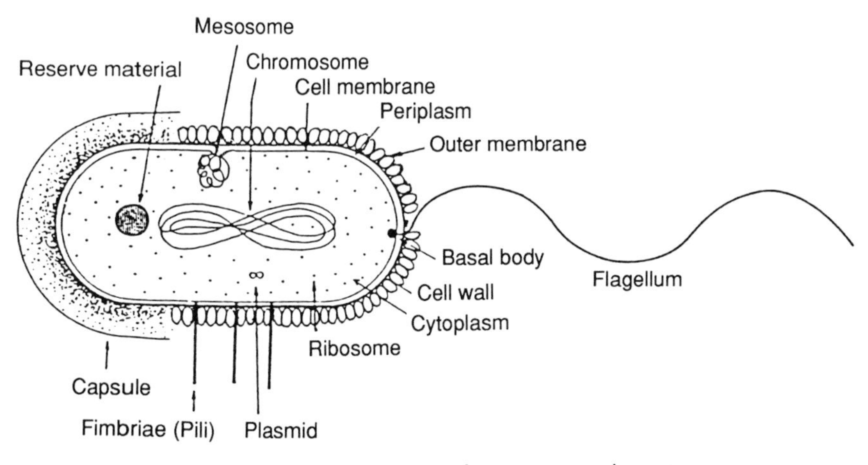


**Fig. 2.1** Schematic structure of gram-negative bacteria.

mycelium 0.5–2.0 μm in width and the formation of chains of spores 0.5–2.0 μm in diameter at the tip of aerial hyphae.

Mycoplasmalike organisms are pleomorphic and usually spherical or ovoid in shape, ranging in size from 0.3 to 2 μm in diameter. Conspicuous morphological variations such as filamentous, spiral, or small spherical bodies (60-100 nm) are observed in diseased plant tissue, depending on the stage of reproduction and environmental conditions.

Spiroplasmas are helically coiled filaments measuring 0.12 μm in width and 2–4 μm in length. The smallest viable unit is a two-turn elementary helix. The most frequently dividing parental helixes are those with approximately four turns, yielding two elementary helixes under favorable cultural conditions. The morphology of spiroplasmas may be altered under suboptimal growth conditions or by inadequate techniques for electron microscopy, resulting in helical filaments, spherical bodies, or *brebs*.

The gross morphology and structure of gram-negative bacteria are shown in Fig. 2.1.

## 2.2 Structure of the Cell Envelope

The cell envelope of gram-negative bacteria is composed of the cell wall, the cell membrane, and the outer membrane. The outer membrane is absent in the gram-positive bacteria.

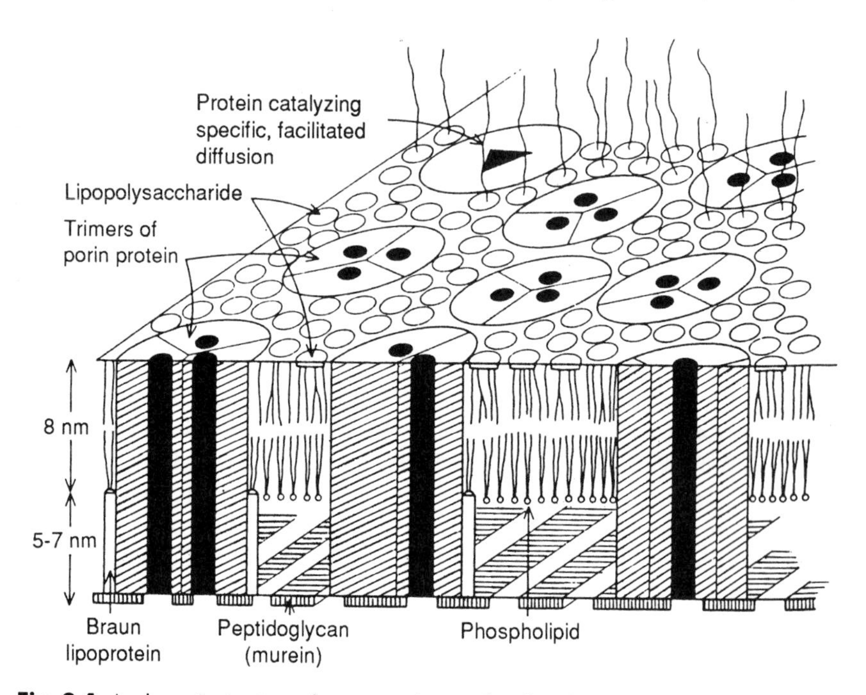


**Fig. 2.4** A schematic structure of outer membrane of *Escherichia coli* and *Salmonella typhimurium* [Nikaido, H. and T. Nakae, (1979). A tentative structure of the cell wall of *Escherichia coli* and *Salmonella typhimurium. Advance in Microbiol Physiology* **20,** 163–260.]

these two bacteria, with 72% DNA homology. The OmpF genes of *P. syringae* have 33% homology with OmpA genes of *E. coli* as far as the terminus-encoding region is concerned.

The gram-negative bacteria carry up to $10^5$ copies of porin per cell on their outer membrane. The porins are wide water-filled channels that allow passive nonspecific diffusions of hydrophilic solutes. The permeability to hydrophobic solutes is small. The more specific porins are also inducible depending on growth conditions. They include the proteins responsible for specific permeation of maltose and maltodextran, anions, iron-siderophores, and vitamin $B_{12}$.

### *Lipopolysaccharides*

Lipopolysaccharides (LPS) cover about 40% of the outer surface of enteric bacteria. LPS are amphipathic molecules with the hydrophobic lipid A moiety and the hydrophilic polysaccharide moiety (Fig. 2.5). The lipid A that is common to all LPS is composed of a D-glucosamyl-$\beta$-D-

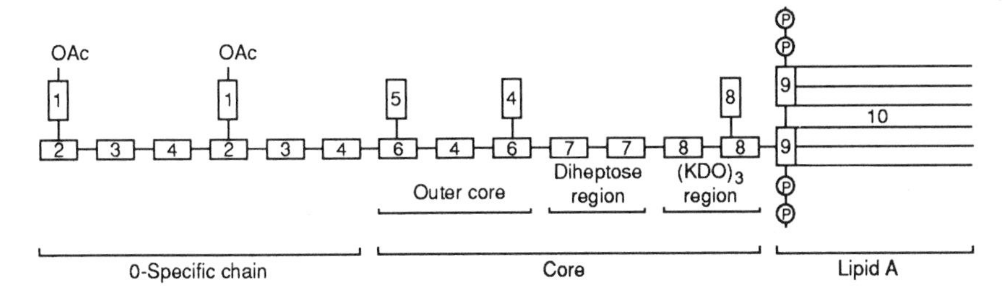


**Fig. 2.5** Schematic illustration of the lipopolysaccharide molecule of *Salmonella typhimurium.* 1, abequose; 2, D-mannose; 3, L-rhamnose; 4, *N*-acetyl-D-glucosamine; 5, D-galactose; 6, D-glucose; 7, heptose; 8, 2-keto-3-deoxyoctonic acid (KDO); 9, glucosamine; 10, fatty acids (1 dodecanoic (12), 4 tetradecanoic (14) and 1 hexadecanoic (16) acid residues).

glucosamine backbone and five to seven saturated fatty acid chains of 12–16 carbon atoms. The polysaccharide moiety is subdivided into core and 0-antigen side chains. The core consists of a nonrepeating series of sugar residues. The structure of 0-antigen exhibits considerable variations among species and strains, and the differences are of taxonomic importance. The LPS of some plant pathogenic bacteria have been suggested to have a major role in host recognition. In *P. solanacearum,* for example, the avirulent and hypersensitive-reaction positive strains have LPS that are missing the 0-antigen consisting of rhamnose, 2-amino-2-deoxyglucose, and xylose at the mole ratio of 4 : 1 : 1.

### 2.2.3 CELL MEMBRANE

The cell membrane is a unit membrane located inside the cell wall and incorporates various intrinsic proteins. The cell membrane consists of a lipid bilayer of 50 to 75% protein and 20 to 35% lipid on a dry weight basis. The lipid of bacteria except archaeobacteria is the fatty acid-glycerol ester type. The cell membrane contains various enzymes for energy-yielding metabolism such as cytochromes, cytochrome oxidase, dehydrogenases, ATPase, protein synthetases, and permeases and plays an important role in respiration, active transport, flagellar rotation, or segregation of nuclear material at cell division.

The energy for active transport and flagella rotation is provided by the proton motive force, or proton electrochemical potential difference across the membrane. The proton motive force is generated by translocation of protons from the matrix to periplasm in coupling with respiration chain and/or ATP hydrolysis by ATPase.

In some bacteria, such as *Clavibacter, Streptomyces,* and *Bacillus,* a part of the cell membrane invaginates into the cytoplasm to form complex membrane-infoldings called *mesosomes.* Mesosomes are associated with respiratory activity, nuclear division, septum formation, spore formation, and secretion of hydrolytic enzymes.

Cells of mycoplasmalike organisms and spiroplasmas lack a true cell wall and are bounded only by a single lipid bilayer membrane. The lipids of *Mycoplasma* consist mostly of phospholipids, glycolipids, and neutral lipids, which are all located in the membrane structure. In *Spiroplasma citri,* the major component of the membrane is an amphiphilic protein *spiralin* of molecular weight 26,000. Glycoprotein located on the surface of the cell membrane has an important role in host–parasite interaction. The helical nature of spiroplasma is associated with the contraction of fibrils found on the inner surface of the cell membrane.

### 2.2.4 PERIPLASM

Periplasm is a matrix of polypeptide and saccharides. It contains various enzymes including plant tissue degrading enzymes such as cellulases and pectinases. Some enzymes located in periplasm function as scavengers that process the conversion of nontransportable metabolites to transportable ones. The periplasmic oligosaccharides such as $\beta$-1,2-D-glucan derived from the outer membrane are responsible for maintaining the osmotic pressure of periplasm equivalent to that of cytoplasm. The oligosaccharides also function for generating the net negative charges to bacterial cells through Donnan equilibrium resulting from their anion residues.

### 2.2.5 GRAM REACTION

The gram reaction is attributed to the structure of the cell envelope. The term *gram reaction* has been used either to indicate the reaction of bacteria to Gram staining or to an alkaline solution. Gram reaction may have also been used to imply distinct taxonomic groups. To avoid this confusion, the term *gram reaction positive, negative,* or *variable* can be used to describe the results of staining. In contrast, *gram-type positive, negative,* or *zero* can be used to indicate the classification of bacteria in taxonomically relevant groups. Archaeobacteria are an example of gram-type zero because they have a cell wall totally different from peptidoglycan.

Gram staining was discovered empirically in 1884 by Christian Gram and used as a diagnostic procedure. The taxonomic value of the procedure was, however, recognized in the subsequent finding of a relationship between the gram reactions and various characteristics such as isoelectric point, tolerance to triphenylmethane dyes, alkali, and antibiotics, and acid-fast staining. Hence, gram-negative and gram-positive bacteria became considered to be distantly related taxonomic groups with different evolutionary backgrounds.

The principal steps of Gram staining are as follows: (1) heated fixed smears of bacteria on a slide are stained with crystal violet, (2) the stain is washed off with water, (3) a dilute iodine solution is added, and (4) the slide is briefly washed with alcohol or acetone. Gram-positive bacteria remain a deep blue-black, whereas gram-negative bacteria are completely decolored. A counterstain of safranin or dilute carbol fuchsin may be applied to differentiate clearly the two types of reaction.

The gram reaction can be also examined by a simple test with 3% potassium hydroxide solution. When bacterial cells are heavily suspended in a drop of KOH solution on the slide, gram-negative bacteria become very viscid, whereas gram-positive ones remain unchanged.

The majority of plant pathogenic bacteria are gram-negative bacteria. The relatively few species of gram-positive plant pathogenic bacteria belong to the genus *Arthrobacter, Bacillus, Clavibacter, Curtobacterium, Nocardia, Rhodococcus,* and *Streptomyces.*

## 2.3 External Structure

### 2.3.1 FLAGELLUM

Most bacterial plant pathogens are motile by means of flagella except for *E. stewartii* and several of the coryneform bacteria. The flagellum (pl. flagella) has a helical shape with various wavelengths. Because flagella are too narrow to be resolved by ordinary light microscopy, they can only be seen after flagellar staining or with the use of electron microscopy. The flagellar filament consists of 11 rows of identical subunits of the protein *flagellin* and has a tubular structure 12–19 nm in diameter, with an axial hole approximately 6 nm in diameter. The base of the flagellum forms a slightly wider hook-shaped structure penetrating the cell envelope and connecting with a basal body 30–40 nm in diameter. The hook acts as a universal joint, transmitting rotational motion to the filament. The basal bodies of gram-negative bacteria consist of three ringlike structures around a central rod, but those of gram-positive bacteria consist of two ringlike structures. These rings are tightly bound to parts of the cell envelope (Fig. 2.6). Bacteria swim by the passive action of the flagella driven by rotary motors coupled to the proton motive force. The motor proteins are located on cell membranes and rotate flagella clockwise and counterclockwise. Switching rotational direction is controlled by two proteins produced by chemotaxis genes *cheY* and *cheZ*.

The arrangement as well as the number of flagella are characteristics of taxonomic importance: monotrichous, lophotrichous, and peritrichous flagellations are distinguished (Fig. 2.7). These flagella arrangements are, however, not absolute. Species of the marine bacteria (*Vibrio*) have both types of polar flagellation and peritrichous flagellation.

Except in the case of soft rot bacteria (*Erwinia* spp. and *Pseudomonas* spp.), bacterial cells in diseased plant tissues are generally nonmotile.

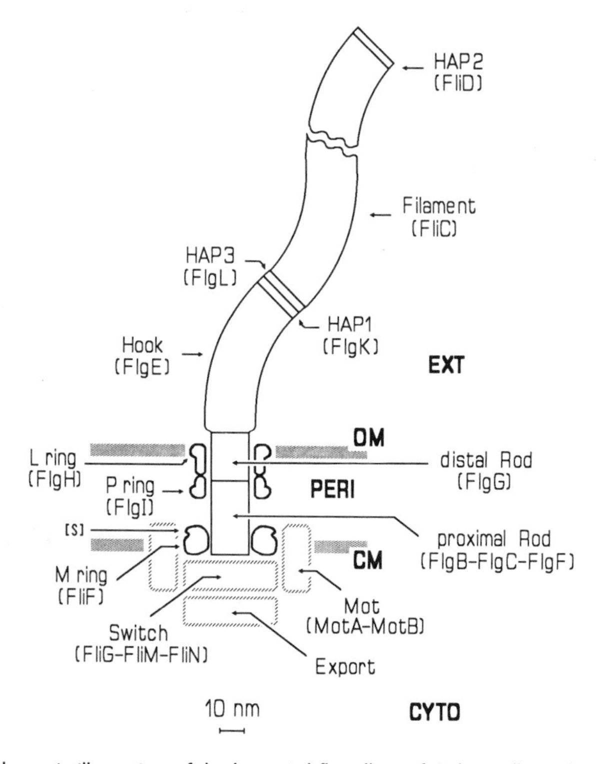


**Fig. 2.6** Schematic illustration of the bacterial flagellum of *Salmonella typhimurium.* [Jones, C. J. and R. M. Macnab, (1990). Flagellar assembly in *Salmonella typhimurium:* analysis with temperature-sensitive mutants. *J. Bact.* **172,** 1327–1339.]

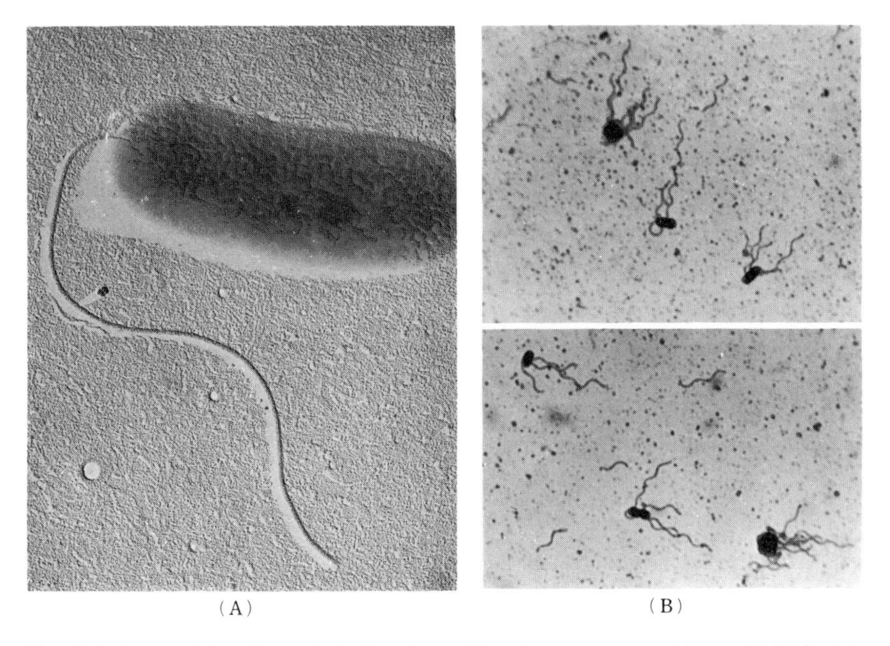


**Fig. 2.7** Bacterial flagella. (A) Polar flagellum of *Xanthomonas campestris* pv. *citri.* (B) Peritrichous flagella of *Erwinia mallotivora* [M. Goto, (1981). "New Bacterial Plant Pathology", Soft Science Co. Ltd. Tokyo.]

However, motile cells arise in the nonmotile population some time after bacteria are released from lesions into a water film on the plant surface.

*P. andropogonis* and *Bdellovibrio* spp. form sheathed flagella, in which the sheath is a membranous continuation of part of the outer membrane.

## 2.3.2 PILUS OR FIMBRIA

Many filamentous appendages are found on the surface of the majority of gram-negative bacteria. They are called pili or fimbriae. These structures range from 3 to 14 nm in diameter and 0.2 to 20 $\mu$m in length. They consist of a protein called *pilin* or *fimbrin* with a molecular weight of 16,600.

Fimbriae are classified into six groups according to their functions and morphology. *Sex pili* act as a specific bridge for gene transfer in conjugation. Fimbriae act as the adhesive organelles with the ability to agglutinate red blood cells and to form pellicles at the surface of liquid culture media. Fimbriation is controlled by phase variation acting as an on-and-off switch in individual cells at the level of transcription. On the other hand, environmental factors such as temperature and nutrition determine the number of fimbriae per cell.

### 2.3.3 EXTRACELLULAR POLYSACCHARIDES

Bacteria often produce extracellular polysaccharides (EPS) outside the cell envelope. The small and dense layer with a definite boundary is called the *capsule*. The water soluble and less adherent layer without a definite boundary is referred to as the *slime layer*. The former structures can be observed under the microscope after capsule staining, whereas the latter structure is difficult to recognize by staining. The liquid culture of the slime-producing bacteria become viscid, and the colonies on an agar plate are domed, umbonate, and convex with a smooth glistening surface.

The EPS comprise an exceedingly diverse variety of strongly hydrophilic homo- and heteropolysaccharides. They have many important physiological functions for maintaining bacterial population in natural habitats, for example, protection of bacterial cells from desiccation, phagocytosis of protozoa and attack of other microorganisms, receptor sites in phage absorption, adhesion of bacterial cells to solid surface, selective uptake of ions, and host–parasite interactions.

The EPS with defined chemical structure include xanthan of *Xanthomonas campestris* (Fig. 2.8), succinoglycan and cyclic(1,2)-β-D-glucan of *Agrobacterium* spp. and *Rhizobium* spp., an EPS of *C. michiganensis* subsp.

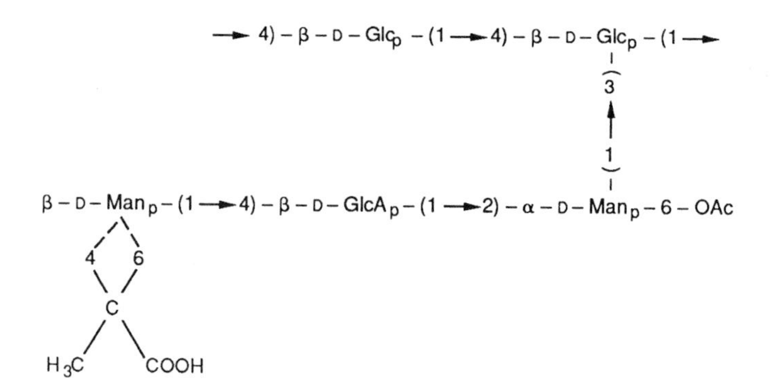


**Fig. 2.8** Structure of extracellular polysaccharide xanthan produced by *Xanthomonas campestris* [Melton, L. D., Mindt, L., Rees, D. A., and Sanderson, G. R. (1976). Covalent structure of the extracellular polysaccharide from *Xanthomonas campestris:* Evidence from partial hydrolysis studies. *Carbohydrate Res.* **46,** 245–257.]

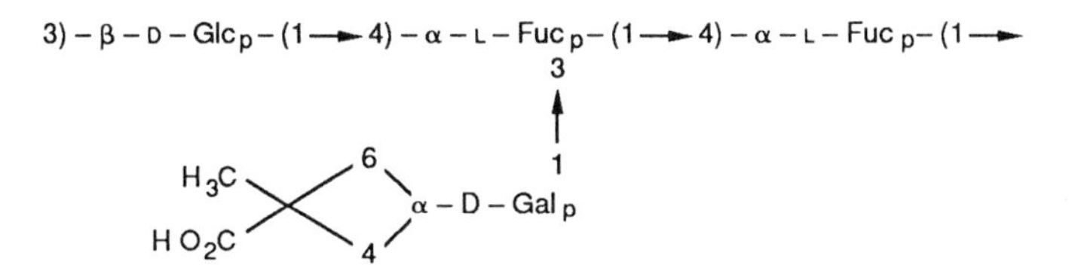


**Fig. 2.9** Structure of extracellular polysaccharide produced by *Clavibacter michiganensis* subsp. *insidiosus.* [Gorin, P. A., Spencer, J. F. T., Lindberg, B., and Lindh, F. (1980). Structure of the extracellular polysaccharide from *Corynebacterium insidiosum. Carbohydrate Res.* **79,** 313–315.]

*insidiosus* (Fig. 2.9), acidic polysaccharides such as alginate and marginalan, as well as the neutral polysaccharide levan of fluorescent *Pseudomonas* spp.

Xanthan has a number of unusual physical properties in solution such as pseudoplasticity, thixotropy, anomalous variation in viscosity with temperature, and the ability to form mixed gels with other plant polysaccharides such as galactomannan, as well as high stability to microbial degradation. These properties give xanthan a number of industrial uses in paints, oil drilling, foodstuffs, and pharmaceutical products.

## 2.4 Internal Structure

### 2.4.1 CYTOPLASM

The cytoplasm contains ribosome granules and various kinds of enzymes, coenzymes, intermediate metabolic products, and inorganic substrates. It is also the basin for protein synthesis and active metabolism. Ribosomes consists of about 40% protein and 60% rRNA and are particles with a sedimentation constant of 70 S. At low concentrations of $Mg^{2+}$, the 70 S ribosomes dissociate into 50 S and 30 S subunits. Within the cell, actual protein synthesis is accomplished on the polysome, which is composed of a large number of ribosomes distributed over the length of mRNA molecules.

### 2.4.2 CHROMOSOME

In prokaryotes the terms *nucleus, genome,* and *chromosome* can be used interchangeably. The nuclear material within the cell is not separated by a membrane as in eukaryote cells, and mitosis does not take place along with cell divisions. There is basically one chromosome per cell, but two or more chromosomes can be seen in a cell in the exponential growth phase because of the lack of synchrony between nuclear division and cell division. The chromosome is highly folded to form a compact mass and attached to the cell membrane at the site of replication.

The bacterial chromosome is a covalently closed loop of double-stranded DNA with a genome size of 900–9,000 kilobase pairs (kb) (in most bacteria it is 2,000–4,000 kb). The genome size of plant pathogenic bacteria seems to be at the upper end of this range, e.g., 5,447 kb in *P. solanacearum* and 5,500 in *P. tolaasii* (Holloway, 1991). It is estimated to be roughly 1 mm in length, or 3,000 to 4,000 in cistrons. Our knowledge of gene mapping of plant pathogenic bacteria is quite limited, i.e., the number of genes mapped on a chromosome are not more than 50, even in *P. syringae* and *E. chrysanthemi* in which gene mapping has been intensively conducted.

### 2.4.3 PLASMIDS

Bacteria often harbor small, extracellular replication units (replicons) that can be stably inherited by their descendants. Such self-replicating entities are called plasmids. Plasmids are often referred to as episome when they are capable of integrating into host chromosome and replicating with it. Bacterial cells generally express new genetic characteristics by acquisition of plasmids irrespective of integration of the plasmid genome into the chromosome. Plasmids can differentially be eliminated in bacterial cells by treatments with mutagenic dye such as acridine orange, antibiotics such as mitomycin C, or high temperature. Such a loss of plasmids is called *curing*.

Plasmids are generally covalently closed circular DNA (ccc-DNA) units with a mass of 4-200 kb. They carry the determinants of additive traits such as sex factors, bacteriocin production, use of unusual substrates, drug resistance, haemolysis, phage resistance, UV resistance, and pathogenesis but usually do not carry the determinants that are vital to bacterial viability.

#### *Transmissible and Nontransmissible Plasmids*

Plasmids can be self-transmissible (conjugative) or non-self-transmissible (nonconjugative). The mobilization of nonconjugative plasmids can be mediated by conjugative plasmids. For example, the nonconjugative plasmid pCR1 of *E. stewartii* is mobilized by a conjugative plasmid of this bacterium. Conjugative gene transfer has been demonstrated in a number of plant pathogenic bacteria with plasmids that include the named groups such as incompatibility group P (IncP) (e.g., RP1, RP4, R751, RK212.1, R68.45) and IncF group (e.g., F'lac$^+$, R100drd56).

### *Drug Resistance Plasmids or R-Plasmids*

Plasmid-borne resistance against antibiotics and heavy metal compounds is known in plant pathogenic bacteria. The advantages of this type of resistance include (1) multiple-drug resistance can be carried in a single plasmid, (2) resistance genes can be amplified when needed and deamplified when not needed, (3) the plasmid can be stored in a minimum portion of the microbial population and regained when needed, (4) the plasmid can serve as a vector to transfer genes, and (5) plasmids serve an evolutionary role in the rearrangement of genetic information (Koch, 1981). Transfer of R-plasmids occurs at high frequency between the plant pathogenic bacteria and saprophytic bacteria as well as animal pathogenic bacteria. Introduced R-plasmids in plant pathogenic bacteria are generally unstable on artificial media and are lost within a short period of time, whereas they can show high stability when the host bacteria survive parasitically or saprophytically in or on plants or both. The presence of R-plasmids has important implications in the field of sanitation as well as the control of bacterial plant disease.

### *Cryptic Plasmids*

The plasmids with unknown phenotypic traits are called cryptic plasmids, resident plasmids, or indigenous plasmids. Many plant pathogenic bacteria can harbor one or more cryptic plasmids, but 11–13 cryptic plasmids have been detected in *Erwinia stewartii.* For elucidating functions of these plasmids, mobilization by other transmissible plasmids, curing or transposon mutagenesis can be employed. Cryptic plasmids are usually stable. Plant pathogenic bacteria live in various habitats, not only in and on host plants but also in soil, water, and insect vectors. The stability of cryptic plasmids may be of value for these bacteria to survive in such secondary habitats.

## 2.4.4 ENDOSPORES

*Bacillus* and *Clostridium* spp. have a life cycle that includes endospore formation. Endospores are dormant resting cells with distinctive properties of great resistance to external environmental factors such as heat, UV irradiation, desiccation, and toxic chemicals. Autoclaving at 121°C for 20 min is necessary to destroy the heat-resistant endospores. For sterilization at 100°C, intermittent heating (Tyndallization) for 30 min for 3 successive days is necessary. With this method, all vegetative cells are killed by heating to 100°C, and surviving endospores are allowed to germinate at room temperature; the resulting vegetative form is destroyed by reheating. The shape and intracellular location of endospores are significant taxonomic traits of spore-forming, rod-shaped bacteria.

### 2.4.5 RESERVE MATERIALS

Reserve materials of prokaryotes includes poly-β-hydroxybutyrate, bolutin (polymers of inorganic phosphate), starch, glycogen, and sulfur. Only poly-β-hydroxybutyrate has been confirmed as the reserve material of plant pathogenic bacteria. It can be readily observed with unstained preparations under the phase contrast microscope as refractile bodies, or stained with 0.3% alcoholic solution of Sudan Black B where it appears as purple black inclusions. This substance is assumed to be used by bacteria as a carbon and energy source under oligotrophic growth conditions.

## 2.5 Pigments

Some plant pathogenic bacteria produce water-soluble or water-insoluble pigments of various colors such as blue, green, yellow, red, and brown under aerobic cultural conditions. Those pigments assist in the identification of bacteria, although many of them remain uncharacterized.

### 2.5.1 FLUORESCENT PIGMENTS OF *PSEUDOMONAS*

The fluorescent pigments produced by *Pseudomonas* had been traditionally called *fluorescein*. However, the term *pyoverdin* has become commonly applied to avoid confusion with chemically synthesized fluorescein *resorcinolphthalein*. Pyoverdins consist of a peptide of 5–8 amino acids and a quinoline derivative chromophore of molecular weight of about 1,000. They are iron-binding and iron-transport substances or siderophores. Different pyoverdins have the different structures of peptide moiety which are responsible for iron-binding functions (Fig. 2.10).

**Fig. 2.10** Structures of pigments produced by plant pathogenic bacteria. 1, Pyoverdin Pa (*P. aeruginosa*); 2, Feruvenulin; 3, Toxoflavin, 4, Reumycin; 5, Rubrifacine, 6, Proferrorosamine A, 7, Indigoidine; 8, Xanthomonadin; 9, Zeaxanthin β-diglucoside.

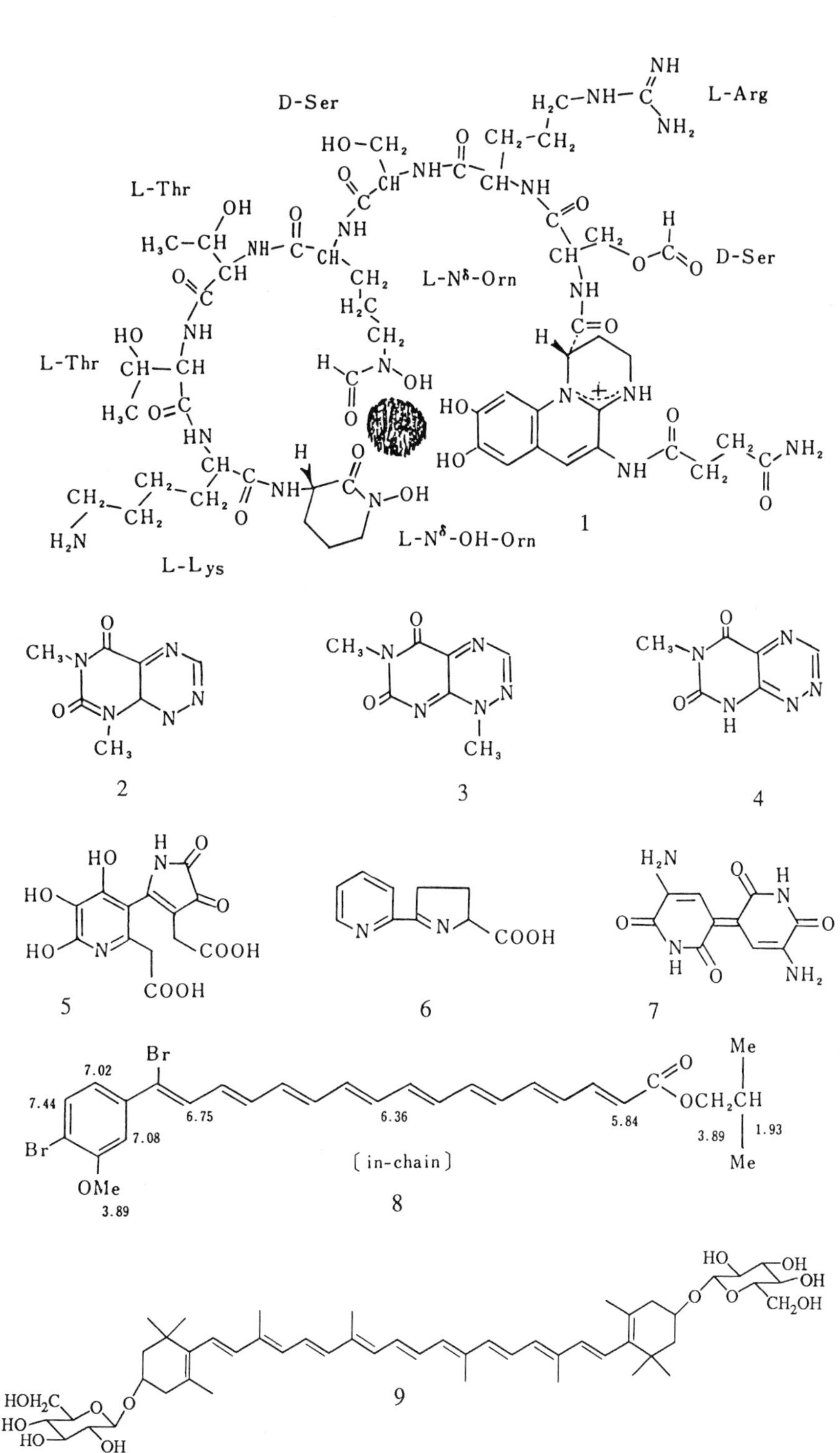
D-Ser
L-Arg
L-Thr
D-Ser
L-N$^{\delta}$-Orn
L-Thr
L-Lys
L-N$^{\delta}$-OH-Orn
1
2
3
4
5
6
7
(in-chain)
8
9

*Pseudomonas glumae* produces two fluorescent pigments called *fervenulin* and *toxoflavin*. The latter is an unstable compound and easily converted to *reumycin* (Fig. 2.10). Those pigments have plant toxicity, inducing chlorosis of leaves as well as inhibition of leaf- and root-elongation in the infected rice seedlings.

## 2.5.2 PIGMENTS OF *ERWINIA* SPP.

*Erwinia chrysanthemi* produces a water-soluble blue pigment *indigoidine* (Fig. 2.10). *E. rhapontici* produces a pink pigment *proferrorosamine A*, which chelates iron converting to *ferrorosamine A* (Fig. 2.10). *E. ruburifaciens* also produces another pink pigment, *rubrifacine*, which is related to indigoidine and assumed to be an oxidation product of nicotine (Fig. 2.10). Rubrifacine inhibits electron transport, and the toxic action of *E. rubrifaciens* to walnut is explained by this function.

*E. herbicola*, *E. uredovora*, and other yellow colony-forming erwinias produce common carotenoid pigments such as *zeaxanthin-β-diglucoside* (Fig. 2.10). These pigments are considered to protect bacterial cells from ultraviolet rays, which induce injuries on either DNA or cell membrane, depending on wave lengths.

## 2.5.3 PIGMENTS OF *XANTHOMONAS*

The yellow water-insoluble pigments of xanthomonads are located exclusively in the outer membrane of cell walls, where they are esterified to a lipidlike moiety. The pigments are brominated aryl-polyene pigments, which have been given the trivial name *xanthomonadins* (Fig. 2.10). Xanthomonadins are differentiated into 15 groups according to the number of bromine atoms, absorption maximum, and mass spectrometric $M^+$ value. These pigments are of great taxonomic significance in the genus *Xanthomonas*, providing a simple and reliable means of identification and differentiation from other yellow gram-negative bacteria.

## 2.5.4 PIGMENTS OF *CLAVIBACTER* AND *CURTOBACTERIUM*

*Clavibacter michiganensis* subsp. *michiganensis* and *Curtobacterium flaccumfaciens* pv. *poinsettiae* produce non-water-soluble yellow pigments (Fig. 2.11), which are carotenoids associated with a protein of cell membrane. The chemical structures and color of the pigments alter from yellow to

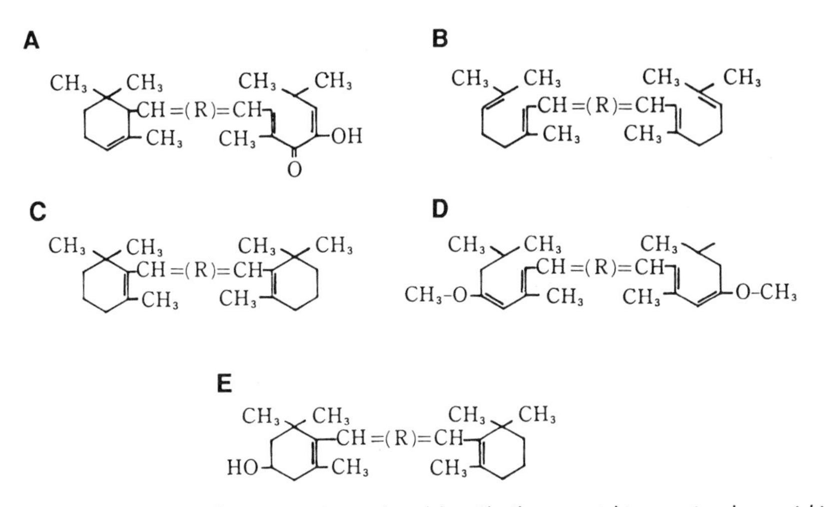


**Fig. 2.11** Structures of carotenoids produced by *Clavibacter michiganensis* subsp. *michiganensis.* A, Canthaxanthin (?); B, Lycopene; C, β-carotene; D, Spirilloxantin; E, Cryptoxanthin. (R) = $C_{20}H_{24}$. [From Saperstein, S., Starr, M. P., and Filfus, J. A. (1954). Alterations in carotenoid synthesis accompanying mutation in *Corynebacterium michiganense. J. Gen. Microbiol.* **10,** 85–92.]

red, depending on the concentrations of thiamine added to the medium. The color of carotenoid pigment produced by *C. flaccumfaciens* pv. *flaccumfaciens* is sharply changed by pH.

# Further Reading


Benz, R. (1988). Structure and function of porins from Gram-negative bacteria. *Ann. Rev. Microbiol.* **42,** 359–393.

Coplin, D. L. (1989). Plasmids and their role in the evolution of plant pathogenic bacteria. *Ann. Rev. Phytopathol.* **27,** 187–212.

Holloway, B. W. (1991). *Pseudomonas* in the late twentieth century. Pseudomonas 1991. *Third International Symposium on Pseudomonad Biology and Biotechnology.* (In press).

Holloway, B. W., and Morgan, A. F. (1986). Genome organization in *Pseudomonas. Ann. Rev. Microbiol.* **40,** 79–105.

Hopkins, D. L. (1989). *Xylella fastidiosa:* xylem-limited bacterial pathogen of plants. *Ann. Rev. Phytopathol.* **27,** 271–290.

Koch, A. L. (1981). Evolution of antibiotic resistance gene function. *Microbiol. Rev.* **45,** 355–378.

Razin, S. (1978). The Mycoplasmas. *Microbiol. Rev.* **42,** 414–470.

Stanier, R. Y., Ingraham, J. L., Wheelis, M. L., and Painter, P. R. (1986). "The Microbial World," 5th ed. Prentice-Hall, Englewood Cliffs, New Jersey.

CHAPTER
**THREE**

# *Taxonomy of Plant Pathogenic Prokaryotes*

Prokaryotes are different from eukaryotes in the following properties: (1) the nucleoplasm is never separated from the cytoplasm by a unit membrane, (2) no basic protein is present in the nucleoplasm, (3) mitosis does not take place in cell division, (4) 70 S ribosomes are dispersed in the cytoplasm but not present in the form of endoplasmic reticulum, (5) streaming of cytoplasm is not observed, (6) culture consists of single cells without forming organs, and (7) mitochondria is absent.

The traditional taxonomy of prokaryotes has largely depended on an artificial classification on the basis of a limited number of phenotypic characters. Now, however, a phylogenetic classification based on the similarity of ribosomal ribonucleic acids is rapidly developing with the background of remarkable advances in molecular biology.

## 3.1 Content of Taxonomy

Bacterial taxonomy encompasses the science or study of classification, identification, and nomenclature of bacteria.

### 3.1.1 CLASSIFICATION

Classification is the grouping of organisms in some orderly fashion. In prokaryotes with simple morphological features, it has not been possible to obtain empirical evidence of a phylogenetic relationship based on a fossil record so that it has been difficult to establish a natural classification as in plants and animals that reflects the process of evolution. Therefore, the traditional taxonomy of prokaryotes on the basis of relationships of certain bacteriological properties has inevitably been an artificial classification. It has developed to satisfy the practical necessity for determining the organisms by binomial nomenclature at the genus and species level.

### 3.1.2 IDENTIFICATION

Identification is a comparative process of determining the taxonomic positions of unknown bacteria with those of known bacteria. The unknown bacterium is determined on the basis of similarity of bacteriological characteristics, whether or not it is the same as a known bacterium or a new one. It is not uncommon to detect phenotypic variations in different strains of the same bacterium. Therefore, a complete identity is rarely obtained even between closely related bacteria when phenotypic properties to be examined are increased, and thus subjective judgment is often required in practical identification.

### 3.1.3 NOMENCLATURE

Nomenclature requires following the rules in the International Code of Nomenclature of Bacteria in naming any bacterial taxa. Names proposed contrary to the rules have no taxonomic standing. Thus, the nomenclature aims at the stability of bacterial systematics by looking over the justifiability of the existing or proposed names, by avoiding or rejecting the use of names that may cause error or confusion, and by avoiding the useless creation of names.

## 3.2 Phylogenetic Taxonomy of Prokaryotes

The phylogenetic taxonomy is based on the concept of the molecular clock. Long-chain molecules such as DNA and RNA undergo various degrees of random mutational alterations along with long evolutional processes from common ancestral origin. The extent of mutational alterations constitutes the molecular clock that may be analyzed by comparing the residuum of the ancestral form in existing molecules.

Since the late 1970s the phylogenetic taxonomy of prokaryotes has undergone a revolutionary development on the basis of remarkable advances in the technologies of molecular biology. The criterion of classification in this system is based on the partial sequence characterization in terms of a "comparative cataloguing" of 5S-ribosomal ribonucleic acids (rRNA) (about 120 nucleotides) or 16S-rRNA (about 1600 nucleotides). The latter in particular is considered to be the most useful chronometer of the evolution of prokaryotes at the level above species. The use of rRNA as the criterion of phylogenetic systematics is based on (1) the ribosome is of ancient origin, (2) it is universally distributed and functionally equivalent among prokaryotes, (3) its changes are not extensive, and the primary structures are well conserved, (4) the presence of extreme conservation and of variability makes it useful in analysis of distant as well as close relationships, and (5) the molecule of 16S-rRNA is large enough for statistical analysis.

The method consists of digesting the 16S-rRNA with a restricted endonuclease or T1 ribonuclease that excises rRNA into oligonuceotides of 6–20 nucleotides, of determining their sequences, and of producing a unique catalog of digestion products from each organism for their comparative similarity.

According to investigations conducted so far, eubacteria and archaeobacteria, like methanogenic bacteria, were genetically split from a common ancestor together with eukaryotes in the same period. The general outline of phylogenetic evolution based on 16S-rRNA comparisons is shown in Fig. 3.1. We can trace the evolution of bacteria back more than 3 billion years. The most ancient bacteria are anaerobic, and aerobic bacteria have arisen many times from them. Nonphotosynthetic bacteria appear to have arisen from photosynthetic bacteria, which are also extremely ancient. *Bacillus* and *Mycoplasma* seem to have arisen from *Clostridium* at the period corresponding to the evolution of actinomycetes.

Although the terminology of phylogenetically defined groupings has not been officially sanctioned, *phylum* (pl. *phyla*) has been proposed instead of *Division* in the traditional taxonomy. Ten phyla have been identified so far, but most of them have no relationship with the phenotypically defined taxa in the present taxonomy. The present *genera* are no more than species within a genus. The important phenotypic properties of higher ranks in the existing taxonomy such as the capacity of photosynthesis or morphological properties such as cocci, bacilli, and spirilla have no validity in phylogenetic systematics. Thus, disagreement is so extensive that it cannot be resolved by minor modification of the present taxonomy, and hence most present taxa stand to lose their taxonomic standing when a proper phylogenetic classification emerges. This is why it is referred to as a revolution in the taxonomy of prokaryotes.

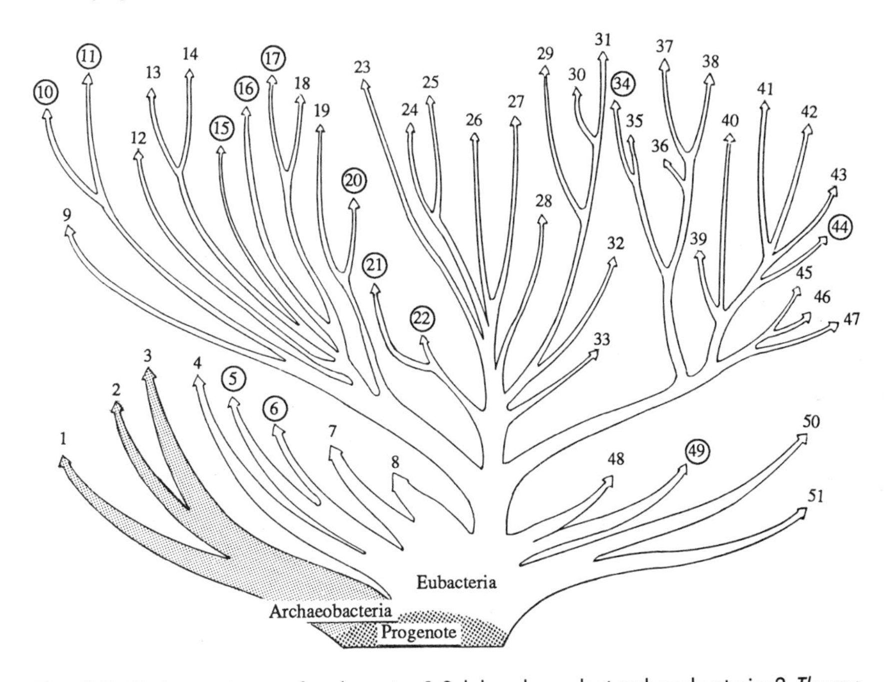


**Fig. 3.1** Phylogenetic tree of prokaryotes. 1. Sulphur-dependent archaeobacteria; 2. *Thermoplasma;* 3. Methanogens/Halophiles; 4. *Planctomyces;* 5. *Chloroflexus;* 6. *Chlorobium;* 7. *Spirochaeta;* 8. *Deinococcus;* 9. *Aquaspirillum itersonii;* 10. *Rhodospirillum rubrum;* 11. *Rhodopseudomonas photometricum;* 12. *Pseudomonas diminuta;* 13. *Rhizobium;* 14. *Agrobacterium;* 15. *Rhodopseudomonas viridis;* 16. *Rhodopseudomonas acidophila;* 17. *Rhodopseudomonas palustris;* 18. *Nitrobacter;* 19. *Paracoccus;* 20. *Rhodopseudomonas sphaeroides;* 21. Chromatiaceae; 22. *Ectothiorhodospira;* 23. *Pseudomonas maltocida;* 24. *Vibrio;* 25. *Photobacterium;* 26. *Escherichia coli;* 27. *Aeromonas;* 28. *Oceanospirillum;* 29. *Pseudomonas fluorescens;* 30. *Serpens;* 31. *Pseudomonas aeruginosa;* 32. *Pseudomonas maltophila;* 33. *Legionella;* 34. *Rhodopseudomonas gelatinosa;* 35. *Sphaerotilus natans;* 36. *Aquaspirillum gracile;* 37. *Pseudomonas acidovorans;* 38. *Pseudomonas testosteroni;* 39. *Aquaspirillum dispar;* 40. *Alcaligenes faecalis;* 41. *Alcaligenes eutrophus;* 42. *Pseudomonas cepacia;* 43. *Chromobacterium lividum;* 44. *Rhodospirillum tenue;* 45. *Nitrosomonas europaea;* 46. *Nitrosolobus multiformis;* 47. *Spirillum volutans;* 48. *Cytophaga/Bacteroides;* 49. Cyanobacteria/Chloroplasts; 50. *Bdellovibrio/Myxococcus/*Sulphur-dependent eubacteria; 51. Gram-positives; ○: Includes photosynthetic bacteria. [Adapted from Stackebrandt and Woese, (1984). Microbiol Sci. **1,** 119–120.]

As phylogenetic systematics are developed, certain phenotypic characteristics with phylogenetic validity can be selected and used as reliable criteria for the new systematics. Murein type, isoprenoid and fatty-acid composition, the presence or absence of whole cell sugars and teichoic acids, end products of glucose fermentation, and the relationship to oxygen are considered to be useful markers in phylogenetic taxonomy.

In plant pathogenic prokaryotes, phylogeny has rarely been investigated in terms of classification, although its usefulness in identification has been demonstrated. In this chapter, therefore, plant pathogenic bacteria are described in the traditional taxonomy represented by *Bergey's Manual of Systematic Bacteriology.*

## 3.3 Concept of Bacterial Species

In plants and animals, *species* is defined as a population of individuals in which the fundamental characteristics are stable, and intraspecies crosses do not cause sterility nor reduce reproduction. In bacteria, the concept of species is inevitably less clear compared with that of higher organisms because of their simple cell morphology and the primitive form of sexual reproduction.

Bacterial species are the taxonomic groups of strains defined on the basis of common phenotypic and genotypic characteristics. In practice, however, the International Code of Nomenclature of Bacteria requires one to designate a strain of the group as the type strain, the representative holding the name of the species. Therefore, the bacteriological characteristics of a species inevitably become different at various degrees around the type strain. These factors makes the concept of bacterial species indistinct and classification unstable.

### 3.3.1 TAXOSPECIES

Taxospecies refers to a group of strains that have a number of phenotypic characteristics in common and can be grouped into distinct phenotypic groups. The bacterial species which are operationally and routinely used in the science, for example, *E. amylovora* or *P. solanacearum,* are all the defined taxospecies. The named group in a formal taxonomy is referred to as a *taxon* (pl. *taxa*).

### 3.3.2 NOMENSPECIES

A nomenspecies is the species with binomial names given in accordance with the rules of nomenclature, irrespective of inappropriateness to a practical nomenclatural system. The names may, therefore, be claimed for another taxa on the basis of some criteria. For example, *X. campestris* is a nomenspecies superseded by *X. campestris* pv. *campestris.*

### 3.3.3 MOLECULAR SPECIES

Molecular species is the term applied to a group of strains with a high degree of nucleic acid (DNA or RNA) homology. The sequence of nucleic acids establishes and preserves the identity of a species. The species in phylogenetic systematics corresponds to this species.

### 3.3.4 GENOSPECIES

A genospecies is a group of strains that can accomplish genetic exchange in some way. Because genetic exchange occurs not only at the species level but also at the genus level or even at the family level, genospecies has not been considered of high practical value.

## 3.4 Methods of Classification

### 3.4.1 CONVENTIONAL TAXONOMY

In conventional taxonomy, strains are classified on the basis of morphological, cultural, physiological, biochemical, and pathological characteristics. This has been the most common approach to bacterial taxonomy, and many of the currently recognized taxa have been established by this method. In conventional taxonomy, however, the investigators used to attempt to select certain definitive properties, either morphological, physiological, biochemical, or pathological, to take precedence over others. The taxa thus established have no phylogenetic relationship.

The taxonomic significance of each definitive property used in conventional taxonomy varies depending on the bacteria. Therefore, minimal standards of phenotypic properties are anticipated with each taxonomic category to which plant pathogenic bacteria belong.

### 3.4.2 NUMERICAL TAXONOMY

In numerical taxonomy, bacteria are classified on the basis of overall similarity of a large number of phenotypic characteristics, and a group of strains with a certain level of similarity is defined as a *taxon*. Thus, this method possesses a higher relationship with natural classification than conventional classification. In practice, more than 50 phenotypic characteristics such as morphological, cultural, physiological, and biochemical properties are examined with individual strains, and the data are programmed for a computer by a binary assignment in which a code of 1 is usually assigned for positive characters and a code of 0 for negative characters. The similarities used in bacterial taxonomy are generally the *similarity index* and the *matching index,* calculated as shown below:

$$\text{Similarity index: } S = \frac{N_s}{N_s + N_d} \times 100$$

$$\text{Matching index: } S_{sm} = \frac{N_s + N_o}{N_s + N_d + N_o} \times 100$$

where $N_s$ is the number of features positive or shared in both strains; $N_d$ is the number of features present or positive in one strain but absent or negative in the other strain; and $N_o$ is the number of features absent or negative in both strains. $N_o$ is not counted in the similarity index.

Similarities are usually shown by a dendrogram or matrix (Figs. 3.2 and 3.3). The clusters grouped in order of similarity values are called *phenon.* The question of what level of similarity is required to designate a species or genus is dependent solely on the subjective decision of the classifier. Some taxonomists propose that genera, species, and strains

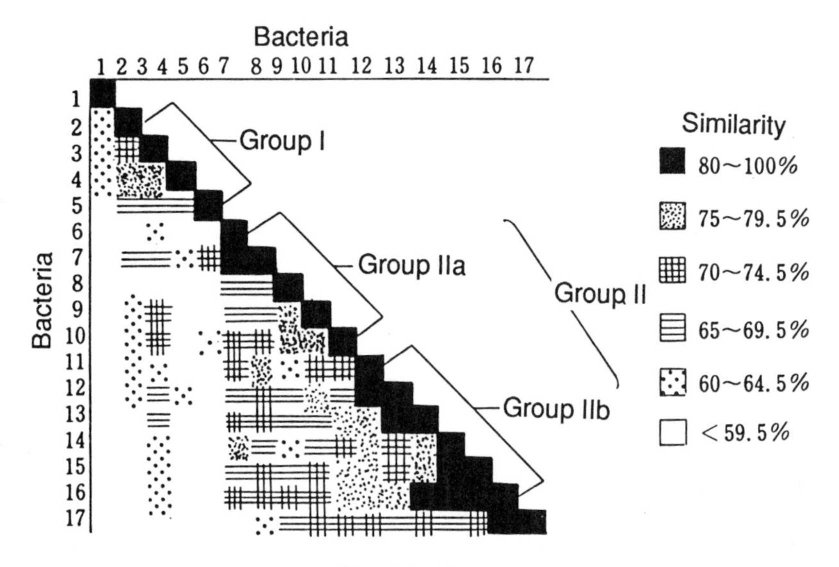


**Fig. 3.2** Matrix.

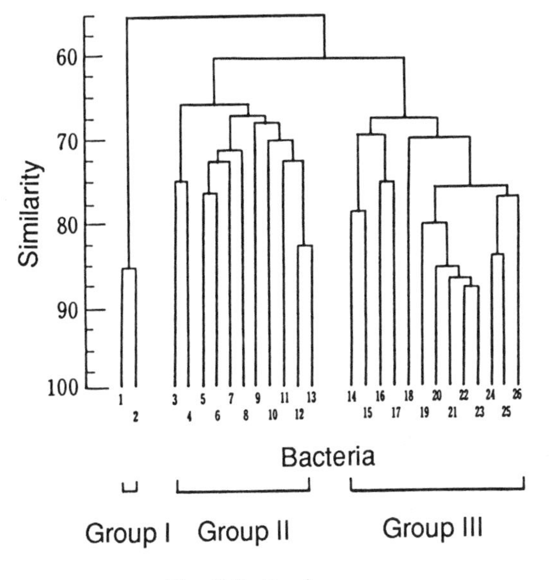


**Fig. 3.3** Dendrogram.

should be defined in terms of S values, e.g., a genus level at 40–60%, a species level at 70–75%, and strain level at 80%.

## 3.4.3 MOLECULAR TAXONOMY

In molecular taxonomy, bacterial strains are clustered into taxa on the basis of DNA parameters such as base composition expressed as guanine plus cytosine content (G + C %), homology, restriction fragment-length polymorphism (RFLP), and gene transfer. The molecular approach attempts to determine the relatedness of chromosomal DNA isolated from two different strains in order to obtain a more direct relationship than conventional and numerical taxonomy that is based on the relatedness of phenotypic properties that reveal only a part of the bacterial genome. Molecular taxonomy is therefore said to be a more rational taxonomic approach toward a better understanding of the ill-defined concept of species and genus.

### *DNA Base Composition*

The G + C content of bacteria varies between 25 and 75%, depending on the taxon. When species are well described, the intraspecific variation of GC content is about ±1% and for genera, the intrageneric variation is about ±5%. When variations are larger than these levels, the taxa are assumed to be heterogeneous. Thus, the GC content has become one of the basic features in the description of new bacterial taxa.

### *DNA–DNA Homology*

A single strand of DNA derived from one strain is radioactively labeled with $^{32}P$ or $^{14}C$ and mixed with an unlabeled single strand DNA from a different strain. The relatedness of base sequences of the two DNA molecules is analyzed from the ratio of hybrid DNA developed in the mixture relative to homologous DNA. Thus, the method elucidates the complementarity of sequences of the four bases (A, T, G, C) between two DNA molecules of different strains. The DNA–DNA homology is suited to classification of taxa below the species level.

There is the possibility that the method of gene diagnosis (Chapter 12, Section 12.1.3), i.e., hybridization between chomosomal DNA and certain probe DNA encoding structural and enzyme proteins, can be effectively applied to the classification of plant pathogenic bacteria in the future. This belief is encouraged by findings that some monoclonal antibodies have revealed the presence of proteins that reflect the relatedness of plant pathogenic bacteria at different taxonomic ranks (Chapter 7, Section 7.4.2) and that some genes associated with pathogenicity, e.g., *hrp* genes (Chapter 8, Section 8.1.3) are well conserved in certain taxonomic groups of bacteria.

### *DNA–rRNA Homology*

In this method, the $^{14}C$-labeled ribosomal RNA from the reference strain is hybridized with DNA from the test bacteria. It is fitted for detecting similarities between distantly related organisms or at the genus or family levels. This is because the genetic loci encoding rRNA are conserved without significant variation during the process of evolution.

### *Restriction Fragment-Length Polymorphism (RFLP)*

Isolated DNA molecules from different strains are cleaved with restriction endonulceases, and the fragments so generated are separated on agarose gels according to size or length. These fragments are hybridized with the independently isolated and labeled DNA probes. The distribution patterns of the fragments which have the sequence homology with the probes can be compared among DNA molecules of different origin. The usefulness of RFLP in bacterial taxonomy is yet to be proven, but it is widely used in the identification of strains at the species or lower levels.

### *Protein Profiles*

Bacteria produce a particular set of cellular and enzyme proteins as gene products. Taxonomy based on the comparison of such protein profiles can, therefore, be referred to as indirect molecular taxonomy. The proteins of two or more strains are simultaneously compared by polysaccharide gel electrophoresis, either in one dimension or two dimensions, with isoelectric focusing in the first dimension. When a number of strains are used, the similarity of the stained gels or zymograms of the proteins can be analyzed by computer.

### *Isozymes*

Enzyme proteins with the same functions may be different in chemical nature. These enzymes are called isozymes and separated by starch electrophoresis. In *Rhizobium* spp., *Escherichia coli, Aeromonas* spp., and *Frankia* spp., a high correlation was noticed between isozyme patterns and DNA homology and RFLP, indicating usefulness in the classification and identification of infrasubspecific subdivisions such as biovars and pathovars.

In plant pathogenic bacteria, the pathological types of citrus canker organism, *X. campestris* pv. *citri,* can be differentiated by the isozyme patterns, indicating its usefulness as a differential marker of infrasubspecific grouping of this bacteria.

### *Gene Transfer*

Gene transfer in bacteria occurs through various mechanisms such as transformation, transduction, lysogenization, conjugation, or phage infection. Successful gene transfers between two taxa have some taxonomic significance, but taxonomic evaluation of gene transfer depends on such factors as stability of transferred genes in the recipient cell, i.e., integration and/or replication. The relative importance of these phenomena in bacterial taxonomy can be considered in decreasing order of gene integration (e.g., transformation, transduction and conjugation associated with chromosome recombination, and phage conversion), gene transmission and expression (e.g., plasmid transfer, phage infection, transfection, or phage lysis), and associated genetic phenomena (e.g., phage adsorption).

### 3.4.4 CHEMOTAXONOMY

The formation of stable cellular components, metabolic profiles, or pigments is all under genetic control and usually is a stable feature of individual strains. For example, the chemical composition of peptidoglycans, particularly sugars and amino acid components, has generally been accepted as an important criterion in the taxonomy of actinomycetes. Various components such as isoprenoid quinons, mycolic acids, polar lipids, fatty acids, polyamines, and cytochromes have also proved to be useful as taxonomic criteria. With the development of relatively fast and reliable analytical devices, attention has increasingly been devoted to the significance of chemotaxonomy.

### 3.4.5 PYROTAXONOMY

Pyrolysis depends upon the general chemical composition of organisms, which is an indirect expression of genes. A small amount (10-100 $\mu$g) of whole cells from colonies grown on a culture plate is directly pyrolysed at high temperature between 200 and 1200°C for a few seconds, and the high molecular weight ions generated are analyzed by mass spectrometry or gas-liquid chromatography. Pyrolysis may be used for relatively quick identification of microorganisms and is becoming known as pyrotaxonomy.

### 3.4.6 SEROLOGICAL TAXONOMY

The most common use of serology in bacterial taxonomy is the infrasubspecific grouping of bacteria or *serovars* according to the antigenic structures of bacterial cells (somatic and flagella antigens). In serological works with polyclonal antibodies, the results often vary depending on the strains, laboratories, and investigators. Such inconsistency is a result of the heterogeneous nature of polyclonal antibodies (Chapter 7, Section 7.2.1). Monoclonal antibodies prepared by an entirely different process react with certain specific antigens. When the proper antibodies are obtained, therefore, they are quite promising in identification of bacteria at the different taxonomic ranks (Chapter 7, Section 7.2.2).

# 3.5 Classification of Higher Ranks of Prokaryotes

Phylogenetic taxonomy has not been developed in the kingdom Procaryotae in the sense of plant and animal kingdom because of the reasons mentioned above. The higher ranks of prokaryotes, therefore, do not necessarily reflect the phylogenetic background in a strict sense, with the possible exceptions of *Proteobacteria* and *Archaeobacteria.*

The higher ranks listed below are mainly based on the descriptions proposed by Murray in *Bergey's Manual of Systematic Bacteriology* (1984). According to his classification, the kingdom Procaryotae is divided into four taxa on the basis of the structural and chemical characteristics of cell envelopes (Fig. 3.4).

## 3.5.1 DIVISION 1. *GRACILICUTES* GIBBONS AND MURRAY 1978

Division 1 consists of the prokaryotes that have a gram-type negative cell envelope consisting of an outer membrane, a peptidoglycan layer, and a unit membrane with fatty acid–glycerol ester type lipids. Endospore is not formed. Usually gram reaction is negative. *Gracilicutes* is divided into the two classes *Proteobacteria* and *Oxyphotobacteria*. The *Proteobacteria* is the only formally designated class on the basis of phylogenetic

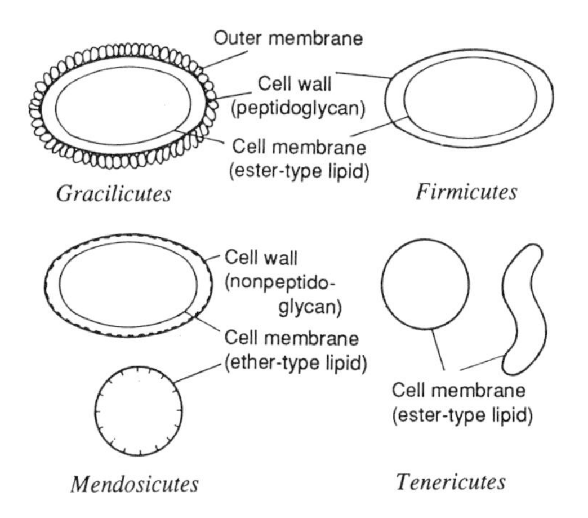


**Fig. 3.4** Classification of divisions of prokaryotes by structure of cell envelope.

principles. It consists of four groups or subclasses called *alpha, beta, gamma,* and *delta.* All gram-negative plant pathogenic bacteria are included in *Proteobacteria* and scattered in four subclasses. Two former classes *Scotobacteria* and *Anoxyphotobacteria* were invalidated.

### 3.5.2 DIVISION 2. *FIRMICUTES* GIBBONS AND MURRAY 1978

Division 2 is the prokaryotes with a gram-type positive cell envelope consisting of a thick peptidoglycan and unit membrane but without an outer membrane. Cells may or may not show branching. Gram reaction is generally, but not always, positive. Some produce endospores. *Firmicutes* is further divided into two classes of *Firmibacteria* and *Thallobacteria. Bacillus* and *Clostridium* are included in *Firmibacteria.* Actinomycetes and related bacteria such as *Streptomyces, Clavibacter, Curtobacterium, Arthrobacter, Rhodococcus,* and *Nocardia* are included in *Thallobacteria.*

### 3.5.3 DIVISION 3. *TENERICUTES* MURRAY 1984

Division 3 is the prokaryotes that lack a cell wall. The cells are enclosed by a unit membrane. They are highly pleomorphic and range in size from large to small (0.2 $\mu$m) deformable form, which is filterable. *Tenericutes* includes the Class *Mollicutes* Edward and Freunt in which plant pathogenic mycoplasmalike organisms and *Spiroplasma* belong.

### 3.5.4 DIVISION 4. *MENDOSICUTES* GIBBONS AND MURRAY 1978

Division 4 is the prokaryotes that have a cell envelope of no conventional peptidoglycan or lack wall material. Cell walls are made purely of protein macromolecules or heteropolysaccharides. Gram reaction is positive or negative. The unit membrane contains ether-linked polyisoprenoid branched-chain lipids. The transfer RNA has unique sequences. *Mendosicutes* includes a class *Archaeobacteria* branched first from a common ancestral progenote in the evolutional process of prokaryotes. No plant pathogenic prokaryotes belong to this division.

Relationships between higher taxa and plant pathogenic bacteria are shown in Table 3.1.

# 3.6 Genera of Plant Pathogenic Prokaryotes

The genera of plant pathogenic prokaryotes and their main characteristics are described below.

## 3.6.1 *PSEUDOMONADACEAE* WINSLOW, BROADHURST, BUCHANAN, KRUMWIEDE, ROGERS, AND SMITH 1917

### Pseudomonas *Migula 1894*

*Pseudomonas* are straight or slightly curved rods, 0.5–1.0 × 1.5–5.0 μm, motile with one to several polar flagella, and gram-negative. They are aerobic, having a strictly respiratory type of metabolism, with oxygen as the terminal electron acceptor (some bacteria can anaerobically grow with nitrate as an alternate electron acceptor). Some bacteria are facultative chemolithotrophs using $H_2$ or CO as energy sources. The mol% G + C of the DNA is 58–70. Type species is *Pseudomonas aeruginosa* (Schroeter 1872) Migula 1900. The species of *Pseudomonas* is divided into five rRNA groups. Many plant pathogenic bacteria are included.

### Xanthomonas *Dowson 1939*

*Xanthomonas* are straight rods of 0.4–0.7 × 0.7–1.8 μm, motile with one polar flagellum, and gram-negative. Colonies are usually yellow with the pigments called "xanthomonadins." Obligately aerobic, having a strictly respiratory type of metabolism, with oxygen as the terminal electron acceptor. Asparagine is not used as a sole source of carbon and nitrogen. Growth is inhibited by 0.1% triphenyltetrazolium chloride. The mol% G + C of the DNA is 63–71. Type species is *Xanthomonas campestris* (Pammel 1895) Dowson, 1939. Mainly plant pathogenic bacteria; but a small number of saprophytic and epiphytic bacteria are also included.

### Xylophilus *Willems, Gillis, Kersters, Van Den Broecke, and De Ley 1987*

*Xylophilus* are straight or slightly curved rods, 0.4–0.8 × 0.6–3.3 μm and gram-negative. They are motile with one polar flagellum and oxidase

**Table 3.1**
Classification of Higher Ranks of Prokaryotes and Affiliation of Plant Pathogenic Bacteria

| Kingdom | Division | Class | Attribute | Plant pathogenic bacteria | |
|---|---|---|---|---|---|
| | | | | Family | Genus |
| | *Gracilicutes* | *Proteobacteria* | Nonphotosynthetic | *Enterobacteriaceae* | *Erwinia* |
| | | | | *Pseudomonadaceae* | *Acidovorax* |
| | | | | | *Pseudomonas* |
| | | | | | *Rhizobacter* |
| | | | | | *Rhizomonas* |
| | | | | | *Xanthomonas* |
| | | | | | *Xylophilus* |
| *Procaryotae* | | | | *Rhizobiaceae* | *Agrobacterium* |
| | | | | Undefined | *Xylella* |
| | | | Photosynthetic without $O_2$ generation | | |
| | | *Oxyphotobacteria* | Photosynthetic with $O_2$ generation | | |

| | | | | | |
|---|---|---|---|---|---|
| *Procaryotae* | *Firmicutes* | *Firmibacteria* | Simple gram-positive bacteria | | *Bacillus* |
| | | | | | *Clostridium* |
| | | *Thallobacteria* | Gram-positive branching bacteria | | *Streptomyces* |
| | | | | | *Arthrobacter* |
| | | | | | *Clavibacter* |
| | | | | | *Curtobacterium* |
| | | | | | *Rhodococcus* |
| | *Tenericutes* | *Mollicutes* | Wall-less prokaryotes | *Spiroplasmataceae* | *Spiroplasma* |
| | | | | Uncertain affiliation | Mycoplasma-like organisms |
| | *Mendosicutes* | *Archaeobacteria* | Unusual walls, membrane lipids | | |

negative. Obligately aerobic, having respiratory type metabolism. Growth is very slow and poor. Reverse to *Xanthomonas* species in that L-glutamate but not calcium lactate is used. The mol% G + C of the DNA is 68–69. Closely related to rRNA subgroup III, which includes *P. avenae* and *P. rubrilineans*. Type species is *Xylophilus ampelinus* (Panagopoulos 1969) Willems, Gillis, Kersters, Van Den Broecke, and De Ley 1987, the causal agent of bacterial necrosis and canker of grape vines, and it is the only species included.

### Rhizobacter *Goto and Kuwata 1988*

*Rhizobacter* are straight or slightly curved, 0.9–1.3 × 2.1–2.5 μm, and gram-negative rods. Motile with polar flagella or lateral flagella or both or nonmotile. Poly-β-hydroxybutyrate granules are formed. Aerobic, having respiratory metabolism. White or yellowish white, plicated, tough or viscid colonies on agar plates. Floccular growth consisting of globular units in liquid medium. Susceptible to 10 μg of vibriostatic agent 2,4-diamino-6,7-diisopropyl-pteridine or 0/129 phosphate. A variety of carbohydrates including starch, glycogen, and dextrin are used as a sole source of carbon. The ubiquinone is Q8. The mol% G + C of the DNA is 66.9–70.6. Type species is *Rhizobacter daucus* (Goto and Kuwata 1988), the causal agent of carrot bacterial gall. The genus includes only one species.

### Rhizomonas *van Bruggen, Jochimsen, and Brown 1990*

*Rhizomonas* are straight or slightly curved rods, 0.43–0.53 × 0.92–1.34 μm, and gram-negative. Motile by one lateral, subpolar or polar flagellum or nonmotile. Colonies are white or yellowish and smooth or wrinkled. Obligately aerobic and have oxidative metabolism. Poly-β-hydroxybutyrate granules are accumulated. A very limited number of carbohydrates are used as a sole source of carbon. Ethanol is not converted to acetic acid. The ubiquinone is Q10. The mol% of G + C is 58–65. Optimum growth temperature is about 28–33°C, and maximum growth temperature is between 36° and 42°C. The type species is *Rhizomonas suberifaciens* van Bruggen, Jochimsen and Brown 1990, the causal agent of corky root of lettuce. The genus includes only one species.

### Acidovorax *Willems, Falsen, Pot, Jantzen, Hoste, Vandamme, Gillis, Kersters, and De Ley 1990*

*Acidovorax* are straight to slightly curved rods, 0.2–0.7 × 1.0–5.0 μm, motile by a single polar flagellum, gram negative, and oxidase positive.

Urease activity varies among strains. No pigment is produced on nutrient agar. Aerobic. Chemoorganotrophic. Some nonplant pathogenic strains show lithoautotrophic growth or denitrification of nitrate. Oxidative metabolism with oxygen as the terminal electron acceptor. Two hydroxylated fatty acid, 3-hydroxyoctanoic acid, and 3-hydroxydecanoic acid are always present; 2-hydroxylated fatty acids are absent, and a cyclopropane-substituted fatty acid is present in most of the strains. The mol% of G + C of the DNA is 62–66. The type species is *Acidovorax facilis* (Schatz and Bovell 1952) Willems, Falsen, Pot, Jantzen, Hoste, Vandamme, Gillis, Kersters, and De Ley 1990. Four plant pathogenic bacteria are currently proposed in this genus: *A. avenae* subsp. *avenae; A. avenae* subsp. *cattleyae; A. avenae* subsp. *citrulli* and *A. konjaci.*

## 3.6.2 *RHIZOBIACEAE* CONN 1938

### Agrobacterium *Conn 1942*

*Agrobacterium* are gram-negative rods, 0.6–1.0 × 1.5–3.0 μm, and motile by 1–6 peritrichous flagella. Aerobic, having a respiratory type of metabolism, with oxygen as the terminal electron acceptor. Abundant extracellular polysaccharides are produced on sugar-containing media. 3-Keto-lactose is produced by *A. tumefaciens* biovar 1 and *A. radiobacter* biovar 1. The mol% G + C of the DNA is 57–63. Type species is *Agrobacterium tumefaciens* (Smith and Townsend 1907) Conn 1942. Mainly occur in soil and oncogenic to plants except *A. radiobacter.*

## 3.6.3 *ENTEROBACTERIACEAE* RAHN 1937

### Erwinia *Winslow, Broadhurst, Buchanan, Krumwiede, Rogers, and Smith 1920*

*Erwinia* are straight rods, 0.5–1.0 × 1.0–3.0 μm, and gram-negative. Motile, with peritrichous flagella except one species (*E. stewartii*). Facultatively anaerobic. Oxidase negative, catalase positive. Acid is produced from sugars such as glucose, fructose, and galactose. The mol% G + C of the DNA is 50–58. Type species is *Erwinia amylovora* (Burrill 1882) Winslow, Broadhurst, Buchanan, Krumwiede, Rogers, and Smith 1920. Mainly plant pathogens, but some bacteria are saprophytic and/or epiphytic. *E. herbicola* has been isolated from plants as well as human and animal hosts.

### 3.6.4 IRREGULAR, NONSPORING GRAM-POSITIVE RODS

**Clavibacter** ***Davis, Gillaspie, Vidaver, and Harris 1984***

*Clavibacter* are gram-positive, pleomorphic rods, which are often arranged at an angle to give V formations at the cell division. Nonmotile. Obligately aerobic. Oxidase negative. Cell wall peptidoglycan contains 2,4-diaminobutyric acid. Rhamnose but not arabinose are found in cell wall. Mycolic acids are not found. Nonhydroxylated fatty acids consist predominantly of *anteiso*- and *iso*methyl branched chains. Straight-chain saturated acids are minor components. Respiratory quinones are menaquinones. Polar lipids comprise diphosphatidylglycerol, phsophatidylglycerol, and glycolipids. The mol% G + C of the DNA is 70± 5. Type species is *Clavibacter michiganensis* Davis, Gillaspie, Vidaver, and Harris 1984. The species includes several subspecies such as subsp. *michiganensis,* subsp. *sepedonicus,* and subsp. *nebraskensis.* In addition, *C. iranicum, C. rathayi, C. tritici, C. xyli* subsp. *xyli,* and *C, xyli* subsp. *cynodontis* are also included in this genus.

**Arthrobacter** ***Conn and Dimmick 1947***

Both forms of rod and coccoid occur in the life cycle of *Arthrobacter.* Gram-positive. The cell wall peptidoglycan contains lysine as the diamino acid. Some dividing cells are arranged at an angle to give V-formation. Obligately aerobic having respiratory type of metabolism. Little or no acid is produced from sugars in peptone media. The mol% G + C of the DNA is 59–70. Type species is *Arthrobacter globiformis* (Conn) Conn and Dimmick 1947. *A. ilicis,* the causal agent of bacterial blight of american holly is included in this genus.

**Curtobacterium** ***Yamada and Komagata 1972***

*Curtobacterium* are small irregular, gram-positive rods. Motile by lateral flagella or nonmotile. Obligately aerobic. Acid is produced slowly and weakly from carbohydrates. The cell wall peptidoglycan contains D-ornithine. Mycolic acids are not present. Cells multiply by bending-type cell division. Generally gelatin is hydrolyzed and DNase is produced. The mol% of G + C of the DNA is 68.3–75.2. Type species is *Curtobacterium citreum* (Komagata and Iizuka 1964) Yamada and Komagata 1972. Four pathovars of *C. flaccumfaciens,* i.e., pv. *betae,* pv. *flaccumfaciens,* pv. *oortii,* and pv. *poinsettiae* are included in this genus.

## 3.6.5 NOCARDIOFORMS

**Rhodococcus** ***Zopf 1891***

*Rhodococcus* are rods to branched substrate mycelium, gram-positive and partially acid-alcohol fast. Aerobic. The cell wall peptidoglycan contains major amounts of *meso*-diaminopimelic acid, arabinose, and galactose. The polar lipids contain diphosphatidylglycerol, phosphatidylethanolamine, and phosphatidylinositol mannosides. The quinons are dehydrogenated menaquinones with either eight or nine isoprene units. The mol% of G + C of the DNA is 63–72. Type species is *Rhodococcus rhodochrous* (Zopf 1889) Tsukamura 1974. *R. fascians,* the causal agent of fasciation of sweet pea, is included in this genus.

## 3.6.6 STREPTOMYCETES

**Streptomyces** ***Waksman and Henrici 1943***

*Streptomyces* are extensively branched vegetative hyphae of 0.5–2.0 μm in diameter. The aerial mycelium at maturity forms chains of three to many spores. The cell wall contains L-diaminopimelic acid and saturated, *iso-*, *anteiso*-fatty acids. The quinone is hexa- or octahydrogenated menaquinones with nine isoprene units. The polar lipids contain diphosphatidylglycerol, phosphatidylethanolamine, phosphatidylinositol, and phosphatidylinositol mannosides. Form discrete and lichenoid, leathery or butyrous colonies (1–10 mm in diameter), which vary in color. Aerobic. Highly oxidative and little acids are accumulated in media. The mol% G + C of the DNA is 69–78. Type species is *Streptomyces albus* (Rossi-Doria 1891) Waksman and Henrici, 1943. *S. scabies, S. acidiscabies, S. ipomoeae,* etc. are included in this genus.

## 3.6.7 *SPIROPLASMATACEAE* SKRIPAL 1983

**Spiroplasma** ***Saglio, L'hospital, Lafleche, Dupont, Bove, Tully, and Freundt 1973***

The cell wall is absent in *Spiroplasma.* Pleomorphic from helical to spherical or ovoid. The helical form, 100–200 nm in diameter and 3–5 μm in length, occurs in logarithmic phase of growth. Motile with flexional, twitching, or rotatory movement but not by flagellum. Facultatively anaerobic. Colonies frequently become diffused because of motility. Nondiffused colonies are 200 μm or less and exhibit a typical umbonate appearance. Chemoorganotrophic, requiring cholesterol for growth. The mol% G + C of the DNA is 25–31. Type species is *Spiroplasma citri* Saglio, L'hospital, Lafleche, Dupont, Bove, Tully, and Freundt 1973, the causal agent of citrus stubborn disease. In addition, plant pathogens of *S. kunkelii* and *S. phoeniceum* are included in this genus.

## 3.6.8 UNDEFINED AFFILIATION

### Xylella *Wells, Raju, Hung, Weisburg, Mandelco-Paul, and Brenner 1987*

*Xyella* are usually straight rods, 0.25–0.35 × 0.9–3.5 μm, with long filamentous strands under some cultural conditions, and gram-negative. Nonmotile. Strictly aerobic. Oxidase negative. Two types of colonies, smooth opalescent and umbonate rough, are formed. Nutritionally fastidious, growing on yeast extract agar medium containing cystein and charcoal or glutamate-peptone agar medium containing serum albumin. Some strains may be slowly grown on ordinary nutrient agar. Colonies develop to 0.6 mm and 1.5 mm in diameter after 10 and 30 days, respectively. The mol% G + C of the DNA is 51–53. Found in xylem. Transmitted by grafting and leaf hopper. Type species is *Xylella fastidiosa* Wells, Raju, Hung, Weisburg, Mandelco-Paul, and Brenner 1987, the causal agent of Pierce's disease of grapevine, phony disease of peach, periwinkle wilt, leaf scorches of almond, plum, elm, sycamore, oak, mulberry, and maples, and citrus leaf blight.

### *Mycoplasmalike Organisms*

Mycoplasmalike Organisms (MLO) are polymorphic organisms with no cell wall and 0.2–0.8 μm in diameter. Morphologically resemble Mycoplasma, but often filamentous and branched forms are found. Observed in the sieve tubes of plants showing yellow-dwarf symptoms and in the salivary glands of insect vectors. Symptoms disappear or become less severe by tetracycline treatments. Uncultivable on artificial media. Until identity becomes possible by cultural techniques, the organisms are relegated to the uncertain aggregation of organisms termed MLO.

## 3.6.9 OTHER BACTERIA

Several other genera include bacteria reported as plant pathogenic bacteria.

### **Acetobacter *and* Gluconobacter**

*Acetobacter* and *Gluconobacter* were reported as the causal agents of "pink disease" of pineapple and "browning and rot" of apple and pear fruits. They are opportunistic pathogens, having no capacity to affect plant organs other than fruits.

### **Bradyrhizobium**

*B. japonicum* are the symbiotic root nodule bacteria of soybean. The yellowing toxin "rhizobitoxin" produced by *B. japonicum* is also produced by a plant pathogenic bacteria *P. andropogonis*. Furthermore, the mechanisms of host specificity in rhizobia provide important implications in host recognition of plant pathogenic bacteria. Therefore, rhizobia are often discussed in parallel with plant pathogenic bacteria in terms of the host–bacterium interactions.

### **Serratia**

*Serratia* are members of gram-negative enterobacteria with peritrichous flagella. Usually form pink or red colonies. *S. proteamaculans* and *S. marcescens* have been reported as plant pathogens. The former had been identified as the causal agent of bacterial leaf spot of *Protea cynaroides,* but existing cultures have no pathogenicity. Instead, *P. syringae* has been reported from Australia as the pathogen of bacterial leaf spot of king protea (*Protea cynaroides*). *S. marcescens* has been reported as the pathogen of "crown rot" of alfalfa that is caused by multiple infection of this bacterium and a pseudomonad.

### **Enterobacter**

*E. cloaceae* has been reported to cause brown discoloration of papaya fruits through opportunistic infection. *Erwinia herbicola,* common epiphyte, is sometimes referred to as the synonym of *Enterobacter agglomerans.*

### **Bacillus *and* Clostridium**

*Bacillus* and *Clostridium* are gram-positive, sporing rods with or without peritrichous flagella. *Bacillus* is aerobic or facultative anaerobic, whereas *Clostridium* is strictly anaerobic. These bacteria cause various types of plant diseases such as potato rot in storage and tobacco leaf rot in the drying process (*Bacillus* spp. and *Clostridium* spp.), tomato seedling rot

(*Bacillus* spp.), soybean rot (*Bacillus* spp.), white stripe of wheat (*B. megaterium* pv. *cerealis*) and wetwood syndrome of poplar and elm (*Clostridium* spp.).

**Nocardia**

Only one species (*N. vaccinii*) has been reported as a plant pathogen. However, the identity of this bacterium should be reconfirmed because there is no information available on this bacterium except the original paper.

***Phloem-Limited Bacteria***

Phloem-limited bacteria cause yellow-dwarf diseases of plants, including clover rugose leaf curl. Bacterial cells are 0.2–0.3 × 1.0–2.0 μm in size. In cell structure they resemble *Xylella fastidiosa,* but they are different in that parasitism is restricted to phloem. They are uncultivable and persistently transmitted by the vector of plant-hoppers.

Bacteria living exclusively in xylem or phloem were once referred to as "rickettsia-like organisms" because of their morphological features and fastidious nature of nutrition. Since their unrelatedness to *Rickettsia* has been clarified by DNA analysis, today they are referred to as xylem-limited bacteria or phloem-limited bacteria, or simply fastidious bacteria.

## 3.7 International Code of Nomenclature of Bacteria

The International Code of Nomenclature of Bacteria was revised in 1976 from the previous edition, which originated in the International Botanical Code dating from May 1, 1753. The most important changes in the new code are described below.

### 3.7.1 PRIORITY

The priority of bacterial names was established as of January 1, 1980, under rule 24 of the new code. For names published prior to January 1, 1980, names were retained only for those listed in the Approved Lists of Bacterial Names and those agreed upon as to the validity for their retention. Names not verified by the above categories were determined as nomenclaturally invalid and available for reuse for the naming of new taxa.

### 3.7.2 PUBLICATION OF NEW NAMES

A new name or a new combination for an existing taxon must be published in the *International Journal of Systematic Bacteriology* (*IJSB*) to be a validly published name. For taxa described elsewhere, validations must be made by listing in the *IJSB*.

### 3.7.3 NOMENCLATURAL TYPE

Before publication of the name of a new taxon, a culture of the proposed type strain must be deposited in a permanently established culture collections from which it will be available. The designations assigned to it by the culture collections must be given in the published description. Nomenclatural types of species or subspecies are the specifically designated strains and distinguished as holotype, monotype, lectotype, and neotype, depending on the status of the original description and designation.

### 3.7.4 INFRASUBSPECIFIC SUBDIVISION

Infrasubspecific subdivision refers to the designation of taxa below the rank of subspecies and is not covered by the rules of the code. For the infrasubspecific taxa, the suffixes *-var* or *-form* are recommended to replace *-type* to avoid confusion with the term *type,* which should be used strictly to mean nomenclatural type. The term *strain* refers to the descendants derived from a single colony in pure culture. The term *clone* refers to descendants derived from a single parent cell. A *culture* of bacteria is a population of bacterial cells in a test tube.

## 3.8 Recent Trends in the Taxonomy of Plant Pathogenic Bacteria

Taxonomy of plant pathogenic bacteria is the discipline that changes unceasingly in response to advances in various fields of science, such as molecular biology, biochemistry, genetics, bacteriology, and plant pathology. Therefore, it has inevitably been in a state of flux.

The instability in the taxonomy of plant pathogenic bacteria often originates in the indefinite concept of species in which pathogenicity or host-specificity cannot be conclusively appraised. Consequently, one extreme of the argument relies on the concept of new host–new species as seen in Wernham's classification of genus *Xanthomonas,* where host-specificity is the primary criterion for speciation.

Another extreme relies on the concept of Adansonian taxonomy, in which pathogenicity is not more than a bacteriological property. In this system, bacteria sharing high phenotypic similarities should be lumped into a species irrespective of differences in pathogenicity. It could be said that taxonomy of plant pathogenic bacteria has historically swung between these two extremes.

The concept of *pathovar* (Dye et al., 1980), that is, the unique taxonomic rank specific to plant pathogenic bacteria, was born in such a background. Since the pathovar system was approved by the International Society of Plant Pathology in 1980, serious questions have been raised on its eligibility, and rearrangements of the system to some extent are being discussed today.

To establish the taxonomic stability of plant pathogenic bacteria based on the sensible speciation, pathogenicity, or host-specificity should be fully elucidated from the viewpoints of molecular biology and biochemistry, clarifying their taxonomic significance in relation to other genetic features. Recent advances in molecular biology have made it possible to assume that approximately 100 genes or gene clusters are involved in pathogenesis. If this is the case, pathogenicity may be assessed by a number of biochemical phenotypes in Adansonian taxonomy, and the importance of pathogenicity in the classification of plant pathogenic bacteria could be reevaluated. Thus, current studies on molecular structures and biochemical functions of hypersensitive reaction and pathogenicity (*hrp*) genes, avirulence (*avr*) genes, and/or disease specific (*dsp*) genes should provide substantial implications not only for the molecular elucidation of host–parasite interactions but also for the taxonomy of plant pathogenic bacteria.

## Further Reading

Bergey's Manual of Systematic Bacteriology, Volume 1 (1984) (Krieg, N. R., and Holt, J. G., ed.); Volume 2 (1986) (Sneath, P. H. A., Mair, N. S., Sharpe, M. E., and Holt, J. G., ed.); Volume 3 (1989) (Staley, J. T., Bryant, M. P.,

Pfennig, N., and Holt, J. G., ed.); and Volume 4 (1989) (Williams, S. T., Sharpe, M. E., and Holt, J. G., ed.). Williams & Wilkins, Baltimore.

Bradbury, J. F. (1986). "Guide to Plant Pathogenic Bacteria." CAB International, U. K.

Dye, D. W., Bradbury, J. F., Goto, M., Hayward, A. C., Lelliott, R. A., and Schroth, M. N. (1980). International standards for naming pathovars of phytopathogenic bacteria and a list of pathovar names and pathotype strains. *Rev. Plant Pathol.* **59,** 153–168.

Krawiec, S. (1985). Concept of a bacterial species. *Int. J. System. Bacteriol.* **35,** 217–220.

Lapage, S. P., Sneath, P. H. A., Lessel, E. F., Skerman, V. B. D., Seeliger, H. P. R., and Clark, W. A., ed. (1975). "International Code of Nomenclature of Bacteria and Status of the International Committee of Systematic Bacteriology and Status of the Bacteriology Section of the International Association of Microbiological Societies." International Association of Microbiological Societies by the American Society for Microbiology, Washington, D. C.

Schleifer, K. H., and Stackebrandt, E. (1983). Molecular systematics of prokaryotes. *Ann. Rev. Microbiol.* **37,** 143–187.

Stackebrandt, E., and Woese, C. R. (1984). The phylogeny of prokaryotes. *Microbiol. Sci.* **1,** 117–122.

CHAPTER
FOUR

# Physiology

Nutritional and metabolic traits of plant pathogenic prokaryotes are extremely complex. For example, some plant pathogenic bacteria can saprophytically grow in plant residues and even in soil. In contrast, *Mollicutes* and xylem- and phloem-limited bacteria are parasites of both plants and insects and quite difficult to grow on artificial media. Thus, knowledge of nutritional requirements and metabolism of plant pathogenic bacteria is essential not only for accomplishing their cultivation in artificial media but also for understanding their host–parasite interactions and behaviors in natural ecosystems.

## 4.1 Nutrition

Plant pathogenic bacteria are all chemoheterotrophs, which obtain their energy from the metabolism of carbohydrates, amino acids, or other organic carbon compounds, which also serve as principle carbon sources.

### 4.1.1 EUTROPH AND OLIGOTROPH

The eutrophs are a group of bacteria that grow well on full-strength rich media, whereas the oligotrophs can grow only on media that are highly diluted, for example, $10^{-2}$ to $10^{-3}$ strength. The oligotrophs are very sensitive to various organic compounds of culture media, but the inhibitory effect of peptone is particularly pronounced. Therefore, the growth of oligotrophs is greatly suppressed on undiluted nutrient agar.

The oligotrophs are common in soil bacteria; colonies counted on rich media are much smaller in number compared to those on a diluted weak media.

All known plant pathogenic bacteria have been eutrophs. However, *Rhizobacter daucus* and *Rhizomonas suberifaciens,* the recently disclosed pathogens of bacterial gall of carrot and corky root of lettuce, respectively, have oligotrophic traits. The discovery of these bacteria implies that if highly diluted media are used for study, more plant pathogenic oligotrophs may be isolated from plant roots disordered by unknown causes.

### 4.1.2 NITROGEN SOURCES

Most plant pathogenic bacteria can utilize inorganic nitrogen sources. Some xanthomonads and non-spore-forming Gram-negative bacteria require nitrogen in an organic form such as amino acids.

#### *Inorganic Nitrogen*

Ammonium salts and nitrates are commonly utilized as sources of nitrogen. Some strains of *Pseudomonas solanacearum* and *P. caryophylli* carry out denitrification, that is, they produce gas from nitrate under anaerobic conditions.

#### *Organic Nitrogen*

Organic nitrogen is commonly provided in the form of amino acids, peptone, beef extract, and yeast extract. Peptone is an excellent nutrient because it contains almost all of the amino acids needed for bacterial growth, and it also provides a good buffering effect to the medium. Beef extract and yeast extract are rich sources of vitamins and are used in combination with peptone to accelerate bacterial growth.

The capacity of plant pathogenic bacteria to grow on single amino acids as the sole source of carbon is poor in comparison with the saprophytic bacteria. This may result from the absence of amino acid permeases or of some enzyme required for metabolism of the amino acid, or from the toxicity of amino acids.

Amino acids such as alanine, glutamate, aspartate, and leucine are widely and effectively utilized by plant pathogenic bacteria. However, a balanced mixture of various kinds of amino acids is preferable for optimum growth. Sulfur-containing amino acids such as methionine or cysteine are required for the growth of some plant pathogenic bacteria such as *Xanthomonas campestris* pv. *translucens*, *X. campestris* pv. *oryzae*, and *X. campestris* pv. *citri*. These bacteria lack the ability to synthesize sulfur-containing amino acids from inorganic sulfur compounds such as $MgSO_4$.

### 4.1.3 CARBON SOURCES

Plant pathogenic bacteria utilize for energy-yielding metabolism various organic compounds such as amino acids, organic acids, monosaccharides, disaccharides, trisaccharides, polysaccharides, primary alcohols, polyalcohols, and glycosides. The carbon compounds most generally utilized by bacteria include organic acids of the tricarboxylic acid (TCA) cycle (e.g., malate, succinate, citrate, and fumarate), sugars (e.g., glucose, fructose, galactose, mannose, and sucrose) and sugaralcohols (e.g., mannitol and sorbitol).

In *P. solanacearum* and *X. campestris* pv. *citri*, utilization of some carbohydrates such as lactose, trehalose, and mannitol often varies with the strain or pathovar within a species. Such differences are useful in differentiation of strains into biovars, and as markers in studying ecological behavior.

### 4.1.4 SALTS

The inorganic elements usually required by bacteria are K, P, S, Ca, and Fe in addition to the nitrogen and carbon sources mentioned above. Phosphates are needed as components of nucleic acids and adenosine triphosphate, and sulfates for synthesis of sulfur-containing amino acids. The phosphates are not only important as nutrients, but through their buffering action they prevent adverse change in pH during metabolism of medium constituents. Such microelements as Co, Cu, Zn, Mo, and Mn are also needed for bacterial growth. In most cases, however, supplement of these constituents to the artificial media is unnecessary because chemicals generally contain these compounds as trace elements.

Salt concentration exerts a significant effect on bacterial growth through its effect on the osmotic pressure of the medium. When non-halophilic bacteria such as plant pathogenic bacteria grow in an environment of high salt concentration, the free amino acids glutamate and proline selectively accumulate in gram-negative cells and proline in gram-positive cells in order to regulate osmolarity to a favorable level.

This selective accumulation of free proline in cells is a universal resistance reaction of prokaryotes and eukaryotes to the osmotic stresses accompanying dry conditions.

### 4.1.5 GROWTH FACTORS

The organic compounds, which are the precursors of various coenzymes and stimulate bacterial growth at very low concentrations, are collectively called growth factors. Growth factors of bacteria include nicotinamide, nicotinic acid, thiamine, pantothenic acid, biotin, folic acid, pyridoxine, riboflavin, vitamin $B_{12}$, thioctic acid, inositol, choline, pimelic acid, and *p*-aminobenzoic acid. The most common growth factor of plant pathogenic bacteria is nicotinic acid and its amide. These substances are components of the coenzymes nicotinamide adenine dinucleotide (NAD) and nicotinamide adenine dinucleotide phosphate (NADP), which are involved in the enzymatic dehydrogenation reaction.

### 4.1.6 OTHER COMPONENTS

To grow spiroplasmas, serum, usually horse serum or fetal calf serum, is incorporated into the pleuropneumonia-like organism (PPLO) medium at high concentrations. The serum provides cholesterol and fatty acids, which are absolute requirements for spiroplasmas. Albumin molecules in the serum also play a role as carriers of fatty acids. A large amount of sorbitol or sucrose is added to adjust the osmotic pressure to the same level as that of the cytoplasm, which in the case of *S. citri* is about 15 atm. *Xylella fastidiosa,* the causal agent of Pierce's disease of grapevine, can also grow on media with similar compositions as those for spiroplasmas except the substances for osmolarity. In addition, various compounds such as activated charcoal, ACES (2-[2-amino-2-oxoethyl) amino] ethanesulfonic acid) buffer or ferric pyrophosphate may be added to adjust the growth conditions of the bacterium.

### 4.1.7 SIDEROPHORES

Iron is a biologically important metal because it is a constituent of cytochrome and other heme or nonheme proteins as well as a cofactor for a number of enzymes. When grown in an iron-deficient medium, aerobic and facultatively anaerobic gram-negative bacteria produce low molecular weight chelating agents specific for ferric ion called *siderophores* (Greek: "iron bearer"). In addition to the ligand, bacteria also produce on their cell envelope a receptor protein for the specific transport of $Fe^{3+}$-siderophore complex into cells where $Fe^{3+}$ undergoes a reductive separation from the ligand (Fig. 4.1).

A wide variety of siderophores with different structures have been identified in bacteria. They are classed chemically as either catechols or hydroxamic acids in addition to the ferric citrate system, but there is an increasing number of siderophores of unknown structure or mode of transport (Fig. 4.2).

Siderophore production is of interest in plant pathology for two reasons: (1) a possible association with virulence, and (2) action as the antibiotics in the interactions between deleterious rhizobacteria and plant growth-promoting rhizobacteria in the rhizoplane–rhizosphere of plants. The fluorescent pigment pyoverdin$_{pss}$, which functions as a

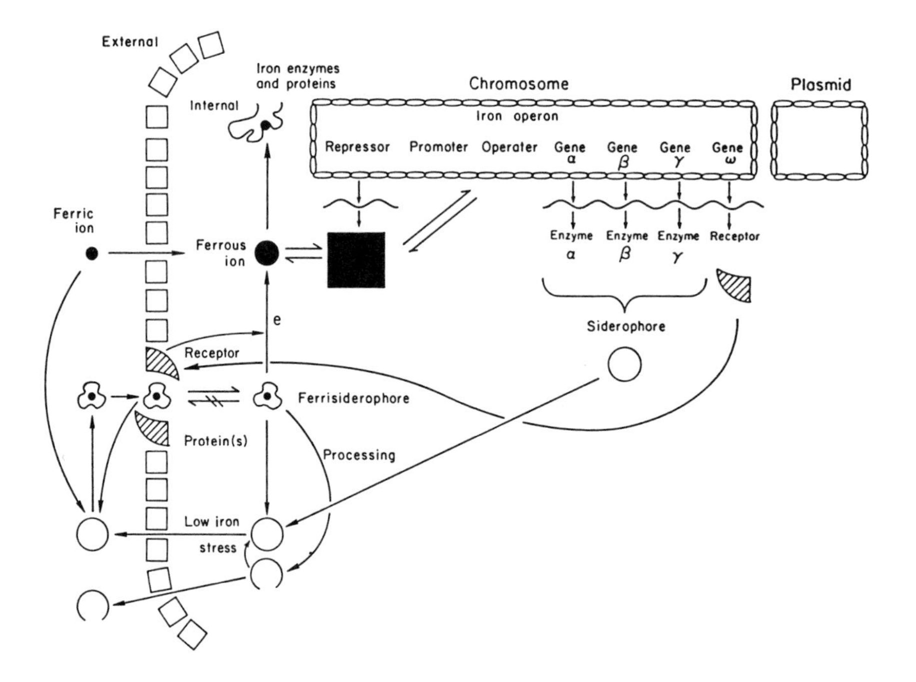


**Fig. 4.1** Schematic illustration of low and high affinity iron assimilation in bacteria [Neilands, J. B. (1982). Microbial envelope proteins related to iron. *Ann. Rev. Microbiol.* **36,** 285–309. Reproduced, with permission, from the Annual Review of Microbiology, Vol. 36, © 1982 by Annual Reviews Inc.]

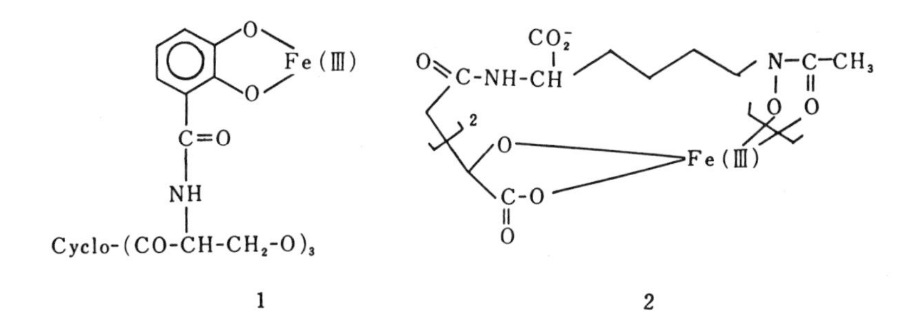


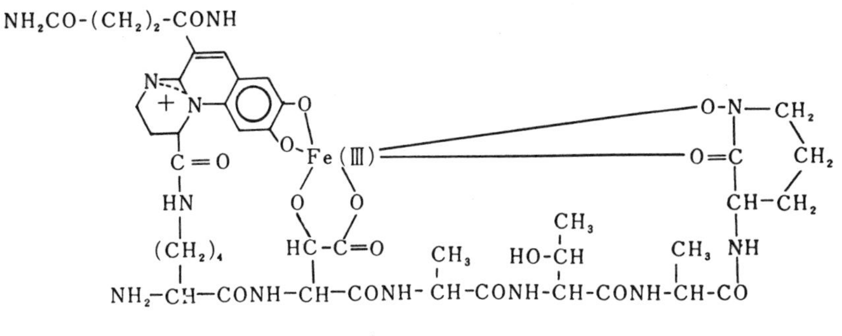


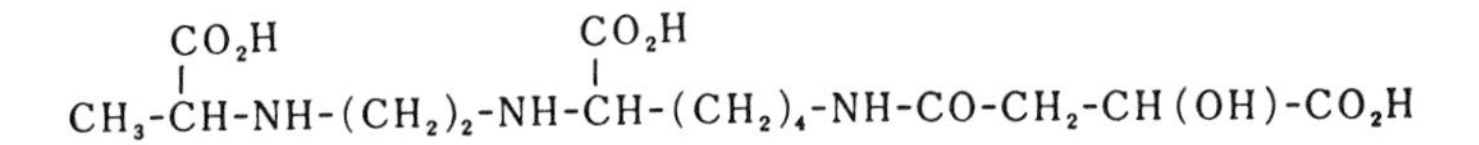


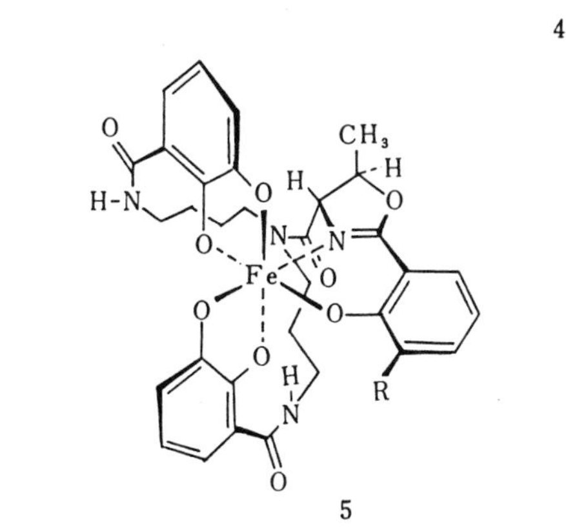


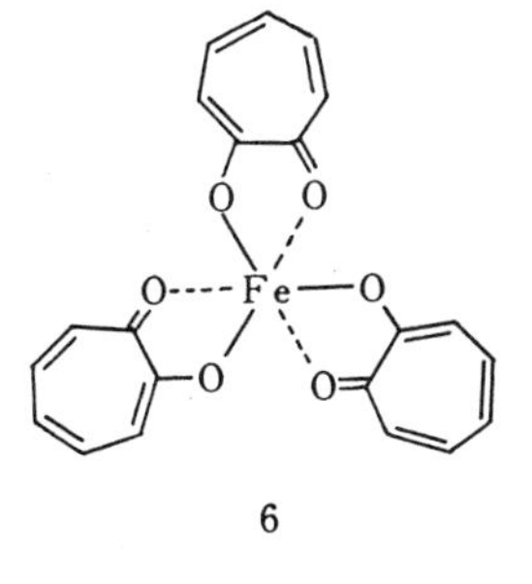


**Fig. 4.2** Structures of siderophores. 1, enterobactin (enterobacteria); 2, aerobactin (enterobacteria); 3, pseudobactin (*Pseudomonas* B10); 4, rhizobactin (*Rhizobium meliloti*); 5, R = H: parabactin (*Paracoccus denitrificans*); R = OH: agrobactin (*Agrobacterium tumefaciens*); 6, tropolone (*Pseudomonas plantarii*).

siderophore of *P. syringae* pv. *syringae*, has no direct relationship with pathogenicity and/or virulence of this bacterium (see Chapter 6, Section 6.3.1).

In mutants of *E. chrysanthemi* induced under iron-limited growth conditions, on the other hand, correlation was found between the disappearance of outer membrane proteins and loss of pathogenicity to saintpaulia and bacteriocin sensitivity. On the basis of this observation, a hypothesis has been presented that the low-iron-inducible outer membrane proteins are the receptors of siderophore and the availability of iron is essential for growth of the bacterium *in planta*.

In bacterial seedling blight of rice caused by *P. plantarii*, a siderophore-like substance, troporone, causes yellowish blight of leaves as well as inhibition of root elongation at concentrations of 3–25 ppm. This toxicity is suppressed by the application of iron (see Chapter 8, Section 8.1.2).

Siderophores play an important role in the antagonism between different microorganisms. Specific strains of the *P. fluorescent-putida* species complex exert their plant growth-promoting activity by depriving native microflora which rely on iron nutrition. The extracellular siderophore of these strains is a yellow-green fluorescent pigment called pseudobactin. The mechanism of the siderophore effect is explained as follows: these strains rapidly colonize plant roots and remove iron in the root zone, making it unavailable to other rhizoplane microorganisms, some of which are deleterious to plant growth and thereby create a more favorable environment for root growth.

Siderophores also play an important role in synergism between different microorganisms. The growth of *A. tumerfaciens* is noticeably stimulated by coexistence with a soil bacterium, *Azotobacter vinelandii*. The mechanism is explained as follows: *A. vinelandii* produces a siderophore which alters the insoluble iron compounds in soil to soluble ones and provides them to *A. tumefaciens* existing in a nearby habitat. *A. tumefaciens* cannot independently grow where no soluble iron compounds are available, because the siderophore, agrobactin, produced by this bacterium does not have the capacity to make iron compounds soluble.

## 4.2 Growth Curve

Growth curves indicating population changes are obtained by plotting the logarithm of the number of viable cells against time. They differ depending on bacterial species, media, temperature, and pH, but generally exhibit the same shapes under identical culture conditions. There are four principal phases: lag phase, exponential growth phase, maximum stationary phase, and death phase.

### 4.2.1 LAG PHASE

In general, when a culture medium is inoculated with bacteria, a certain period of time elapses before exponential growth takes place. This is the lag phase, in which the cells increase in size as a result of the accumulation of various cellular components, and enzyme induction takes place before the onset of active metabolism. The presence of a lag phase and its length are largely dependent on the parent culture used as inoculum.

### 4.2.2 EXPONENTIAL GROWTH PHASE

In the exponential growth phase, also known as the logarithmic growth phase, bacterial growth rate is constant, and the log of the number of viable cells plotted against time gives a linear curve. The *doubling time* of some plant pathogenic bacteria is shown in Table 4.1. The doubling time of fast-growing bacteria that form macroscopic colonies on plates 24 hr after inoculation is about 25 to 30 min. In bacteria forming visible colonies 2 to 3 days after inoculation, doubling times range between 60 and 90 min.

The rate of growth in the exponential phase can be expressed either by the growth rate constant or by the doubling time. If the number of bacterial cells increased from $M_0$ to $M_1$ during the period of time from $T_0$ to $T_1$, the *generation time* and *growth rate constant* can be calculated from the equation,

$$M_1 = 2^n M_0$$

$$\text{Log}\, M_1 = n \log 2 + \log M_0$$

$$n = \frac{\log M_1 - \log M_0}{\log 2} \qquad k = \frac{n}{T_1 - T_0} \qquad g = \frac{1}{k} = \frac{T_1 - T_0}{n}$$

where $n$ is the number of generations, $k$ is the growth rate constant, and $g$ is the generation time.

The temperature characteristic of bacterial growth can be expressed by the following Arrhenius equation, which was originally developed to determine the effect of temperature on chemical reactions,

$$k = Ae^{-u/RT}$$

$$\ln k = \ln A - u/RT$$

**Table 4.1**
Doubling Time (or Generation Time) of Plant Pathogenic Bacteria

| Bacteria | Doubling time (min) | Culture method |
|---|---|---|
| *Erwinia amylovora* | 78 | Stationary culture |
| *E. carotovora* subsp. *carotovora* | 25–30 | Shaking culture |
| *E. chrysanthemi* pv. *zeae* | 25 | Shaking culture |
| *E. herbicola* | 25–30 | Shaking culture |
| *E. milletiae* | 25–30 | Shaking culture |
| *Pseudomonas solanacearum* | 66 | Shaking cuilture |
| *P. syringae* pv. *syringae* | 73 | Shaking culture |
| *P. syringae* pv. *apii* | 80 | Shaking culture |
| *Xanthomonas campestris* pv. *citri* | 90 | Shaking culture |
| *X. campestris* pv. *phaseoli* | 134 | Stationary culture |
| *X. campestris* pv. *oryzae* | 90 | Shaking culture |
| *X. campestris* pv. *pruni* | 92 | Shaking culture |
| *Rhodococcus fascians* | 88 | Stationary culture |
| *Curtobacterium flaccumfaciens* pv. *flaccumfaciens* | 80 | Stationary culture |
| *Clavibacter michiganensis* subsp. *michiganensis* | 115 | Stationary culture |
| *Agrobacterium tumefaciens* | 78 | Stationary culture |
| *A. rhizogenes* | 121 | Stationary culture |

where $k$ is the specific reaction rate constant, $R$ is the gas constant, $T$ is the absolute temperature in Kelvin, $A$ is referred to as the frequency factor, and $u$ is the activation energy of the reaction. When the logarithms of specific growth rate constants of bacteria, $r$ instead of $k$, are plotted against the reciprocal of the absolute temperature ($1/T$), the temperature dependence of bacterial growth is shown by a linear relationship within a certain range of temperatures. However, it is claimed that this modification of the Arrhenius equation results in a curve rather than a straight line. As an alternative, a linear relationship between the square root of the growth constant ($r$) and temperature ($T$) in the following equation is proposed as being applicable to the growth of a wide range of bacteria.

$$\sqrt{r} = b(T - T_0)$$

where $b$ is the slope of the regression line, $T$ is the temperature, and $T_0$ is a conceptual temperature of no metabolic significance. In *E. amylovora*, an

Arrhenius plot showed a linear relationship between doubling time and temperatures between 9° and 18°C: the higher the temperature, the shorter the doubling time. No significant difference is observed, however, between 18° and 30°C. The temperature of 18°C is of special interest because occurrence of blossom blight sharply declines below this temperature.

### 4.2.3 MAXIMUM STATIONARY PHASE

In this phase, the viable count of bacteria is at its maximum and remains constant. In the later stages of the exponential phase, the growth rate gradually reduces because of the development of an unfavorable environment in which there is a deficiency of available oxygen, exhaustion of required nutrients, and change in pH. The reduced growth in population is counterbalanced by death occurring at an equivalent rate, so that the viable count remains unchanged.

The maximum viable cell number of the culture in the stationary phase is called *M* (maximum) concentration, and this value is generally constant for a bacterial species under defined cultural conditions. Although the population size is around $10^9$ cells/ml regardless of the species, it increases several to tenfold in shaken or aerated cultures. The maximum stationary phase continues from several hours to a few days, depending on the species or cultural conditions.

### 4.2.4 DEATH PHASE

In the death phase, the viable cell count decreases with time. As the environment of the culture becomes increasingly adverse, the count of dead cells exceeds the count of viable cells to a greater extent. The death rate finally reaches a maximum constant value, resulting in exponential death of the culture. In the death phase the bacterial cells become smaller in size or abnormally elongated or uneven in staining properties.

## 4.3 Lethal Dilution Effect

Bacteria can survive, in general, for a long period of time without causing phenotypic mutations when they are suspended in distilled or pure water at high concentrations. This procedure is, therefore, applied to bacteria as a preservation method. When bacteria are suspended in pure water at a concentration of $10^7$ cells/ml or lower, however, the patterns of survival curves become quite different from those at concentrations of $10^8$ cells/ml or higher.

Plant pathogenic bacteria are divided into three groups by the ability to survive in highly diluted suspensions: (1) those that rapidly die within 24 hr, (2) those that retain the initial populations for 24 hr, and (3) those that multiply, recovering a population of $10^6$ cells/ml within 24 hr when they are diluted to a density lower than this level (Fig. 4.3). Goto (1985) called the quick extinction of bacteria in a highly diluted suspension in distilled and deionized water the "lethal dilution effect."

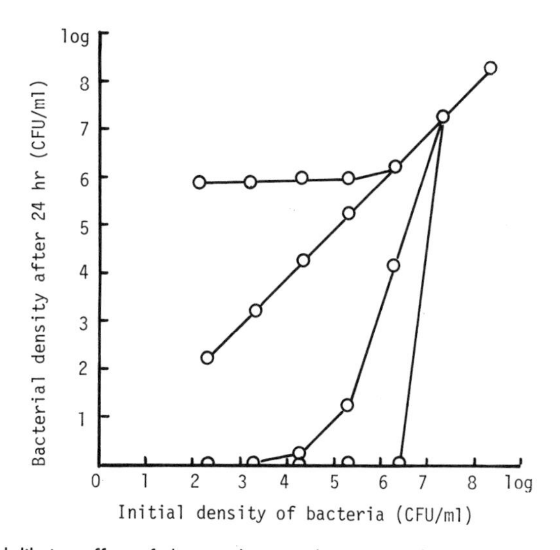


**Fig. 4.3** Lethal dilution effect of plant pathogenic bacteria in distilled water (28°C). Bacteria of type 1a: *P. cissicola, P. fluorescens, P. syringae* pv. *eriobotryae, E. mallotivora, E. milletiae, X. campestris* pv. *begoniae*, pv. *campestris*, pv. *citri*, pv. *nigromaculans*, pv. *oryzae*, pv. *phaseoli*, pv. *pruni*, pv. *vesicatoria*. Bacteria of group 1b: *P. syringae* pv. *lachrymans*, pv. *phaseolicola*, pv. *tabaci*. Bacteria of group 2: *P. aeruginosa, P. andropogonis, P. avenae, X. campestris* pv. *tardicrescens, E. carotovora* subsp. *carotovora, E. chrysanthemi* pv. *zeae, A. tumefaciens, C. michiganensis* subsp. *michiganensis C. flaccumfaciens* pv. *oortii*. Bacteria of group 3: *P. solanacearum*. [Goto, M. (1985). The role of extracellular polysaccharides of *Xanthomonas campestris* pv. *citri* in dissemination and infection: A review. Abstracts of Fallen leaf lake conference "On the genus *Xanthomonas*." p. 15.]

The effect is weakened by the presence of salts; the populations become stable by the addition of phosphate buffer or KCl at a concentration of 0.1 *M*. Extracellular polysaccharides produced by bacteria also have a high protective effect: bacteria of group 1 were completely relieved from the lethal dilution effect by the presence of "xanthan" at concentrations between 0.05 and 0.5 mg/ml.

## 4.4 Chemotaxis

Certain motile bacteria are capable of making a behavioral response to a chemical concentration gradient, which is known as *chemotaxis*. The bacteria are either attracted toward chemical substances (positive chemotaxis) or repelled from these substances (negative chemotaxis). Attractants include sugars, amino acids, alcohols, and oxygen. When oxygen is an attractant, it is called aerotaxis.

Chemotaxis is regulated by the following steps: (1) recognition of the specific chemotactic substance by a receptor protein located on the cytoplasmic membrane, (2) transfer of the stimuli to the basal part of the flagella by diffusible signal proteins, and (3) activation of flagella motors by the proton motive force, (4) induction of motility of the bacterial cell toward the attractant (see Chapter 2). In general, the optimal conditions for chemotaxis consist of growth temperature 5°C lower than the optimum temperature, pH 6–7, and the presence of $10^{-4}$–$10^{-5}$ *M* EDTA.

*E. amylovora* shows positive chemotaxis to aspartate, fumarate, malate, maleate, malonate, oxaloacetate, and succinate, but not to any of the sugars that are effective attractants for other bacteria such as *Salmonella typhimurium* and *P. aeruginosa*. Response of *E. amylovora* to these attractants is inhibited by substitution of an amide group for a carboxy group or addition of a carbon atom between carboxy groups as well as the presence of malate, suggesting that the bacterium has a single chemoreceptor site for all these attractants.

Chemotaxis has been examined with several plant pathogenic bacteria in relation to disease resistance. Some studies concluded that chemotaxis toward leaf extracts or guttation fluid correlated well with the resistance of cultivars, or chemotaxis significantly affects infection ratio. However, other studies are in disagreement with these results, it being difficult at present to draw the general concept on the role of chemotaxis in host–parasite interactions.

# 4.5 Degradation of Macromolecules

A wide variety of enzyme-catalyzed chemical transformations occurs in bacterial cells to generate chemical energy in the form of inorganic or organic compounds, as well as to obtain nutrients in degradation metabolism (catabolism).

## 4.5.1 DEGRADATION OF PECTIN

The ability to produce pectic enzymes *in vitro* is not necessarily predictive of plant pathogenicity, because not all of the pectin-degrading bacteria are plant pathogens. This is mainly based on the fact that the type of pectinases and their secretion capacity are different depending on the bacteria.

### *Pectic Substances*

Pectic substances consist basically of polymers of galacturonic acid joined in $\alpha$-1,4-glycosidic linkage; they are subdivided into three categories, depending on the degree of polymerization and extent of esterification of the carboxyl groups. Protopectin of plant tissues is the intact, relatively water-insoluble high molecular weight substance consisting of about 1,000–2,000 galacturonic acid units, with between 50 and 80% of the carboxyl groups esterified with methanol, together with small amounts of galactose, arabinose, rhamnose, xylose, and acetyl groups. Pectic acid is defined as polygalacturonic acid free of esterified carboxyl groups. Pectinic acid is the form of pectic acid with a low number of esterified carboxyl groups, and pectin is the water-soluble polymer of about 200 galacturonic acid units with 50–80% esterification of the carboxyl groups. Link pectin is referred to as the pectic acid in which the carboxy groups were nearly 100% methylated.

Polygalacturonate chains are associated in the middle lamella with other polygalacturonate chains or with protein through ionic bonds involving the divalent cations $Ca^{2+}$ or $Mg^{2+}$. These pectic substances, together with other macromolecular cell wall components, vary in their chemical and physical properties, depending on the plant species, organ, tissue or age, which determines their resistance to enzymatic degradation.

### *Pectic Enzymes*

Pectic enzymes are classified on the basis of (1) mode of cleavage of $\alpha$-1,4-glycosidic bonds of polygalacturonate, (2) preference for either pectic acid or pectin, and (3) point of cleavage in the polymeric chain (EC numbers in parenthesis indicate the names of the enzymes recommended by the International Enzyme Committee).

*Pectin Methylesterase (PME) (EC3.1.1.11)*

This enzyme is widely distributed in various plants and microorganisms; the enzyme hydrolytically removes methyl groups from esterified carboxyl groups of pectic substances (Fig. 4.4). Although the enzyme is not directly involved in splitting of polygalacturonic acid chains, it

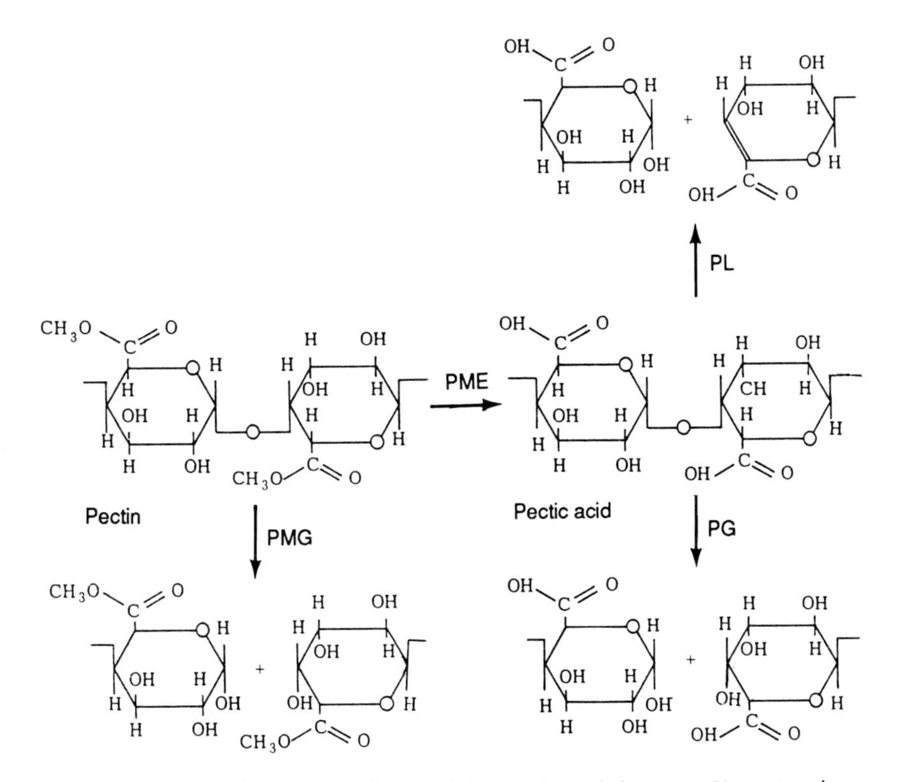


**Fig. 4.4** Enzymatic degradation of pectin. PME, pectin methylesterase; PL, pectate lyase; PMG, polymethylgalacturonase; PG, polygalacturonase.

alters the property of the chains so that they become more susceptible to the action of hydrolase-type pectinases.

*Pectic Hydrolase*

These enzymes bring about hydrolytic cleavage of $\alpha$-1,4-glycosidic bonds (Fig. 4.4). In general, the activity of these enzymes is depressed by the presence of $Ca^{2+}$ and strongly inhibited by phenolic substances, especially their oxidation products (quinones).

*Polyglacturonase (PG)* This enzyme attacks pectic acid in preference to pectin, cleaving $\alpha$-1,4-glycosidic bonds at random, producing galacturonic acid or oligogalacturonic acid as the end products (endo-PG; EC3.2.1.15), or at the terminal position, releasing galacturonic acid units (exo-PG; EC3.2.1.67). (Fig. 4.4). The former induces a rapid decline in viscosity of the reaction mixture with a slight increase in reducing activity, whereas the reverse occurs in the latter.

*Polymethylgalacturonase (PMG)* This enzyme hydrolyses pectin in preference to pectic acid. Two enzymes, endo-PMG and exo-PMG, are differentiated.

*Lyase or Transeliminase*

This enzyme splits $\alpha$-1,4-glycosidic bonds at C-4 and simultaneously eliminates a proton from C-5 to give an unsaturated bond between C-4 and C-5 of the reaction product (Fig. 4.4). The reaction products have maximum absorption at 235 nm, which enables assay of enzyme activity to be carried out spectrophotometrically. Enzyme activity has an alkaline pH, optimum at or above 8.0 and is enhanced in the presence of $Ca^{2+}$. In plant tissues these enzymes are inhibited by oxidized phenols (quinones) or indole acetic acid.

*Pectate Lyase (PL) or Polygalacturonic Acid Transeliminase (PATE)* This enzyme splits pectic acid in preference to pectin; it is subdivided into endo-PL (EC4.2.2.2) and exo-PL (EC4.2.2.9). The end products of their activity are saturated and unsaturated galacturonic acid or oligogalacturonic acid.

*Pectin Lyase (PNL) or Pectin Transeliminase (PTE)* This enzyme cleaves pectin in preference to pectic acid and is subdivided into two enzymes, endo-PNL (EC4.2.2.10) and exo-PNL.

*Oligogalacturonate Lyase (OGL) or Oligogalacturonate Transeliminase (OGTE) (EC4.2.2.6)* This enzyme cleaves the oligogalacturonic acids into unsaturated deoxy-keto-uronic acid or D-galacturonic acid.

Pectic substances are degraded by these enzymes into galacturonic acid and deoxy-keto-uronic acid. These compounds are converted to the intermediate metabolite, 2-keto-3-deoxygluconate (KDG), which is further converted to pyruvate and glyceraldehyde-3-phosphate (Fig. 4.5).

### *Regulation of Pectic Enzyme Activity*

Biosynthesis of pectin-degrading enzymes is induced by 5-keto-4-deoxyuronate (DK1), 2,5-diketo-3-deoxygluconate (DKII), and/or 2-keto-3-deoxygluconate (KDG), which are the intermediate metabolites of

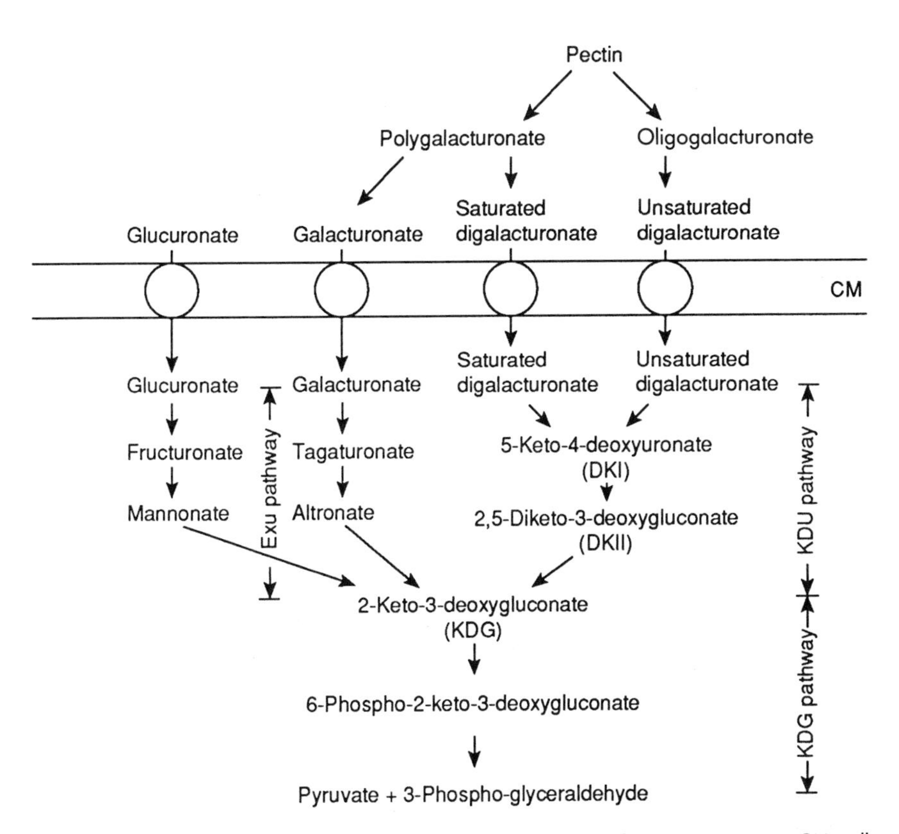


**Fig. 4.5** Degradative pathway of pectin and uronides. ○, uronide transport system; CM, cell membrane.

pectin catabolism. Induction of pectinases is subject to catabolite repression by glucose and to self-catabolite repression. The relationship between pectic enzymes and pathogenesis is described in Chapter 8.

### 4.5.2 DEGRADATION OF CELLULOSE

Cellulose consists of long chains of glucose residues linked by $\beta$-1,4-glycosidic bonds. The degree of polymerization is estimated at about 600–1,000 in wood pulp cellulose and about 10,000 in cotton fiber cellulose. Crystalline cellulose consists of microfibrils of average diameter of 35 Å, each containing 40 parallel chains linked through hydrogen bonds. Native cellulose fibers are composed of these crystalline cores and amorphous cortex in the ratio 7 : 1.

Several enzymes are involved in microbial degradation of native cellulose. The production of cellulases is induced by cellobiose or sophorose (generated from cellulose by cellulase and $\beta$-glucosidase-associated transglucosidase activity) and is under the control of catabolite repression by glucose. Cellulases are competitively inhibited by cellobiose and carboxymethyl cellulose (CMC) and inactivated by halogens, heavy metals, and detergents. Exocellobiohydrolase is more susceptible than Cx cellulase to chemical inhibitors. A polymeric leuco-anthocyanin found in the unripe fruit of persimmon is also a powerful but nonspecific inhibitor of cellulases.

#### *Exocellobiohydrolase (EC3.2.1.91)*

Exocellobiohydrolase is also termed $\beta$-1-4-glucancellobiosylhydrolase or avicelase. It shows exo-type activity, splitting off terminal glucose dimers from glucan chains in crystalline forms of cellulose such as avicel. It is distinguished from carboxymethylcellulase by its high activity toward crystalline cellulose.

#### *Cx Cellulase or Endoglucanase (EC3.2.1.21)*

This enzyme is also called carboxymethylcellulase (CMCase) and is widely distributed in microorganisms. It splits $\beta$-1,4-glycosidic linkage in random fashion to produce cellobiose, cellotriose, or cellotetraose as the end products.

***Beta-glucosidase (EC3.2.1.21)***

$\beta$-glucosidase hydrolyses cellobiose, cellotriose, or cellotetraose as the end products of Cx cellulase or exocellobiohydrolase and releases glucose. Three glucosidases are recognized, one of which is cytoplasmic (aryl-$\beta$-glucosidase I), the others being periplasmic (aryl-$\beta$-glucosidase II and III). $\beta$-glucosidases of *E. chrysanthemi* are differentiated into the protein that is encoded on the *clb* genes and cleaves cellobiose, arbutin, and salicin and the protein that is encoded on the *arb* genes and cleaves only aromatic $\beta$-glucosides.

***Exo-1,4-β-D-glucosidase (EC3.2.1.74)***

This enzyme cleaves $\beta$-1,4-D-glucan chains and oligosaccharides releasing glucose. The activity of the enzyme toward cellobiose is very low.

In enzymatic degradation of cellulose, a synergistic effect has been observed between Cx cellulase and exocellobiohydrolase. This fact implies that cellulose molecules are degraded stepwise as follows: (1) exocellobiohydrolase cleaves cellobiose subunits, (2) Cx cellulase cleaves $\beta$-glucosidic bonds at random in the middle of cellulose molecules, releasing cellodextrins, (3) $\beta$-glucosidases hydrolyze cellobiose and cellodextrins into glucose.

Many plant pathogenic xanthomonads produce Cx cellulase. In host plants attacked by these bacteria, the cellulose of the host cell wall is partly degraded, showing a considerable degree of depolymerization. However, none of them can grow on native cellulose as a sole source of carbon because of the absence of exocellobiohydrolase.

The role of Cx cellulases in pathogenesis has been studied in *X. campestris, E. chrysanthemi,* and *P. solanacearum* by cloning their genes. These studies demonstrated that Cx cellulases are not essential for the development of pathogenesis, i.e., they are more advantageous to saprophytic life than parasitic life. In *E. chrysanthemi,* the enzymes are produced in the logarithmic phase and accumulated in the periplasmic space, secreting 35% outside the bacterial cell.

## 4.5.3 DEGRADATION OF STARCH

Starch is a mixture of two main components, amylose and amylopectin, of which amylose is usually present in a smaller amount. Amylose consists of long chains of 300 to 1,300 $\alpha$-1,4-linked glucose residues, whereas amylopectin consists of an undefined number of chains containing 20 to 24 α-1,4-linked glucose units, which are linked to each other through α-1,6-bonds to give a dendroid structure. There are several enzymes involved in the degradation of this complex substrate. Some plant pathogenic bacteria, particularly xanthomonads, are active starch hydrolysers, but little is known about their mode of action.

***Alpha-amylase; Amylo(1,4)-dextrinase (EC3.2.2.1)***

This enzyme hydrolyzes the internal α-1,4-linked glucose units with the formation of maltose and a small amount of glucose from amylose. From amylopectin it produces the same end-products and, in addition, dextrin (α-limit dextrin) with 4–7 glucose units joined by α-1,6 linkages. The first reaction results in a rapid loss of iodine-staining capacity, with little formation of reducing sugars which later slowly increase.

Amylases of *Bacillus* spp. have been intensively studied from the viewpoint of characterization, classification, and genetic regulation of production. Some bacterial plant pathogens such as xanthomonads produce α-amylase. In addition to α-amylase, several other enzymes are involved in microbial degradation of starch, e.g., β-amylase (EC3.2.1.2), glucoamylase (EC3.1.1.3), isoamylase (EC3.2.1.68), and limit dextrinase (EC3.2.1.41). However, none is known to occur in plant pathogenic bacteria.

## 4.6 Respiration and Fermentation

Complex macromolecules like polysaccharides, proteins, and fats are degraded into their constituent sugars, uronides, amino acids, fatty acids, and glycerol and finally converted into pyruvate through various pathways of intermediate metabolism. When pyruvate is oxidized through the tricarboxylic acid (TCA) cycle into $H_2$ and $CO_2$, with the generation of chemical energy in the form of adenosine triphosphate (ATP), the process is called *respiration*. In respiration, organic and inorganic compounds are oxidized, releasing electrons which are accepted by molecular oxygen (aerobic respiration) or inorganic compounds such as sulfate, nitrates, and carbonates (anaerobic respiration).

*Fermentation* is the metabolic process whereby organic compounds serve as both electron donors and electron acceptors and end products such as alcohols or lactate are formed through reduction and decarboxylation of pyruvate.

### 4.6.1 GLUCOSE METABOLISM

Glucose is converted to the key metabolic intermediate, pyruvic acid, through various catabolic pathways including the Embden-Meyerhof-Parnas pathway (EMP), pentose-phosphate pathway, and Entner-Doudoroff pathway. The plant pathogenic bacteria of the genus *Erwinia* produce acid from glucose under both aerobic and anaerobic conditions, whereas other genera such as *Pseudomonas* and *Xanthomonas* do so only under aerobic conditions. The Hugh and Leifson test to examine this capacity is, therefore, a valuable diagnostic procedure in systematic plant bacteriology.

In xanthomonads and pseudomonads, the pentose-phosphate pathway has a minor role, and 80–90% of glucose is converted to gluconate or 2-keto-gluconate and transported into bacterial cells. These compounds are then metabolized through the Entner-Doudoroff pathway (Fig. 4.6).

In *Erwinia,* in contrast, glucose is transported into cells through the active transport system called *phosphoenolpyruvate sugar phosphotransferase system* (PTS) and converted to fructose-1,6-diphosphate, which is cleaved ultimately into two molecules of pyruvates through the EMP or pentose-phosphate pathways. The Entner-Doudoroff pathway does not have any importance (Fig. 4.6). Although *Escherichia coli* is primarily a glucose fermenter, it may oxidize glucose in the presence of a coenzyme *pyroquinoline quinone* (PQQ). However, *E. cypripedii* and *E. milletiae* can also convert glucose into 2,5-diketo-gluconate and 2-keto-gluconate in the absence of PQQ.

### 4.6.2 ELECTRON TRANSPORT SYSTEM

The pyruvate produced in the various pathways is cleaved into three molecules of $CO_2$ and five pairs of hydrogen atoms by decarboxylation and dehydrogenation.

Hydrogen atoms released in the TCA cycle are transferred to nicotinamide-adenine dinucleotide (NAD), flavine-adenine dinucleotide (FAD), and to cytochromes. Hydrogen atoms become hydrogen ions by passing electrons to cytochromes. The electrons are passed through the electron transport chain, which consists of cytochromes such as b, c, a, and chytochrome oxidase, to the terminal electron acceptors such as $O_2$, $NO_3^-$, $SO_4^{2-}$, and $CO_3^{2-}$. Cytochromes of gram-negative bacteria generally consist of b, c, o, $a_1$, and d, whereas those of gram-positive bacteria b, c, a, $a_3$, and o. Cytochrome c is of taxonomic importance and can be

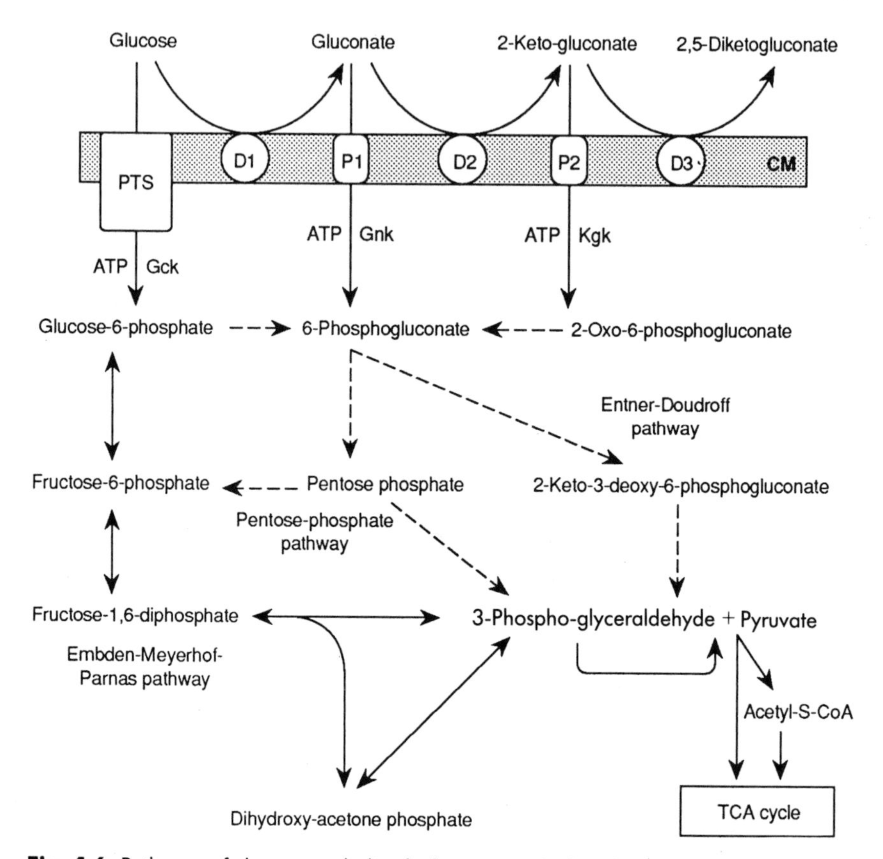


**Fig. 4.6** Pathways of glucose catabolism by bacteria. PTS, phosphoenolpyruvate sugar phosphotransferase system; D1, glucose dehydrogenase; D2, gluconate dehydrogenase; D3, 2-keto-gluconate dehydrogenase; P1, active transport system of gluconate; P2, active transport system of 2-keto-gluconate; ATP, adenosine triphosphate; Gck, glucose kinase; Gnk, gluconate kinase; Kgk, 2-keto-gluconate kinase; CM, cell membrane.

tested empirically by means of the oxidase test. *Erwinia* is negative in this test because of the absence of this enzyme.

The energy released in the process of electron transport along the electron transport chain is stored in the form of adenosine triphosphate (ATP). The phosphorylation generating the high-energy compound ATP is coupled with electron transport and is accordingly termed *oxidative phosphorylation.*

### 4.6.3 METABOLISM OF URONIDES

Uronide metabolism, exemplified by the pathway for the breakdown of galacturonic and glucuronic acid in *E. chrysanthemi*, is shown in Fig. 4.5. There is an active transport system common for galacturonate and glucuronate, and the former acts as the inducer for the transport system of both substrates. Induction of the transport system is very sensitive to catabolite repression by glucose, and the effect is strongly reversed by cyclic AMP.

## 4.7 Biosynthesis of Macromolecules

Bacteria synthesize various macromolecules of the cell from simpler compounds taken from the external environment and chemical energy in the form of adenosine triphosphate (anabolism).

### 4.7.1 BIOSYNTHESIS OF EXTRACELLULAR POLYSACCHARIDES

In general, biosynthesis of extracellular polysaccharides (EPS) is accomplished through the catalytic activity of glycosyltransferases from precursors such as uridine diphosphate glucose (UDPG) or adenosine diphosphate glucose (ADPG). Initiation of biosynthesis by these enzymes usually requires the presence of a small fragment of the polysaccharide end product as a primer in the reaction mixture. The EPS produced by *E. stewartii* and *E. amylovora* consist of glucose, galactose, and glucuronic acid. The former EPS contains these components in the ratio of 10 : 5 : 1, and the latter, which has the trivial name "amylovorin" in the ratio of 1 : 7 : 2, with a small amount of mannose. Thus, the EPS of both *E. amylovora* and *E. stewartii* are synthesized through the activity of galactosyltransferase, which transfers galactose from UDP-galactose to the EPS primer. *A. tumefaciens* synthesizes cyclic β-D-(1,2)-glucan through the processes of 235 kDa protein + UDP-glucose → protein β-D-(1,2)-glucan + UDP and protein β-D-(1,2)-glucan → protein + cyclic β-D-(1,2)-glucan.

Biosynthesis of bacterial EPS is affected by the osmotic pressure of the environment. *A. tumefaciens* produces abundant cyclic β-D-(1,2)-glucan in the medium with low osmotic pressure, whereas it remarkably reduces its production in media supplemented with sodium chloride or mannitol at a concentration of 0.5 *M*. When *X. campestris* is grown under conditions of limited nitrogen and excess carbon, the EPS xanthan is actively biosynthesized during the late exponential phase to early stationary phase where the viscosity of the culture media is still low. The rate of xanthan biosynthesis decreases as the viscosity of the media increases. The energy required for the biosynthesis of xanthan amounts to between 60 and 90% of that required for cell production under these conditions. Such a high ratio of energetic demand is supplied by the effective turnover of ATP.

Cell wall peptidoglycan is synthesized from UDP-*N*-acetylglucosamine (UDP-GlcNAc) and another precursor nucleotide with the general formula of UDP-*N*-acetylmuramyl-X-D-isoglutamyl-Y-D-alanyl-D-alanine (UDP-MurNAc-X-D-isoGlu-Y-D-Ala-D-Ala). The amino acids X and Y linked to UDP-MurNAc are different, depending on the bacterial species.

## 4.7.2 BIOSYNTHESIS OF PROTEIN

### *Assimilation of Ammonia*

Ammonia assimilation and synthesis of glutamate, glutamine, and aspartate occur in bacteria through one of the following reactions: (1) asparagine synthetase, (2) NADP-dependent glutamate dehydrogenase (GDH), (3) glutamine synthetase (GS), and (4) NADP- (or NAD-) dependent glutamate synthase or glutamine-oxoglutarate amino transferase (GOGAT).

Ammonia assimilation reactions resulting in the formation of glutamic acid and glutamine depend on the concentration of ammonia available in cells. GDH and GS pathways operate efficiently when ammonia concentrations are high, whereas GOGAT, coupled with GS, occurs at low concentrations of ammonia. The pathways of ammonia assimilation differ depending on the bacteria.

In *Erwinia,* the major components of the amino acid pool are glutamate, alanine, aspartate, and glycine, regardless of the three intrageneric clusters of "*amylovora,*" "*carotovora,*" and "*herbicola*" with other amino acids at very low levels. A consistent increase in the content of glutamate, aspartate, glycine, and lysine results from growth under conditions of ammonia excess, suggesting a role in the maintenance of electron neutrality. On the other hand, conditions of ammonia limitation lead to high pool concentrations of arginine.

### *Protein Synthesis*

Synthesis of a protein is brought about by coordinate functions of several adjacent genes such as promoter, operator(s), structural gene(s), and a terminator. Such a collection of adjacent genes coding for a single messenger-RNA molecule under the control of a single promoter is called an *operon*. The operon is regulated by *repressors* that are the product of regulatory genes (Fig. 4.7). The promoter is the region of DNA to which RNA polymerase binds and starts transcription of mRNA on a specific strand of its DNA template. The operator is the specific sequence that binds repressor protein encoded on regulator genes and controls the synthesis of mRNA. The terminator is the region of DNA where RNA polymerase stops adding nucleotides to the elongating RNA chain and dissociates from the DNA template, releasing the mRNA molecule.

The enzymes and structural proteins of bacterial cells are synthesized on the cytoplasmic ribosomes in association with mRNA molecules. The multifunctional transfer-RNA (tRNA) molecule plays a specific role in protein synthesis through recognizing the nucleotide sequence of the mRNA, linking corresponding amino acids, and binding to ribosomes.

The amino acid sequence of the protein is determined by the nucleotide sequence of DNA. Amino acids are activated by the action of amino acid–activating enzymes and ATP. The activated amino acids are linked to tRNA, forming amino-acyl-tRNAs through the action of aminoacyl-tRNA transferases. A tRNA molecule is equipped with a sequence of three bases at its distal end (anticodon). The sequence is complementary to that of mRNA (codon) and specific to each amino acid to be attached. An aminoacyl-tRNA molecule attaches to a particular region known as the "A site" of the 50 S subunit of the ribosome, and its anticodon determines the site (codon) to be bound on mRNA (recognition).

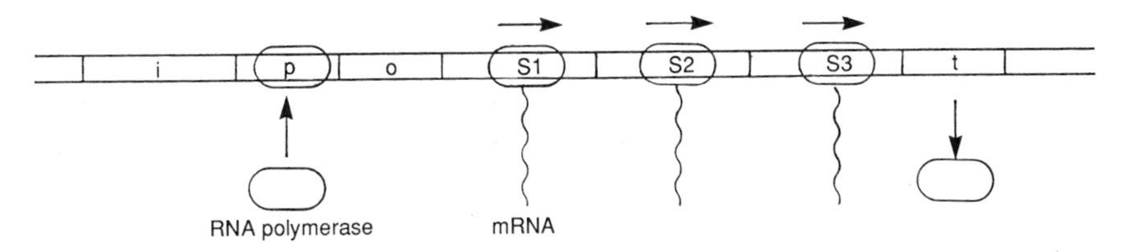


**Fig. 4.7** Schematic illustration of an operon and its associated regulatory genes. i, regulatory gene; p, promoter; o, operator; S1–S3, structural genes; t, terminator.

A peptide bond is then formed between -$NH_2$ of amino-acyl-tRNA and -COOH of the terminal amino acid of the peptidyl-tRNA at the "P-site" of the ribosome, lengthening the peptide chain by one amino acid residue (peptidyl transfer).

After the peptidyl transfer, the free tRNA molecule is displaced from the ribosome and simultaneously the ribosome transfers one codon, resulting in the transfer of the peptidyl-tRNA from the A site to the P site. The amino-acyl-tRNA with the anticodon corresponding to the next codon of mRNA binds to the vacant A site. The repetition of these steps is "translation" of the genetic information of mRNA. The initiation and termination of translation is regulated by the codons on mRNA.

## *Regulation of Biosynthesis and Activity of Enzymes*

### *Induction and Repression*

Enzymes are distinguished as two types: constitutive enzymes, which are produced irrespective of the presence of the substrates, and inducible enzymes, which are produced only in the presence of substrates. Both types of enzymes are under strict genetic control. Where operator genes are nonfunctional or repressor is produced in an inactive form, the enzymes are synthesized constitutively.

The repressors are noncatalytic allosteric proteins and repress the synthesis of enzymes by binding to the operator genes of DNA, blocking the initiation of transcription of the corresponding mRNA molecules. A single repressor often affects the synthesis of several mRNA molecules.

In inducible enzymes, repressor is inactivated by binding to the substrate of the enzyme so that the operon becomes functional and enzyme is produced. When the substrate is exhausted, the repressor is again activated and enzyme production is inhibited. The substrate and substrate analogue that inactivate the repressor are called *inducers* or *effectors,* and the phenomenon is referred to as the *negative control* of enzyme induction (Fig. 4.8). As found in arabinose metabolism of *E. coli,* the repressor may bind to a chromosome, blocking the initiation of transcriptions. In this case, the substrate binds to the repressor and causes conformational change, converting it into an activator and triggering transcription. This is referred to as *positive control* of enzyme induction.

### *End-Product Repression*

When addition of the end product of a biosynthetic pathway to the growth medium stops or slows down the synthesis of an enzyme that is specifically and genetically under the control of regulator genes, the

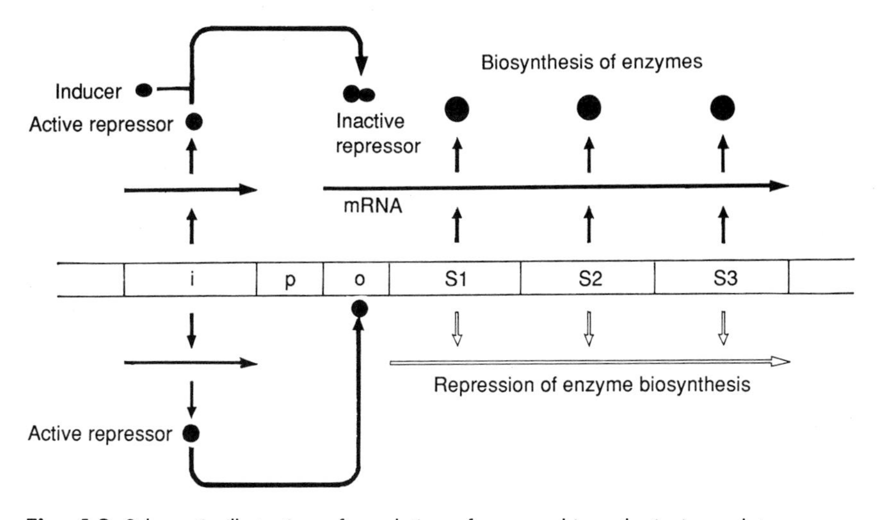


**Fig. 4.8** Schematic illustration of regulation of enzyme biosynthesis. i, regulatory genes; p, promoter; o, operator; S1–S3, structural genes.

phenomenon is called *end-product repression,* and the enzymes involved are called *repressive enzymes.* Enzyme synthesis becomes functional again (derepressed) when the intracellular concentration of the end product falls to a very low level (Fig. 4.9).

*End-Product Inhibition of Enzyme Activity or Feedback Inhibition*

In feedback inhibition, the end product (effector) of a given synthetic pathway prevents the activity of the first enzyme in the pathway, resulting in the arrest of a series of biosynthetic reactions. When the intracellular concentration of the particular end product falls to a low level, the allosteric inhibition also decreases and the rate of biosynthesis increases.

When the enzyme protein binds to a specific small effector molecule at a site other than the active site, the enzyme may change in conformation so that it is unable to bind the substrate at its active site and biosynthesis is stopped. This type of enzyme protein is called an *allosteric protein.* Even if enzyme synthesis is prevented by end-product repression, biosynthetic reactions continue with an increase in end product until the intracellular level of the specific enzyme decreases to a certain level. Therefore, biosynthetic regulation is effectively controlled when both end-product repression and feedback inhibition function synergistically in the same cell.

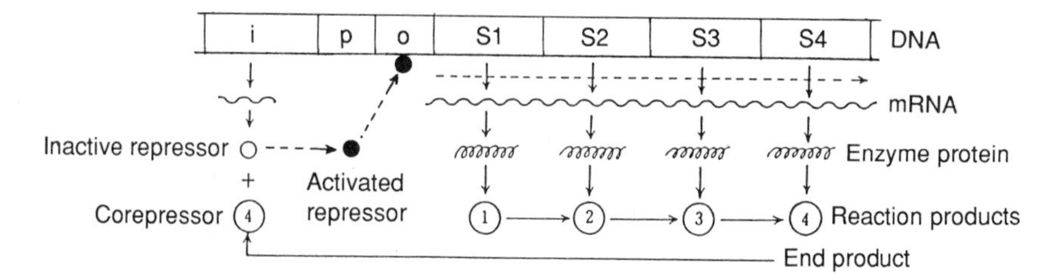


**Fig. 4.9** End-product repression of enzyme synthesis. i, regulatory genes; p, promoter; o, operator; S1–S4, structural genes.

*Catabolite Repression*

When a rapidly metabolizing compound such as glucose is used as the carbon source, the induction of various enzymes may be greatly repressed compared with the induction that occurs when more slowly metabolizing substrates such as glycerol and lactic acid are used. This phenomenon, known as catabolite repression, continues until glucose is exhausted by the bacteria. There is release from catabolite repression when 3′,5′-cyclic adenosine monophosphate (cAMP) is added to the medium because the repression is caused by the low level of cAMP resulting from the rapid metabolism of glucose. Cyclic-AMP binds as an effector to an allosteric protein called *catabolite gene activator protein* (CAP), a dimeric molecule of molecular weight 44,000, which attaches to chromosomal DNA at the catabolite gene activator protein (CAP) site of promoter stimulating the transcription of mRNA. The CAP protein is inactive when not bound to cAMP. Thus, the biosynthesis of enzymes under the control of catabolite repression requires the presence of CAP protein and of cAMP at a level sufficient to activate the CAP protein. In *X. campestris* pv. *campestris,* the strains unable to synthesize the CAP protein show remarkable reduction in pathogenicity but not in the ability to use various carbon sources. The attenuation of CAP-deficient strains is supposed to originate in reduction of the synthesis of xanthan, protease, and pectate lyase that are under the control of catabolite repression.

*Enzyme Induction by an SOS Reaction*

This is the repairing mechanism of the injured DNA regulated by two proteins of RecA and LexA. When DNA molecules are intact, the repressor proteins (LexA, molecular size 22.7 kDa) produced by *lexA* genes inactivate several genes in SOS boxes located near promoters (17 proteins in *E. coli*), completely preventing enzyme production or suppressing it to a very low level. When DNA is injured by ultraviolet (UV) rays or mitomycin C, the signal of the SOS repairing system (single strand DNA) develops its functions, activating the multifunctional protein RecA (37.8 kDa) which has protein- and ATP-degradation capacity in addition to DNA recombination activity.

The RecA protein cleaves LexA protein at the site of -ala-gly, eliminating its function as the repressor. As the concentration of LexA protein diminishes, repression is removed, inducing repair of the injured site of DNA. At the same time, various damage-inducible genes (*din*) are induced and the production of enzymes is initiated (Fig. 4.10).

The *recA* genes have been identified in *E. carotovora* subsp. *carotovora*, *E. chrysanthemi*, and *P. syringae* pv. *syringae*. The *recA* mutants of *E. carotovora* are sensitive to UV irradiation, methyl methansulfonate, and nitroquinoline oxide, and production of pectin lyase and carotovoricin are not induced by mitomycin C. Likewise, the *recA* mutants of *E. chrysanthemi* are sensitive to UV irradiation and deficient in generalized recombination, but their pectic and cellulolytic activities are unaffected.

### *Secretion of Proteins*

Enzyme proteins are secreted through the three steps of (1) inner membrane (cytoplasmic membrane) passage, (2) outer membrane passage, and (3) release from outer membrane (Fig. 4.11). In general, biosynthesis of enzyme proteins is initiated with precursors called *leader*, which consist of 20–30 amino acids. The precursor binds to the transfer protein with ATP and elongates the polypeptide while the formation of

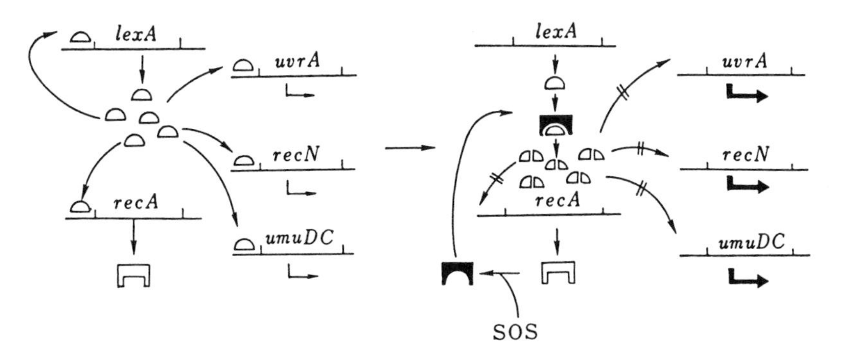


**Fig. 4.10** Schematic model of SOS regulatory system [Walker, G. C. (1985). Inducible DNA repair systems. *Ann. Rev. Biochem.* **54,** 425–457. *lexA,* resistance to UV; *recN,* recombination; *uvrA,* UV repair; *umuDC,* UV matagenesis. Reproduced, with permission, from the Annual Review of Biochemistry.]

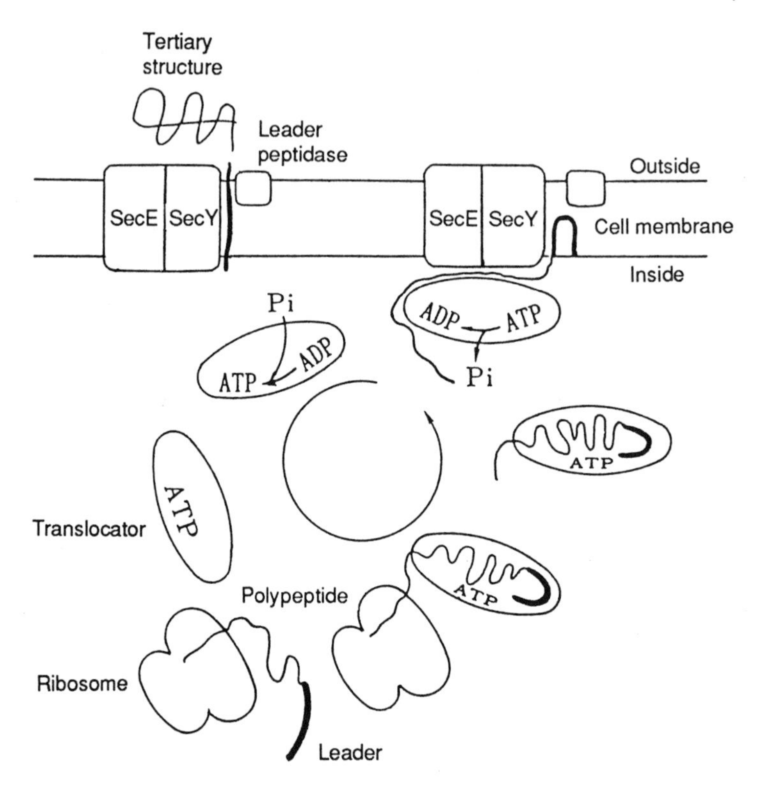


**Fig. 4.11** Schematic illustration of protein export system.

tertiary structures is prevented. The precursor binds to cytoplasmic membrane at the periplasmic side and transfers the polypeptide to periplasm by the membrane transport proteins and proton motive force. The polypeptide may have completed its elongation at this stage or may be transferred while it is being synthesized.

The polypeptide chain secreted into periplasm immediately takes the tertiary structure, and the leader peptides are digested by the peptidases located on the inner membrane, releasing the active form of enzyme molecules. In addition to the peptidase genes, several genes, such as *prlA(secY), secA, secB, secC,* and *secE,* are involved in the series of secretion mechanism. The SecY and SecE proteins are located on the inner membrane and seem to have the most important role in secretion of enzyme protein. Many plant pathogenic bacteria seem to secrete proteins in this way. Some bacteria, however, may form tertiary structures when the protein molecules pass through the outer membrane. The transfer of enzyme proteins through the inner and outer membranes is under the different regulation of chromosomal genes.

## Further Reading

Beguin, P. (1990). Molecular biology of cellulose degradation. *Ann. Rev. Microbiol.* **44,** 219–248.

Bouvet, O. M. M., Lenormand, P., and Grimont, P. A. D. (1989). Taxonomic diversity of the D-glucose oxidation pathway in the *Enterobacteriaceae. Int. J. Syst. Bacteriol.* **39,** 61–67.

Csonka, L. N. and Hanson, A. D. (1991). Prokaryotic osmoregulation: Genetics and physiology. *Ann. Rev. Microbiol.* **45,** 569–606.

Leong, J. (1986). Siderophores: their biochemistry and possible role in the biocontrol of plant pathogens. *Ann. Rev. Phytopathol.* **24,** 187–209.

Lessie, T. G., and Phibbs, P. V., Jr. (1984). Alternative pathways of carbohydrate utilization in pseudomonads. *Ann. Rev. Microbiol.* **38,** 359–387.

Miller, R. V., and Kokjohn, T. A. (1990). General microbiology of *recA:* environmental and evolutionary significance. *Ann. Rev. Microbiol.* **44,** 365–394.

Neilands, J. B. (1982). Microbial envelope proteins related to iron. *Ann. Rev. Microbiol.* **36,** 285–309.

Randall, L. L., Hardy, S. J. S., and Thom, J. R. (1987). Export of protein: a biochemical view. *Ann. Rev. Microbiol.* **41,** 507–541.

Walker, G. C. (1985). Inducible DNA repair systems. *Ann. Rev. Biochem.* **54,** 425–457.

CHAPTER
**FIVE**

# *Lysis of Bacteria*

Plant pathogenic bacteria in nature remain in a pure state only in plant tissues at the initial stage of disease development. The lesions are soon invaded by various microorganisms such as bacteria, yeasts, and fungi. Thereafter, pathogens have to live as a member of such complex microbial populations, although pathogens may be independent from others in very limited peripheral areas of lesions. The ability of pathogens to survive in such a mixed population may be significantly affected by synergism or antagonism with other organisms.

An example of synergism can be seen in *Xanthomonas campestris* pv. *oryzae,* which increases its viability and longevity in irrigation water through coexistence with blue-green algae. *Agrobacterium tumefaciens* effectively grow in soil by using $Fe^{3+}$ that is solubilized by other soil bacteria. In general, however, our knowledge on synergism between pathogens and other organisms with respect to the survival capacity of the former is very limited.

*Antagonism* is the growth inhibition of pathogens by other organisms as the result of parasitism, predation, nutritional competition, or the effects of harmful metabolites or antimicrobial substances. This chapter deals with the lysis of pathogenic bacteria mainly resulting from antagonism in a broad sense.

## 5.1 Bacteriophage

In earlier era of the 20th century, plant bacteriologists attempted to use bacteriophages (or simply phages) as a therapeutic agent for plant disease control. All of these efforts were, however, abandoned without any success. The same was also true with trials to control human and animal diseases. Currently, phages are mainly used for typing bacterial strains and for analyzing the ecological behavior of pathogenic bacteria.

### 5.1.1 MORPHOLOGICAL AND CHEMICAL COMPOSITION

Phages are morphologically classified into three categories: tadpole, spherical, and filamentous. The heads of tadpole-shaped phages are actually icosahedral and measure about 50–100 nm in diameter and the tails 15–170 × 10–20 nm. The size of spherical phages is generally 20–26 nm in diameter and that of filamentous phages is 800–1,400 nm in length and about 4–7 nm in width. The typical structure of a tadpole-shaped phage is shown in Fig. 5.1.

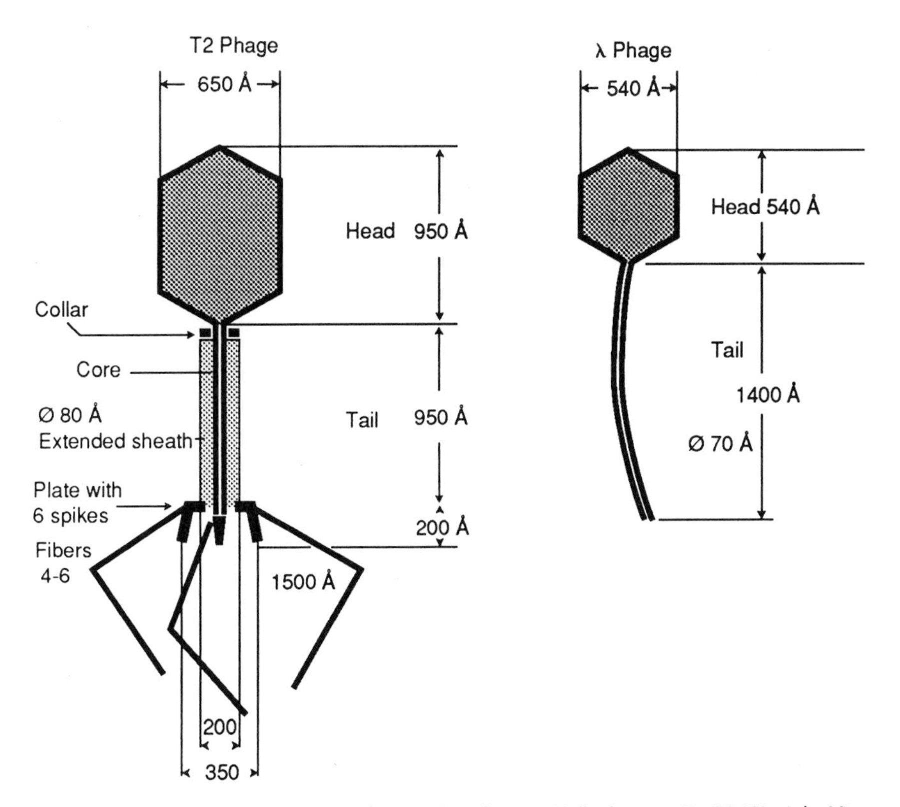


**Fig. 5.1** Shapes and dimensions of T2 and λ phages. (Kellenberger, E., (1962). *Adv. Virus Res.* 8.)

Most phages of plant pathogenic bacteria have DNA, but RNA phages are also known. The base composition of DNA varies with the phages, although most phage nucleic acids consist of the usual bases of thymine, adenine, cytosine, and guanine. The phage Xp-12 of *X. campestris* pv. *oryzae* has 5-methylcytosine instead of cytosine. Two filamentous phages, Xf and Xf2 of the same bacterium differ in length, percentages of bases in the DNA, and the amino acid composition of the coat protein. Two coat proteins of Xf2 have been isolated, one with a molecular mass of 4,740 and another of 56,000 Da. The filamentous phage Cf of *X. campestris* pv. *citri* is characterized by a single-stranded DNA, length of 1008 ± 99 nm, small and clear plaques, and a very narrow host range.

Some RNA phages are characterized by male (F) specificity, and small spherical shapes with an average diameter of 25 nm. The ds-RNA lipid-containing phage $\phi$6 of *P. syringae* pv. *phaseolicola* is about 60 nm in diameter, approximately three times the average size. The RNA molecule of this phage is unique among phages in that it has three segments in a particle.

## 5.1.2 LYTIC CYCLES

### *Infection*

The first step of phage infection is the adsorption of phage particles to receptor sites on the bacterial surface (Fig. 5.2). The collision of phage particles and bacterial cells is purely physical and is determined by the concentrations of both the phage and bacterial cells. Phage adsorption to

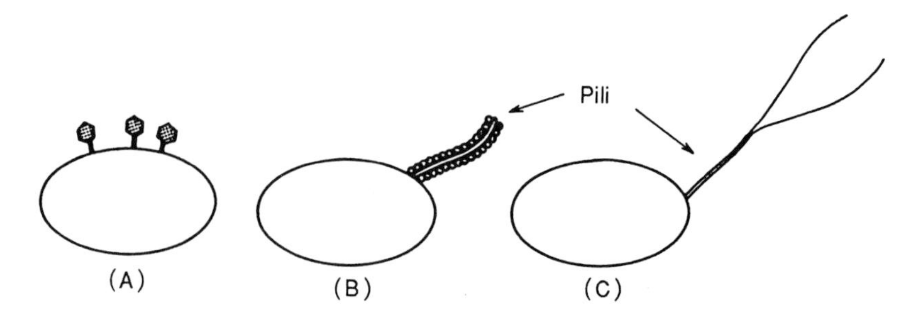


**Fig. 5.2** Schematic illustration of attachment of bacteriophage. (A) Tadpole phage; (B) F-pilus specific spherical RNA phage; (C) filmentous phage [Goto, (1981). "New Bacterial Plant Pathology." Soft Science Co. Ltd., Tokyo.]

bacterial cells does not occur at bacterial concentrations lower than $10^4$ to $10^5$ colony-forming units (CFU)/ml, so that the phage population remains constant until the bacterial population reaches that level. Phage particles adsorb with their tail to the receptors on the bacterial cell envelope. The tail fibers have the role of probing the adsorption sites and the spikes to fix phage particles to the receptors. As the lysozyme located in the tail lyses the bacterial cell wall, the tail sheath contracts so that the core with an internal hole is penetrated, injecting phage-DNA into the bacterial cell. The coat protein of the phage particles, called "ghost," remains outside the bacterial cells.

The RNA phages and filamentous phages adsorb to pili. Spherical RNA phages attach to the side of conjugative pili, which are plasmid-borne. Filamentous phages attach to the tip of pili, forming end-to-end adsorption. The pili are then depolymerized so that the phage particles come in contact with the cell surface and are taken into the bacterial cell. The *Mycoplasma* phages adsorb to the specific receptor protein on the unit membrane.

### *Reproduction*

Phage progeny in the reproduction process do not acquire infectivity until mature phage particles are assembled in the infected host cell. The noninfective form is called the *vegetative phage.* In the early stage of phage reproduction, infective phage particles cannot be recovered from bacterial cell lysates. This stage is called the *eclipse phase.* Replication of phage nucleic acid, formation of structural proteins, and biosynthesis of enzymes required for synthesis of these viral components as well as the assembly of phage particles are performed on a regular time schedule according to the phage genes. Some genes start working immediately after infection, while others are transcripted late in the viral growth cycle. For example, genes that start working immediately include enzymes necessary to switch cellular synthesis from host to viral genes and genes that are transcripted late include lysozymes needed for cell lysis.

The phage head is likely assembled through the process in which viral nucleic acid is first pushed into prohead protein coat at the portal vertax and then is pulled in as a consequence of ATP hydrolysis.

### *Release*

Bacterial cells that contain the mature phage particles usually burst to release the phage progeny into the environment. This state is termed *lysis,* which is caused by the hydrolysis of hydrogen bonds of the cell wall peptidoglycan layer due to the phage-encoded lysozyme. This enzyme is produced in the cell at the end of the latent period of lytic infection. Along with mass lysis of the infected bacterial cells, turbidity sharply drops in liquid cultures, and *plaques* are formed on agar plates (Fig. 5.3). Under certain cultural conditions, the morphology of plaques is specified according to the phage-bacteria system and can be used as the genetic marker of phages.

The lytic cycle of virulent phages normally is completed within an hour after infection (Fig. 5.4). The time-course of phage production in single bacterial cells can be demonstrated by the *one-step growth experiment.* The experiment can be accomplished either by a single-infection method or a multiple-infection method, depending on the number of phage particles to be adsorbed onto a host cell. The number of phage particles released from a single cell is called the *average burst size.*

Filamentous phages, however, generally are released from host cells at the ratio of several hundred per cell, without inducing lysis of the host. Therefore, the turbidity of phage–bacteria mixtures does not change in

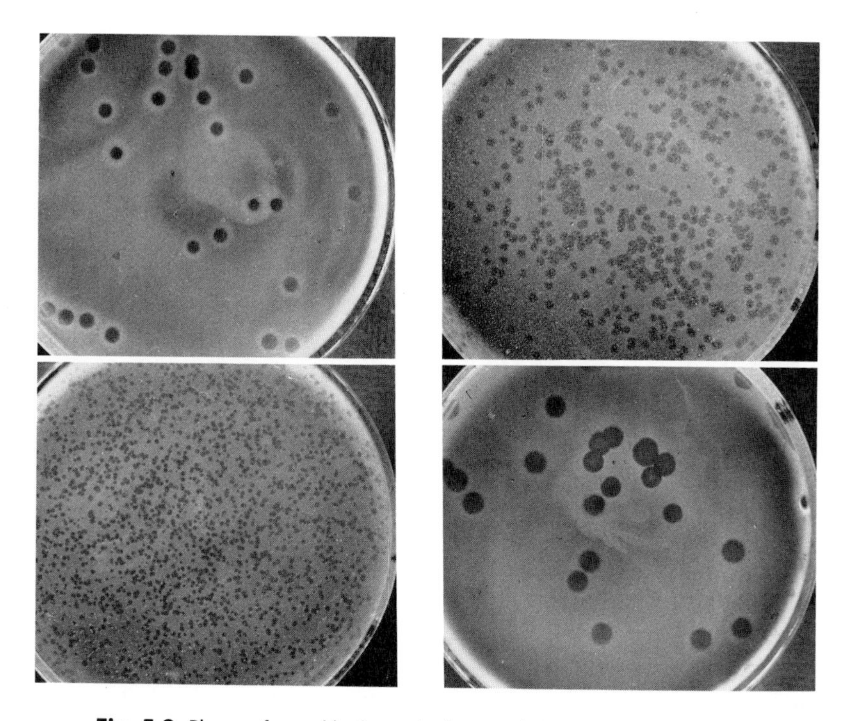

**Fig. 5.3** Plaques formed by bacteriophages of *X. campestris* pv. *oryzae.*

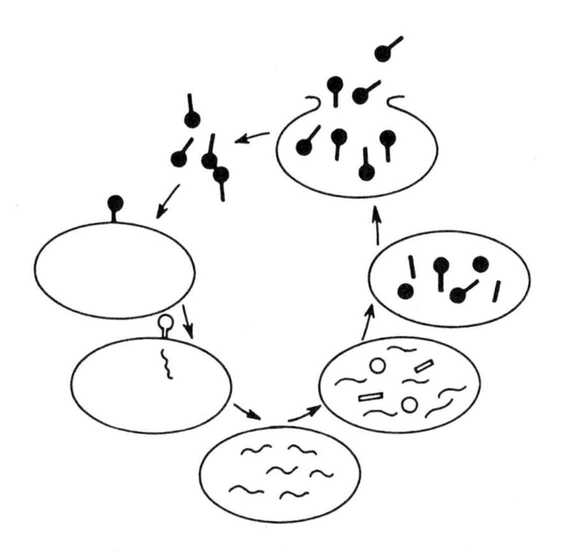

**Fig. 5.4** Lytic cycle of virulent phages. ∿, phage DNA; ●—, phage particle; ○ ▭, coat protein.

the infections due to filamentous phages. The filamentous phage of *X. campestris* pv. *citri* has unique properties in that it is integrated into the host chromosome, forming the state known as prophage.

### *Temperate Phage and Lysogeny*

Temperate phages differ from virulent phages in that phage DNA injected into the cell is integrated into the host chromosome, and thereafter it replicates synchronously with the host genome. The phage genome in this state is called *prophage,* and phages that have the capacity to undertake a prophage state are called *temperate phages.* Such relationships occurring between temperate phages and bacteria are called *lysogeny,* and the bacteria that contain prophage are called *lysogenic bacteria.*

Lysogenic bacteria successively transfer their prophage DNA to the daughter cells in cell divisions without releasing any free phages. The prophage is, however, occasionally released from the host chromosome in some host cells and initiates the lytic cycle as in virulent phages. These cells are finally lysed and liberate free phage particles. This phenomenon is called *spontaneous induction.* Thus, the cultures of lysogenic bacteria usually contain a small number of phages. This state is different from *carrier* strains, which can be subcultured to maintain a low population of

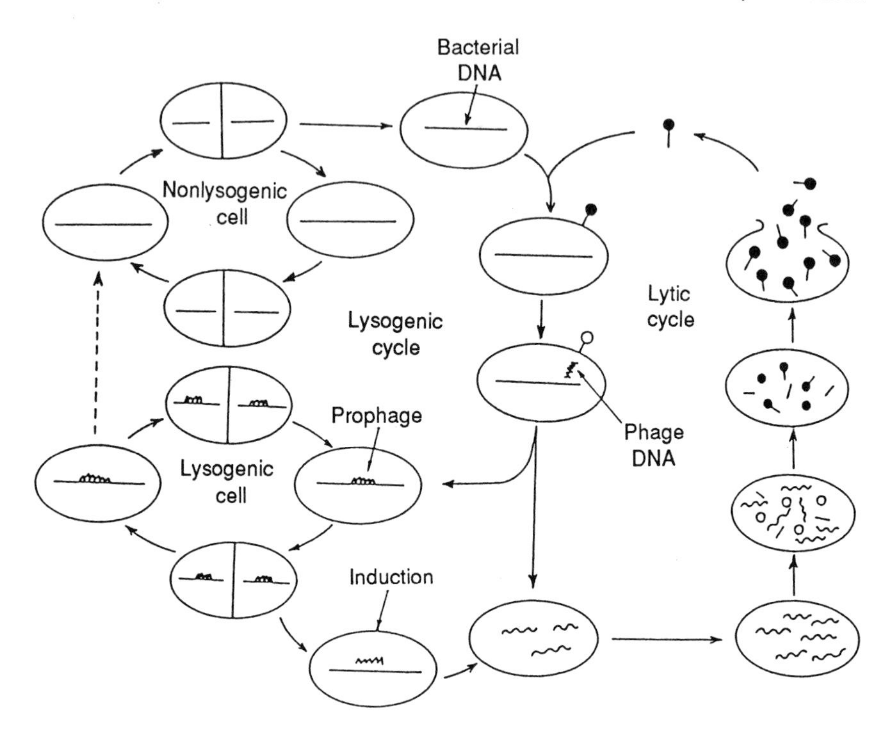


**Fig. 5.5** Schematic illustration of lysogeny.

virulent phages. Induction of prophages into productive phages can be generated by ultraviolet irradiation or mitomycin C treatment (Fig. 5.5).

Lysogenic bacteria carrying prophages become immune to the same or related phages. Lysogeny is a common feature of bacteria. It has been postulated that all bacteria can be shown to be lysogenic if adequate indicator strains are available. *X. campestris* pv. *citri* and *X. campestris* pv. *campestris* are characteristic in this respect because almost all strains are lysogenic and liberate several temperate phages, differing in inactivation temperature or host range patterns.

### 5.1.3 PHAGE MUTATIONS

Phage mutations may be detected by changes in plaque morphology, host range, UV sensitivity, growth temperature sensitivity, etc. Plaque type mutants are readily detectable on plates; the size of plaques, which reflects rapidity of lysis, and different clarity, which reflects heterogeneity in lysis, are common features of mutant plaques.

The average number of plaques per plate may vary when an equal amount of a phage sample is plated with different bacterial strains. The variation of plaque counts is due to differences in the *efficiency of plating* (EOP) of each strain. The efficiency of plating is thus defined as the ratio of plaque count on a given strain to that on a standard or homologous strain. When phage is propagated on strains with a low EOP, the phage progeny acquires an increased EOP on that particular strain. However, the altered efficiency of the phage returns to the original efficiency after a single passage through the homologous strain. This variation in EOP by passage of a phage through different bacterial strains is referred to as *host-controlled modification.*

### 5.1.4 USE OF BACTERIOPHAGE IN BACTERIAL PLANT PATHOLOGY

#### *Identification of Bacteria by Phage Sensitivity or Phage-Typing*

Successful application of phage technique in identification of bacteria requires that the phages have high specificity for species or the group of strains to be examined. The host specificity of phages varies greatly, ranging from strain specific phages to nonspecific polyvirulent phages which can attack bacteria belonging to different taxa such as species, genera, or even families. In phage-typing, therefore, the host specificity of the phages to be employed should be carefully checked beforehand.

The host specificity of phages may vary depending on the source of isolation. In general, phage strains obtained from diseased plant materials have a high specificity or narrow host range and form large plaques, whereas those from soil or sewage have a broad host range for many different bacteria and form small plaques. The host ranges of phages can also vary with the bacteria. For example, phages of *P. solanacearum*, and *E. carotovora* subsp. *carotovora* have narrow host ranges, attacking only a limited number of strains. Therefore, a number of phage strains are necessary to phage-type many bacterial strains of different geographic origins. In contrast, the phages of *X. campestris* pv. *citri* apparently have a broad host range within this pathovar, i.e., only two phages, Cp1 and Cp2, can infect some 98% of the bacterial strains isolated in Japan.

When phage-susceptibility correlates with pathogenicity or certain biochemical properties, these phages can be used as the markers for quick diagnosis of pathological races or biovars. Strains of *P. solanacearum,* which are characterized by lactose oxidation and pathogenicity to tobacco, can be distinguished by certain phages. Also, a lactose-utilizing strain of *X. campestris* pv. *malvacearum* and a mannitol-utilizing strain of *X. campestris* pv. *citri* can be quickly identified by phages. The plum strain and two races of the cherry strain of *P. syringae* pv. *morsprunorum,* which are distinguished only by their ability to colonize their homologous hosts from small inocula, can be readily identified in the laboratory by their reaction to certain phages. Correlations have also been observed between phage sensitivity and host-plant origin of bacterial strains in *E. chrysanthemi* and *X. campestris* pv. *citri.*

### *Use in Ecological Studies*

The use of phages in ecological studies of bacterial plant pathogens has the advantage that the pathogenic bacteria can be selectively detected from a complex population of microorganisms in nature. The principle is that phages added to samples harboring their host bacterium greatly increase in number in a short period of time. The propagated phages can then be readily separated from contaminating microorganisms either by centrifugation or by chloroform treatment. However, the phage technique may not be effective where the population of host bacterium in the samples is not at a sufficient level. Phage–bacteria interactions take place at bacterial population levels higher than $10^4$ to $10^5$ CFU/ml. Therefore, selective enrichment of the host bacterium may be required when the number of bacteria is lower than this level. Successful enrichment is actually dependent on the relative densities and/or growth rates of the host bacteria and saprophytes.

The phage technique has been effectively used to detect infected seeds in kidney bean, soybean, pea, and rice. For this purpose, the presence of pathogenic bacteria can be directly detected from seeds or indirectly from young seedlings. In bacterial canker of tomato (*Clavibacter michiganensis* subsp. *michiganensis*), the latter is more effective for detecting infected/infested seeds taken from diseased fruits. A forecasting system of bacterial leaf blight of rice was devised from the population dynamics of phages which appear in irrigation water in advance of disease appearance.

### *Use in Studies of Pathogenicity*

Phage-resistant mutation often pleiotropically occurs with a mutation in pathogenicity. Such pleiotropic mutations have been reported in various bacteria such as *X. campestris* pv. *oryzae, X. campestris* pv. *pruni, P. syringae* pv. *morsprunorum,* and *E. amylovora.* In *X. campestris* pv. *oryzae,* mutation in phage susceptibility occurs simultaneously with mutations in colony morphology, EPS biosynthesis, and virulence. In *X. campestris* pv. *citri,* colony forms and patterns of susceptibility to other virulent and temperate phages are altered through lysogenization by different temperate phages.

### *Use for Disease Control*

Practical use of phages for control of bacterial plant diseases in the field has not been successful. When some control was achieved, this was brought about by inoculation with a mixture of the phage and the bacterium, or by plant or seed treatment with phage before challenge with bacteria. In practice, however, the pathogenic bacteria in plant tissues are in a dense mass and frequently surrounded by abundant extracellular polysaccharides, which prevent effective adsorption of phage particles. Another obstacle is the complexity of phage–bacterium interrelationships in nature, due to the diversity of bacterial strains that differ in phage susceptibility.

## 5.2 Bacteriocins

Bacteria produce proteinaceous antagonistic substances that are lethal to other strains of the bacteria. These substances are called *bacteriocins,* and the ability of bacteria to produce bacteriocins is called *baceriocinogeny* or *bacteriocinogenicity.* Other antagonistic substances that bacteria produce include bacteriophages, antibiotics, lysozymes, and metabolites. Bacteriocins differ from these substances in the following ways: (1) the antimicrobial spectrum is very narrow, being limited to the strains of the same or closely related species; (2) the effect to sensitive bacteria is lethal; (3) in sensitive bacteria, various metabolic pathways are inhibited; (4) bacteriocins are specifically adsorbed onto receptor sites on the bacterial cell surface; (5) bacteriocinogeny is usually attributed to plasmids; (6) a bacterium is immune to its own bacteriocins; and (7) production of bacteriocins is a lethal process for the producing bacterium.

Bacteriocinogenic strains are somewhat similar to lysogenic strains in the following ways: (1) bacteriocin-producing activity may be found only in a portion of the bacterial population; (2) the majority of cells of a culture may be induced with mitomycin C or ultraviolet irradiation to

**Table 5.1**
Comparison of Bacteriocin and Temperate Phage

| Characteristics | Bacteriocin | Temperate phage |
|---|---|---|
| Multiplication | — | + |
| Formation of scattered plaques | — | + |
| Sensitivity to trypsin | S(R)[a] | R |
| Sensitivity to ultraviolet ray | R | R |
| Heat stability | R(S)[a] | S |

[a] ( ) Reaction of R-type bacteriocin.

produce bacteriocin; and (3) bacteriocin production is a lethal process for the producer. They differ in the several characteristics of their products, bacteriocins and temperate phages (Table 5.1). Characteristics 1 and 2 are particularly useful for distinguishing these two antagonistic substances.

The first record of bacteriocins and their use in plant pathogenic bacteria dates to 1954, when Okabe detected bacteriocinogenic strains of *P. solanacearum* and used them to identify pathogenic groups of the bacterium. Since then, bacterial strains that produce bacteriocins have been detected in many phytopathogenic bacteria examined.

## 5.2.1 DETECTION OF BACTERIOCINS

In general, bacteriocin-producing strains are detected at various ratios when as many strains as possible are tested against each other. Production of bacteriocins may be affected at various degrees by cultural conditions such as the components of media, culture age, colony forms, and growth temperatures. Therefore, different methods should be tried for successful detection of bacteriocins.

Usually, bacteriocins can be detected on agar plates at high frequency. To detect them in liquid culture media, the test bacterium is grown in a nutrient medium for 1–2 days and then centrifuged. The supernatants thus obtained are tested for bacteriocin activity on agar plates. Alternatively, the cells are grown overnight in the same medium after being induced by mitomycin C treatment or UV irradiation and then centrifuged.

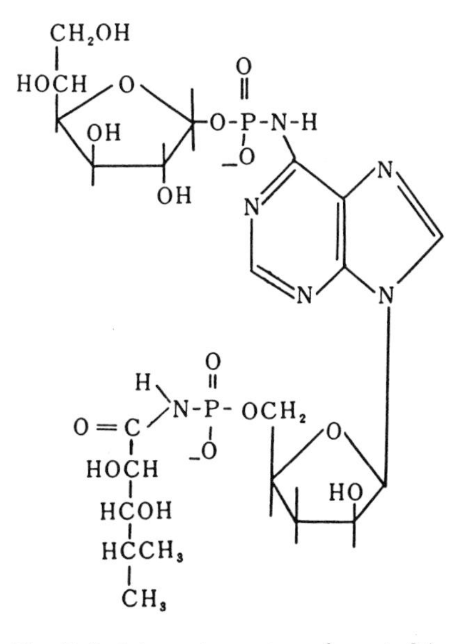


**Fig. 5.6** Schematic structure of agroin 84.

## 5.2.2 PROPERTIES OF BACTERIOCINS

Bacteriocins kill or inhibit the growth of sensitive bacteria without themselves being replicated. Although the majority of bacteriocins are proteinaceous, the bacteriocin *agrocin 84* produced by a soil bacterium, *Agrobacterium radiobacter* strain K84, is a substituted nucleotide adenosine (Fig. 5.6).

In general, bacteriocins that are inactivated by trypsin treatment are heat stable and nonsedimentable by ultracentrifugation. These are called S-type bacteriocins. In contrast, those that are resistant to trypsin treatment and heat liable are called R-type bacteriocins. These bacteriocins are proteins with high molecular weights and have morphological characteristics similar to phage tails so that they are considered to be a kind of defective phage.

## 5.2.3 MODE OF ACTION OF BACTERIOCINS

Bacteriocins adsorbed to the receptor on the surface of bacterial cells eliminate electric charges of inner membrane stopping the biosynthesis of adenosine triphosphate. Consequently, the biosynthesis of DNA,

RNA, protein, and phospholipids is inhibited, stopping growth of bacterial cells and eventually causing their death. In bacteria with phospholipases in the outer membrane, adsorption of bacteriocin disorganizes the regulation systems of the enzymes, inducing degradation of the cell membrane and lysis of bacterial cells. The lethal effect of many bacteriocins is effectively brought about by a single-hit process, i.e., one particle of bacteriocin is sufficient to kill one bacterium.

Bacteriocin producers are generally immune to the bacteriocins that they produce, and the immunity is highly specific for the homologous bacteriocins. However, the immune cells have the capacity to adsorb bacteriocins so that the immunity is down when bacteriocins attach to the immune cells at high concentrations.

In contrast, bacteriocin-resistant mutants, which often develop in cultures, do not have the ability to adsorb bacteriocins because of the alteration or absence of specific receptor sites. Therefore, resistance is quite stable and complete.

### 5.2.4 BACTERIOCIN PLASMIDS

The genetic determinants of agrocin 84 of *A. radiobacter* strain K84, syringacin of *P. syringae* pv. *syringae,* and bacteriocin of *P. cepacia* have been confirmed on plasmids. Other bacteriocins of plant pathogenic bacteria are also assumed to be plasmid-borne, although no confirmative evidence is available yet. The bacteriocinogenic plasmids are generally transferred independently of the chromosome to other strains, conferring on them bacteriocin-producing ability and specific immunity to the bacteriocin.

### 5.2.5 BACTERIOCINS OF PLANT PATHOGENIC BACTERIA

Bacteriocins have been detected for many plant pathogenic bacteria. However, those that have been isolated in a chemically pure state and characterized are limited to a few bacteriocins such as agrocin 84, syringacin 4A, a bacteriocin of *P. solanacearum,* and carotovoricin *Er* of *E. carotovora* subsp. *carotovora.*

*A. radiobacter* strain K84 carries a large plasmid and a small plasmid, and the latter (45 kilobase pairs) encodes both determinants for production of and immunity to agrocin 84. Susceptibility of *A. tumefaciens* to

**Table 5.2**
Functions of Plasmids of *Agrobacterium radiobacter* Strain K84 and *A. tumefaciens*

| Bacteria | Plasmid | Function |
|---|---|---|
| *A. radiobacter* K84 | Small | Production of agrocin 84; immunity to agrocin 84 |
| | Large (Ti) | Catabolism of nopaline; conjugation |
| *A. tumefaciens* | Large (Ti) | Pathogenicity; sensivity to agrocin 84; catabolism of opines; conjugation; replication; exclusion of Ap1 phage |

agrocin 84 is determined by the Ti plasmid, which confers pathogenicity (Table 5.2). These three plasmids are transmissible by conjugation.

Carotovoricin *Er* produced by *E. carotovora* subsp. *carotovora* exhibits a similar shape to that of phage tail, consisting of the core, contractile sheath, and tail fibers. This bacteriocin is coinduced together with pectin lyase (PNL) by DNA-injuring treatment with mitomycin C, nalidixic acid, UV irradiation etc. Thus, the *recA*-dependent SOS reaction seems to be involved in the induction of carotovoricin *Er*.

Agrocin 84 is an unusual substituted adenine nucleotide and seems to be transported into sensitive strains of *A. tumefaciens* by one or more plasmid-encoded binding proteins located on the cell envelope. Syringacin 4A is trypsin resistant and heat sensitive with a phage-tail shape. In contrast, a bacteriocin produced by *P. solanacearum* has been described as the protein of low molecular weight (65,000), heat labile, and sensitive to trypsin.

### 5.2.6 USE OF BACTERIOCIN IN BACTERIAL PLANT PATHOLOGY

Bacteria can be classified into groups by the production of or sensitivity to bacteriocins. These procedures are called *bacteriocin-typing* and are used for epidemiological purposes. As an epidemiological marker of bacteria, bacteriocin-typing is different from phage-typing in that productivity is generally more stable and reliable than sensitivity.

Bacteriocin-typing of phytopathogenic bacteria has been attempted in *P. solanacearum, E. carotovora* subsp. *carotovora, A. tumefaciens,* and others. Bacteriocin production patterns have also been used to differentiate *P. cepacia* strains of clinical and plant origin.

*A. radiobacter* strain K84, which produces bacteriocin agrocin 84, has been used successfully as a biological control agent of crown gall disease. Agrocin 84 has been postulated to play a role in the successful use of this strain in biological control, but competition with the pathogen for attachment sites on the host cell wall and for nutrients is also considered to be involved in the mechanisms.

## 5.3 Bdellovibrio

The bdellovibrios are vibrioid bacteria with the unique property of being the parasite of other bacteria. Literally, *bdello-* is derived from the Greek noun *bdella,* meaning "leech," and *bdellovibrio* means "a leech-like vibrio" (Fig. 5.7). Bdellovibrios are widely distributed in nature and frequently detected in soil, sewage, and water from different sources. Their population in soil may reach $1 \times 10^3$ to $7 \times 10^4$ CFU/g.

In this bacteria, the parasitic, or host-dependent, form and saprophytic, or host-independent, form may be differentiated. The parasitic form is able to multiply either in living bacterial cells or in a suspension of heat-killed cells of a congenial host or other bacteria but not in laboratory nutrient media. In contrast, the saprophytic form is unable to multiply in living bacterial cells but can propagate in suspensions of heat-killed cells or in nutrient media.

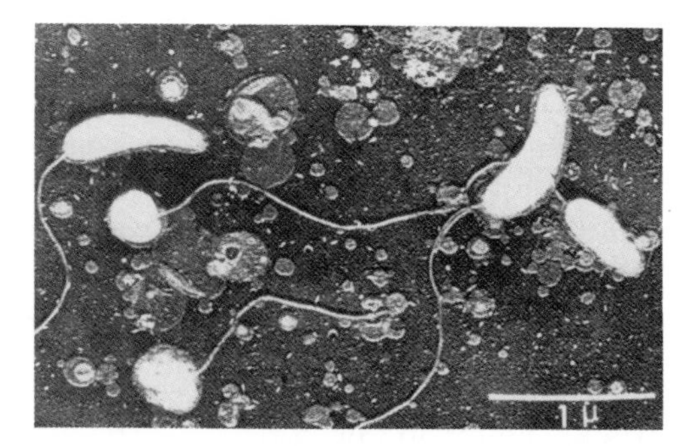


**Fig. 5.7** Bdellovibrios attacking *Xanthomonas campestris* pv. *oryzae* (Courtesy of Dr. S. Wakimoto.)

### 5.3.1 MORPHOLOGY AND CHEMICAL COMPOSITION

Bdellovibrios are taxonomically related to *Aquaspirillum* or *Campylobacter*. Cells are comma shaped gram-negative rods, 0.2–0.5 × 0.5–1.4 μm in size. They are strictly aerobic, having a respiratory type of metabolism and are actively motile by a sheathed flagellum, i.e., they swim 100 times cell length per second. Nutrition is chemoorganotrophic. Their life cycle is biphasic, consisting of a nongrowing predatory phase and an intracellular reproductive phase. Intracellular forms are either spiral shaped nonflagellated developing cells or flagellated matured cells. The mol% G + C of the DNA is 42–51. Protease activities are high, but carbohydrates are not utilized. Metabolism depends on the TCA cycle. The saprophytic form dissociates from the parasitic one at the ratio of $10^{-6}$–$10^{-8}$. The saprophytic form is characterized by high sensitivity to photodynamic action and low viability in natural conditions. The contents of protein and DNA are 60–65% of dry weight. Phospholipids contain phosphatidylethanolamine and phosphatidylserine as well as phosphonosphingolipids that are rarely found in bacteria. On the basis of GC ratio and homology of DNA, three species of *Bdellovibrio bacteriovorus, Bd. starii,* and *Bd. stolpii* have been distinguished. In addition, some unnamed marine strains are known.

### 5.3.2 HOST RANGE

Most bdellovibrios exhibit lytic activity to gram-negative bacteria of *Pseudomonadaceae* and *Enterobacteriaceae*. The bacteria attacked by bdellobibrios include plant pathogenic bacteria of *Agrobacterium, Erwinia, Pseudomonas,* and *Xanthomonas,* as well as *Rhizobium*. There are no bdellovibrios reported for the nonfluorescent pseudomonads such as *P. solanacearum* and *P. caryophylli,* nor for any gram-positive bacteria.

### 5.3.3 PHYSIOLOGY

Parasitic bdellovibrios can multiply on heat-killed cells of the host bacteria in the presence of $Ca^{2+}$ and $Mg^{2+}$. This indicates that the host bacteria can provide all the nutrients necessary for bdellovibrio growth. The saprophytic form is a mutant that has acquired the ability to synthesize certain nutritional elements for itself.

Bdellovibrios cannot metabolize free sugars. Activities of glycolytic enzymes are very low except for phosphoglucoisomerase and phosphoglyceraldehyde dehydrogenase, implying that these enzymes could be involved in the biosynthetic processes rather than energy generation. Bdellovibrios are unique in that 60–80% of adenosine triphosphate trophosphatase (ATPase) is detected in the soluble fraction, and ATPase in membrane particles is not inhibited by dicyclohexylcarbodiimide. Bdellovibrios incorporate into their outer membrane the intact outer-membrane proteins (OmpF) derived from the host bacterium.

About 20–40% of the ribonucleotides of the host cells are degraded into nucleoside-1-phosphate and incorporated by bdellovibrios. The majority of these intracellular nucleotides are used for biosynthesis of DNA and RNA. A part of the nucleotides is further converted to ribose-1-phosphate and free nucleic acid bases. Ribose-1-phosphate is degraded through the pentose-phosphate pathway and used for energy-generating metabolism and for biosynthesis of cellular components other than nucleic acids. The released nucleic acid bases benefit for the growth of bdellovibrios by strengthening the buffering effect of medium.

Bdellovibrios utilize amino acids of the host bacteria as major energy sources. Fatty acids are derived largely from the lipids of the host bacterium with little or no alteration. More than 50% of the carbon of the host cell is converted into cell components of bdellovibrios.

## 5.3.4 HOST–PARASITE INTERACTION

Parasitic interactions between bdellovibrios and host bacteria consist of the four steps of attachment, penetration, multiplication, and release of progeny cells. The series of steps is completed in 3 to 3.5 hr (Fig. 5.8).

### *Attachment*

Primary contact and subsequent attachment of bdellovibrios to their host bacteria are due to purely physical collisions resulting essentially from active motility of the bdellovibrios. The efficiency of attachment does not differ with colony forms or cell wall compositions of the host bacteria. Attached bdellovibrios rotate about 100 times per second along the long axis by the propelling action of flagellum. Attachment of bdellovibrios causes immediate stopping of the host cell motility, regulation of membrane permeability, respiration, and biosynthesis of nucleic acids and proteins.

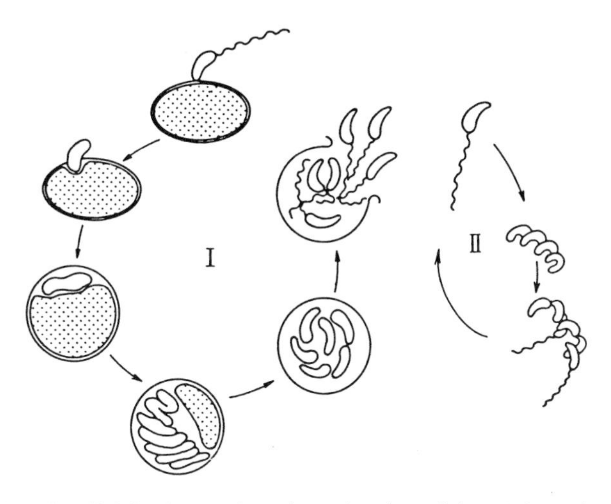

**Fig. 5.8** Life cycle of bdellovibrios. I, host-dependent form; II, host-independent form [Goto, (1981). "New Bacterial Plant Pathology" Soft Science Co. Ltd., Tokyo.]

### *Penetration*

Bdellovibrios physically penetrate the outer membrane of the cell envelope. Pore formation through the peptidoglycan is due to the action of two enzymes, glycanase and peptidase. Where the pore is formed, the unflagellated end of bdellovibrio is attached to the inner membrane of the host cell until penetration is completed. The peptidoglycan layer and inner membrane of the host cell are separated around the pore site by a differential expansion so that the bdellovibrio cell finally enters the expanded host periplasm. The pore is entirely cured when penetration is completed, leaving a slight trace.

The step of penetration takes 5 to 60 min. When penetration is complete, the rod-shaped host cell is converted into a spherical body, or spheroplast. The bdellovibrio cell settles in the periplasmic space and does not enter into the cytoplasm of the host cell. The cytoplasm of the host cell is broken down by proteases produced by bdellovibrio and used as nutrients for its growth.

### *Growth and Multiplication*

At the step of bdellovibrio growth, the role of the host cell wall is no more than a container for the bdellovibrio cell and nutrients, preventing their release into the medium. Growth of the bdellovibrio cell occurs longitudinally and forms a spiral cell, which is several times longer than the original one. The filamentous elongation is mediated by regulatory signals that are derived from the host cell. The long spiral-shaped cell divides into several daughter cells, and each progeny cell develops flagellum, becoming morphologically identical to the parent cell.

#### *Release*

Amino sugars and diaminopimelic acids of the host cell wall peptidoglycan are dissolved by the lytic enzymes, glycanase and peptidase, resulting in degradation of the cell wall and release of matured progeny cells of bdellovibrio. The average number of progeny cells per host cell, average burst size, depends on the capacity of host cells and ranges from 6 to 30. The average burst size with bacterial plant pathogens has been reported to be 6–10 cells.

### 5.3.5 USE OF BDELLOVIBRIOS IN BACTERIAL PLANT PATHOLOGY

There are limited studies on the biological control of bacterial plant diseases with bdellovibrios. In bacterial blight of soybean caused by *P. syringae* pv. *glycinea,* disease development was significantly inhibited by inoculation with a mixture of the pathogen and bdellovibrios at a ratio of 1 : 9 or 1 : 99. However, practical control of plant diseases with bdellovibrios has not been successful as in the case of bacteriophages. The effectiveness largely depends on many unknown factors, e.g., viability and longevity of predators as well as efficiency of predator–prey interactions in and on plants.

## 5.4 Nonobligate Bacterial Predator

Some soil-inhabiting bacteria attack other bacterial cells as bacterial predators in the absence of nutrients such as amino acids and proteins, although they can independently grow where such nutrients are sufficiently supplied. These unique bacteria are called nonobligate bacterial predators. The gram-positive predators generally inhabit nutritionally rich soil, attack a relatively narrow range of bacteria, and are even attacked by other gram-negative predator bacteria. In contrast, gram-negative predators attack a range of both gram-positive and gram-negative bacteria in nutritionally poor soil, having a dominant role in the control of bacterial populations in soil.

*Agromyces ramosus, Ensifer adherens, Cupriavidus necator,* and *Streptomyces* spp. are known as nonobligate bacterial predators at present. Nothing is known, however, about the interactions between these predators and plant pathogenic bacteria in soil.

## 5.5 Protozoa

Protozoa are nonphotosynthetic, motile, unicellular, and eucaryotic protists widely distributed in water and soil. They vary in size, ranging from several micrometers to several millimeters, although the majority of species are microscopic. Protozoa are either naked at the surface or enveloped by a cytoplasmic membrane and contain colloidal cytoplasm which often includes undigested food particles such as algae or bacteria. The cells usually contain one or two nuclei, and in some cases, several. These nuclei often consist of larger, polyploid macronucleus, which are necessary for normal cell division, and a small diploid micronucleus necessary for sexual reproduction.

Protozoa are subdivided into four groups by locomotor organelles such as flagella, pseudopodia, or cilia. Each group has an ameboid stage in the life cycle. Although protozoa are distributed extensively in humid soil enriched with organic matter, they form only a small part of the microbial population of soil. Some protozoa predate bacteria by phagocytosis. The ability to reduce the number of other bacteria by predation in natural soil is limited. Typical periodic fluctuations of populations resulting from the food chain are observed between protozoa and prey bacteria in the pure culture system. Under natural conditions, however, the predatory effect of protozoa, particularly on plant pathogenic bacteria, is difficult to assess because of the complex prey organisms involved and entirely different environments surrounding the predator–prey system.

The interactions between protozoa and plant pathogenic bacteria have been investigated with *P. syringae* pv. *mori, X. campestris* pv. *citri,* and *X. campestris* pv. *campestris.* A protozoa isolated from the culture of *P. syringae* pv. *mori* attacked several other gram-negative plant pathogenic bacteria in addition to the pathogen. In *X. campestris* pv. *campestris,* the bacterial population added in soil quickly declined to the level of $10^5$ CFU/g in contrast to the rapid increase of protozoa. When multiplication of protozoa was inhibited by actidione, the population of the pathogen remained at a high level, suggesting that its quick decline resulted from the predation of protozoa.

## 5.6 Autolysis

When bacteria are placed in unsuitable conditions that drastically disturb their metabolism, spontaneous lysis may occur by their own intracellular glycanase and peptidase which are found in any bacteria synthesizing peptidoglycan. This phenomenon is called *autolysis.* The autolytic enzymes thus produced may cause lysis of other bacteria in close proximity from outside.

Bacterial cells at the exponential phase are most sensitive to autolysis, but those in the stationary phase are tolerant. This may originate in the perfectly balanced and well-controlled system of peptidoglycan-hydrolyzing and peptidoglycan-synthesizing enzymes occurring in the growth period. In such a growth phase, autolysis may be induced by the following environmental factors: (1) sudden removal of oxygen from strictly aerobic bacteria, (2) sudden supply of oxygen to strictly anaerobic bacteria, (3) sudden cooling down to 0°C of bacteria growing at optimum temperature, (4) exposure of psychrophilic bacteria to high temperature, (5) cultivation under nutritional conditions with D-methionine, D-alanine, D-galactose, or glucose depletion (Stolp and Starr, 1965). Autolysis occurring in environments of osmolarity equivalent to that of bacterial cells results in the formation of protoplasts and spheroplasts (L-phase).

## Further Reading

Brock, T. D. (1966). "Principles of Microbial Ecology." Prentice-Hall, Inc., Englewood Cliffs, New Jersey.

Makkar, N. S., and Casida, L. E., Jr. (1987). *Cupriavidus necator* gen. nov., sp. nov.: a nonobligate bacterial predator of bacteria in soil. *Int. J. Syst. Bacteriol.* **37,** 323–326.

Reeves, P. (1972). "The Bacteriocins." Springer-Verlag, Berlin, Heidelberg, New York.

Starr, M. P., and Seidler, R. J. (1971). The Bdellovibrios. *Ann. Rev. Microbiol.* **25,** 649–678.
Stolp, H., and Starr, M. P. (1965). Bacteriolysis. *Ann. Rev. Microbiol.* **19,** 79–104.
Watson, J. D. (1975). "Molecular Biology of the Gene," 3rd ed. W. A. Benjamin, Inc., Menlo Park, California.
Zinder, N. D., and Horiuchi, K. (1985). Multiregulatory element of filamentous bacteriophages. *Microbiol. Rev.* **49,** 101–106.

# CHAPTER SIX

# *Genetics*

Bacteria have a great capacity to adapt to diverse environments through mutation, gene transfer, or metabolic regulations. In this respect, plasmids are particularly important because they transfer themselves or mediate transfer of chromosomal DNA to other bacteria, altering their genetic properties. This chapter deals with the gene transfer, mutation, and genetic studies of some phenotypic characters. In addition, some basic techniques of gene manipulation are described to help understanding of the approaches for genetic analysis as well as biotechnologies in plant pathology. The genetic aspects of enzyme synthesis and regulation were discussed in Chapter 4. Gene expression in relation to pathogenesis will be described in Chapter 8.

## 6.1 Gene Transfer

Genetic materials are transferred from one bacterium to another by transformation, transduction, and conjugation, resulting in genetic recombination in the recipient cell which acquires new characteristics. These mechanisms of gene transfer are an important background of the techniques for gene manipulation.

### 6.1.1 CONJUGATION

The donor cell attaches to the recipient cell, usually with specific appendages called *sex pili,* and transfers its genes by direct cell-to-cell contact. In *Escherichia coli* strain K-12, the ability of the donor strain to produce conjugative pili is determined by the F (fertility) plasmid. This plasmid can be transferred from $F^+$ to $F^-$ cells through DNA replication and converts the latter to $F^+$ by its own $F^+$ determinant. In gram-negative bacteria, a pilus of the $F^+$ cell attaches to the outer membrane protein A (OmpA) of the $F^-$ cell and retracts, resulting in the wall-to-wall contact of both cells. The DNA is transferred through a transmembrane pore that is subsequently formed between the juxtaposed donor and recipient cell envelopes. In gram-positive bacteria, conjugation can occur in the absence of pili.

The process of plasmid transfer during conjugation was well documented in *E. coli*. Transfer of F-plasmid DNA is initiated by formation of a nick at *ori* or origin of transfer by the *traYZ* endonuclease. The termini of the single strand DNA thus formed are attached to cell membrane of the pore and transferred in the 5′ to 3′ direction as shown in Fig. 6.1.

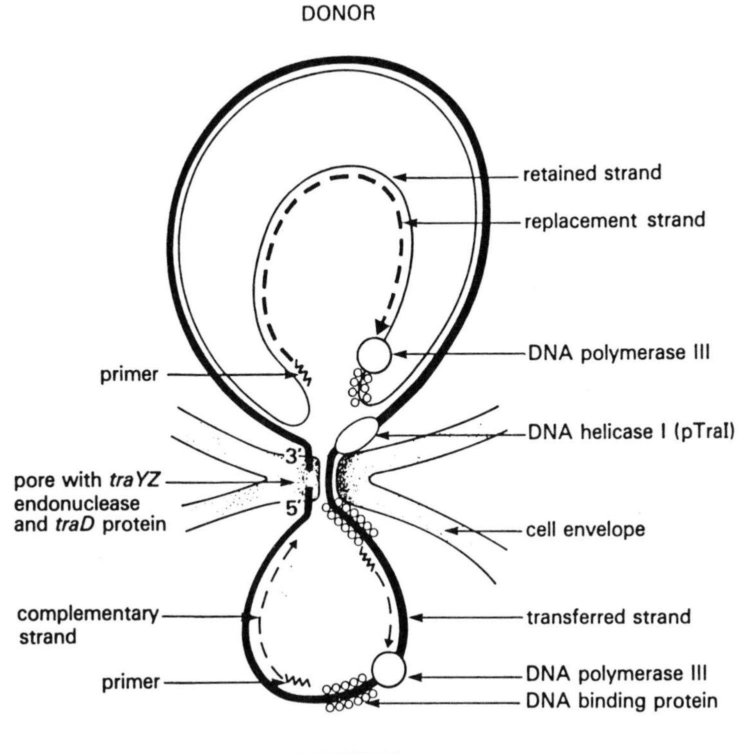


**Fig. 6.1** Schematic model of conjugative transfer of F-plasmid [From Willetts, N. and Wilkins, B. (1984) Processing of plasmid DNA during bacterial conjugation. *Microbiol. Rev.* **48,** 24–41.]

Transferring DNA is protected from winding by helicase and adenosine triphosphatase and from endonucleases by single-strand DNA binding coat protein. As a single strand of plasmid DNA transfers to the recipient cell, a replacement strand is synthesized in the donor cell and a complementary strand in the recipient, forming double strand DNA.

Although a high frequency of plasmid transfer occurs in conjugation, transfer of chromosomal DNA also takes place in a low frequency of $10^{-4}$ to $10^{-7}$. This results from mutant cells, which occur in a population of $F^+$ cells and have the ability to achieve a high frequency of recombination. They are called Hfr (high frequency of recombination) and arise by the integration of the F plasmid into chromosome by crossing over. When Hfr cells are mixed with an excess number of $F^-$ cells, every cell of the former binds to cells of the latter, initiating transfer of chromosome DNA. In conjugation with Hfr, the transfer of chromosome takes place in such a way that replication initiates at *oriT* of the integrated plasmid and terminates at the $F^+$ (sex) determinant.

There is another form of F plasmid called *F prime* (F′) derived from Hfr. When F is released from the Hfr chromosome, it may contain within its circular structure a segment of host chromosome through crossover in the region of exceptional pairing. The F′ plasmid has the property of autonomous replication and can integrate into the chromosome and achieve recombination at high frequency. The well-known plasmid F′lac$^+$ carries chromosomal genes which function in lactose fermentation.

The interrupted mating experiment is an effective approach for determining genetic maps of bacteria. In this experiment, the Hfr strain with such genetic markers as nutritional requirements, drug resistance, and phage susceptibility is grown together with an $F^-$ strain, and the samples are removed from the conjugation mixture at certain time intervals. They are violently agitated to interrupt conjugation and plated on selective media to determine the genes transferred before the interruption. This experiment gives information on the time needed for transfer of particular genes, the frequencies of recombinants and crossing over, and elucidates the arrangement and distance of the marker genes on a chromosome allowing gene mapping.

In *E. amylovora*, *cys*$^+$ and *ser*$^+$ transferred as proximal markers during the first 15 min, *pro*$^+$ started entering at about 75 min, and *lac*$^+$ entered toward the end of the 3 hr mating period, with *ilv*$^+$ as the distal marker. In *E. chrysanthemi*, an Hfr strain with chromosomal integration of an *F′lac*$^+$ plasmid transferred with the linked loci of *leu*$^+$ and *thr*$^+$ as the proximal markers and *lac*$^+$ as the distal marker in the following sequence: origin . . .*leu* . . .*thr* . . .*ade* . . .*lys* . . .*mcu* (multiple carbohydrate utilization) . . .*pel* (pectic acid transeliminase) . . .*his* . . .*try* . . .*gal* . . .*gtr* (utilization of galacturonate) . . .*lac* . . .F.

The F plasmid is disadvantaged for developing a general chromosomal transfer system for plant pathogenic bacteria because its host range is limited to *Enterobacteriaceae.* Instead, the broad host range IncP group plasmids, such as R68, R68.45, PR1, RK2, and FP2, are useful because they have been shown to facilitate chromosomal mobilization in a number of Gram-negative bacteria, including *Pseudomonas, Xanthomonas, Agrobacterium, and Erwinia.* The ability of drug-resistant plasmids to mobilize chromosomes is generally lower than that of F plasmid; it may be improved by the insertion of transposons such as Tn5, Tn10, and Tn501. On the other hand, some drug-resistant plasmids carry $fi^+$ (fertility inhibition) factor, preventing conjugation.

Recently developed new technologies for gene manipulation provide entirely different approaches for gene mapping of bacteria. These technologies include the pulsed field gel electrophoresis (PFG) for separation of large DNA fragments, construction of DNA hybridization probes, restriction enzyme site mapping technique, polymerase chain reaction (PCR), and computational methods for DNA sequencing and DNA data bases. These innovations collectively made it possible to construct physical maps and gene maps in virtually any bacteria, irrespective of the presence of established gene exchange systems. The gene mapping principally consists of construction of genomic restriction maps in which the genetically mapped, cloned genes randomly chosen from a species specific library or even from cloned genes of other bacteria are used as the hybridization probes to PFG-fractionated complete and partial digests of chromosomal DNA.

### 6.1.2 TRANSFORMATION

Transformation may be demonstrated in nature with bacteria such as *Streptococcus pneumoniae,* depending on growth phase or nutritional conditions. In general, however, transformation is accomplished with the artificially extracted and purified DNA of donor bacterium (Fig. 6.2). The transformation process consists of three steps: (1) external binding of the DNA fragment onto the outer membrane, (2) penetration of the DNA fragment through the cell envelope, and (3) gene expression in the state of independent replicon or integrated into the chromosome.

The frequency of recombination derived from transformation may significantly increase by treatment of recipient cells at the exponential phase with 75–100 m*M* $Ca^{2+}$ or $Mg^{2+}$. These cations degenerate the

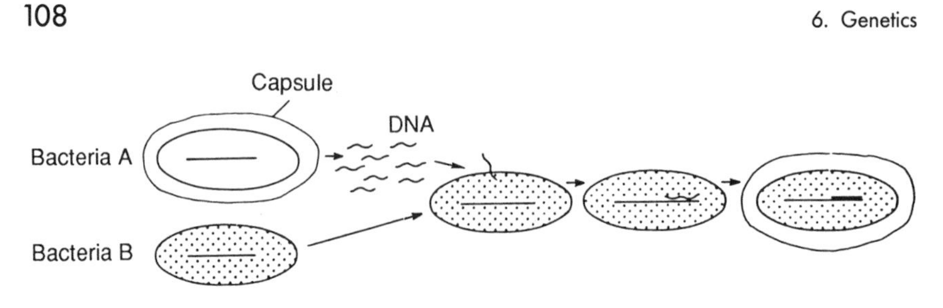


**Fig. 6.2** Schematic illustration of bacterial transformation.

outer membrane proteins or lipopolysaccharides, facilitating binding of the donor DNA to the cell envelope and its penetration into the recipient cell. Frequency of transformation varies depending on other factors such as experimental conditions, bacterial species, the mol% G + C as well as the extent of homology of the donor and recipient DNA, or purity of the extracted DNA. Recent development of the new technology electroporation (see below) has significantly simplified the process of transformation and improved transformation frequency.

In *P. solanacearum* strain K60, transformation frequencies with the genetic markers of *met*$^+$ and *try*$^+$ are in the range of $10^{-5}$ to $10^{-6}$ per recipient. Transformation of *P. syringae* pv. *syringae* strains with plasmid pRO161 and RSF1010 DNA occurs at frequencies of $1 \times 10^{-3}$ to $4 \times 10^{-9}$, depending on the strains, plasmids, and conditions for transformation. In transformation of *A. tumefaciens* strain GV3100 and plasmid RP4 DNA, transformants were obtained at a maximum frequency of $3.5 \times 10^{-7}$.

When purified DNA extracted from bacteriophages is introduced into bacteria, the process is called *transfection*. In *E. carotovora* subsp. *carotovora*, virulent phage DNA showed a higher level of transfection than temperate phage DNA. Transfection of *A. tumefaciens* strains B6S3 and B6-6 with DNA of the temperate phage PS8cc186 yielded a maximum frequency of $2 \times 10^{-7}$ transfectants per total recipient population. The transformation efficiency of *C. michiganensis* subsp. *michiganensis* with the specific phage CMP1 yielded about $2 \times 10^3$ transformants per microgram of plasmid DNA by electroporation.

### 6.1.3 TRANSDUCTION

When prophages are released from host chromosome and initiate independent replication, they may carry a very small portion of the host chromosome, resulting in the assemble of temperate phage particles with the DNA consisting of phage and host-cell DNA. The phage particle is a type of defective phage because it does not contain a full complement of the DNA's part that is left on the host chromosome. When this phage infects other cells, crossing over takes place between a fragment of chromosome of the donor cell and the homologous chromosome of the recipient, resulting in alteration of the corresponding phenotypic properties. Such phage-mediated recombination of the chromosome is called transduction (Fig. 6.3). Because the fragment of host DNA transferred by phage is very small, transduction can be effectively used for analysis of fine structures in gene mapping.

When any chromosomal genes have an equal chance of being incorporated into phage particles, it is referred to as generalized transduction. In contrast, when only certain specific genes can be incorporated into phage particles, it is called specialized or restricted transduction. The frequency of generalized transduction is generally lower than that of specialized transduction.

The temperate phage Erch-12 of *E. chrysanthemi* strain KS612, which was induced by DNA-damaging agents such as mitomycin C, mediated generalized transduction of the chromosomal genes *arg, leu, his, ser, thr, ura* to the recipient strain EC183 of the same bacterium. This study further showed the linkage between *thr-ser,* and *rif* (rifampicin resistance)-*ade,* and also that the recombinant frequency increased 5 to 100 times after UV irradiation. A chromosomal gene of *E. amylovora, trp,* can be integrated into the chromosome of *Salmonella typhimurium* by transduction with the bacteriophage P22.

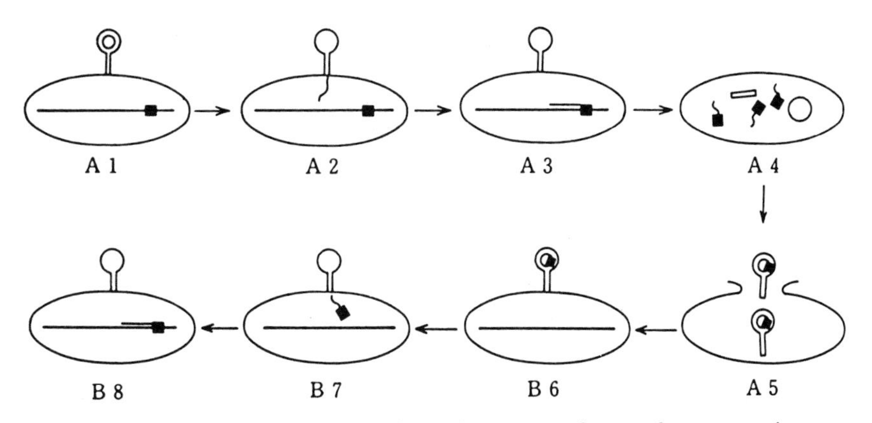


**Fig. 6.3** Schematic illustration of bacterial transduction. 1,2, infection of temperate phage on bacteria A; 3, integration into chromosome as prophage; 4,5, induction of lysis; 6,7, infection of temperate phage on bacteria B; 8, completion of transduction.

### 6.1.4 PHAGE CONVERSION

When bacteria are lysogenized by infection with temperate phages, certain properties may be conferred on the host bacteria solely by the presence of the phage genome. This acquisition of new phenotypic properties as a result of phage infection is called phage conversion. Phage conversion differs from transduction in the following ways: (1) the genes involved are entirely of phage origin; (2) the acquired new property is lost when phage genes are released from the chromosome; and (3) the converting phage particles are completely normal.

Phage conversion in plant pathogenic bacteria has been little documented, although similar phenomena are observed when *X. campestris* pv. *citri, X. campestris* pv. *phaseoli,* and *X. campestris* pv. *begoniae* are lysogenized with *X. campestris* pv. *citri* temperate phages, resulting in the alteration of colony forms and phage susceptibility patterns.

## 6.2 Mutation

The phenotypic characteristics of bacteria may change during adaptation to adverse environmental conditions. These altered properties are reversed when the environment returns to normal. Such noninheritable variation is called *adaptation.* Mutation is the heritable variation involving permanent alteration in the sequence of bases of DNA so that altered characters are passed on to descendants.

Mutation can be differentiated into *spontaneous mutation* and *induced mutation.* Spontaneous mutations usually occur at the rate of around $10^{-8}$ in the normal growth of bacterial cultures without the presence of known mutagenic treatments. These mutants do not become dominant over the population of the parent because the mutants have the same or lower multiplication rate, unless environmental factors are specifically beneficial to the multiplication of the mutants. In contrast, if a selective pressure (e.g., antibiotics) is added to the bacterial population containing mutant cells (e.g., antibiotic resistant mutant), the entire population is quickly replaced by the latter.

The number of mutant cells occurring in a bacterial population is a function of the two parameters *mutation rate* and *mutant frequency.* Mutation rate refers to the probability that any one cell mutates during a defined interval of time or the average number of mutations per cell per generation, and can be expressed as, for example, $1 \times 10^{-8}$ per cell-generation. Mutant frequency refers to the proportion of the mutant or the number of mutant cells contained in the bacterial population at any time. Mutant frequency is a variable parameter affected by mutation rate and efficiency in the selection of the mutants and is generally higher than the mutation rate, ranging between $10^{-4}$ and $10^{-8}$, in some cases $10^{-2}$.

Mutants are essential in the study of bacterial genetics. To obtain mutant clones with various marker properties, the bacterial culture is treated with chemical and physical mutagens, and the induced mutant cells are isolated on selective media. Mutation generated by mutagens is called *induced mutation.*

Many mutations have the capacity of back-mutation or reverse mutation to the original state. Although reverse mutation has its own rate, it generally occurs at the rate of about $1 \times 10^{-8}$ per cell-generation. If two mutations are carried by the same cell, the probability of simultaneous back-mutation of two characters will be $1 \times 10^{-16}$, an almost undetectable rate. Therefore, polyauxotrophs can be used as stable markers in the study of bacterial genetics.

### 6.2.1 MUTATION BY TRANSPOSABLE ELEMENTS

Genetic entities that promote their own transposition are *transposable elements* or *mobile elements.* The transposable elements of prokaryotes were first discovered in *Escherichia coli* in the late 1960s. They randomly insert into many different sites of chromosomal and plasmid DNA, inducing mutations, gene transfer, or recombination. Thus, transposable elements are responsible for the diverse variation of prokaryotes in nature. However, these elements cannot replicate themselves but increase their copies together with the bacterial genome on which they are integrated.

Transposable elements are differentiated into *insertion sequences* (IS) and *transposons* (Tn) mainly on the basis of molecular size and encoded genetic markers (Fig. 6.4). Insertion sequences are small elements of less than 2,000 base pairs (bp) and have only inverted repeats which are essential for recognition of the transposition sites and their insertion. In contrast, transposons are genetic elements of larger than 2,000 bp having two insertion sequences at both ends and antibiotic resistance determinants between them. Therefore, transposons have more benefit than insertion sequences as a tool for studying bacterial genetics because the mutants induced by the former can easily be detected by the drug-resistance markers.

Among a number of transposons, Tn5 that encodes kanamycin-, streptomycin-, and bleomycin-resistant genes is widely used as a tool for mutational analysis of plant pathogenic bacteria. This is mainly based on

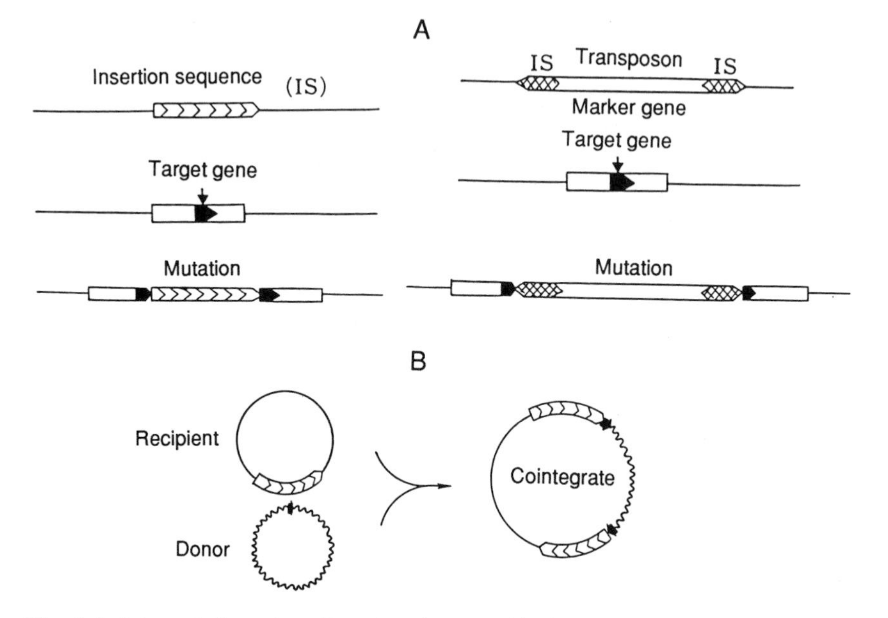


**Fig. 6.4** Schematic illustration of insertion of tranposable elements to target DNA. (A) Insertion of transposable elements involves symmetrical joining of particular nucleotides at the ends of the elements to a target DNA. (B) A new copy of a cointegrate structure is generated by transposition of a transposable element (the black arrow) from a donor plasmid to a recipient plasmid.

the following properties: (1) kanamycin resistance rarely occurs in plant pathogenic bacteria, (2) relatively high frequency of transposition, (3) low insertional specificity, (4) generation of stable mutations, (5) rare occurrence of secondary transposition, (6) infrequent occurrence of genome rearrangements, and (7) a defined physical map (Mills, 1985).

Transposable elements are considered to have a significant role in the mutations of plant pathogenic bacteria in nature. For example, IS51 and IS52 of *P. syringae* pv. *savastanoi,* IS476 of *X. campestris* pv. *vesicatoria,* and IS868 of *A. tumefaciens* integrate into pathogenic genes or avirulence genes inducing mutations in pathogenesis and/or host ranges of these bacteria. The high homology has been detected among these insertion sequences in their nucleotide sequences of inverted repeats, implying their common ancestral background.

## 6.2.2 MUTATION BY CHEMICAL MUTAGENS

In general, chemical mutagens are divided into three groups on the basis of mechanisms involved in the mutagenesis.

### *Nitrites and Nitrosoguanidine*

These mutagens induce chemical changes in the "resting" DNA molecules. The extracted DNA can be treated with chemicals and introduced into recipient cells by transformation, inducing mutations. Nitrates cause transitions of the bases of DNA (replacement of a pair of bases by another pair) resulting from deamination. *N*-methyl-*N'*-nitro-*N*-nitrosoguanidine is the most powerful mutagen widely used in bacterial genetics. This substance induces a number of chemical changes in DNA derived from alkylation of the bases of DNA and causes mutations at high frequency. It induces mutation not only of the target property but also of many other properties. Special care must be taken in handling this chemical because of its strong mutagenic effects.

### *5-Bromouracil, 2-Aminopurine, etc.*

These compounds are the DNA base analogues and readily incorporated into newly synthesized DNA. Bromouracil substitutes for thymine and aminopurine mainly for adenine, causing mutation of bacterial cells resulting from the transitions of the bases of DNA which originate in the tautomeric shift of electron. Efficiency is, however, smaller than that of the nitrites and nitrosoguanidine.

### *Acridine Dyes*

Acridine dyes intercalate between the stacked bases of DNA, expanding the distance between them inducing the insertions and deletions of one or a few base pairs in DNA. Acridine dyes are widely used for eliminating (curing) plasmids, but they are not powerful mutagens of bacterial chromosomes.

## 6.2.3 MUTATION BY PHYSICAL MUTAGENS

### *Ultraviolet Ray*

The ultraviolet ray (UV) is the most widely applied and useful mutagen for bacteria. Effective wavelengths range between 200 and 300 nm, which corresponds to the wavelengths of the maximum absorbance of DNA. The mutagenic activity of UV is mainly attributed to the formation of pyrimidine dimers. The injury of DNA caused by UV-irradiation is repaired by immediate postirradiation of visible light of 300–400 nm. This is called *photoreactivation*.

### *X-Ray, Gamma Ray, and Neutron*

These mutagens may be applicable to bacteria only when no other effective mutagen is available. The effect of irradiations of these rays is so intensive that translocation and/or inversion may occur on the bacterial genome, making it difficult to analyze gene mutation.

## 6.2.4 BACTERIAL MUTATIONS

### *Colony Morphology*

Many plant pathogenic bacteria produce extracellular polysaccharides (EPS) on agar plates, forming slimy and opaque colonies with a smooth surface. These bacteria often dissociate mutants forming butyrous, transparent, and small colonies with a smooth surface (a rough surface in some cases), resulting from loss of the ability to synthesize EPS. In vascular pathogens such as *P. solanacearum* and *X. campestris* pv. *campestris,* such colony mutants generally accompany a complete loss of pathogenicity or marked attenuation. In the majority of parenchyma pathogens, however, noticeable attenuation does not occur through this type of mutation.

### *Biochemical Mutation*

Biochemical mutation includes inherited alterations in the ability to utilize nutrients such as carbohydrates and amino acids, or to produce various pigments. The mutations in carbohydrate utilization have been interesting from the taxonomic viewpoint because the ability to use carbohydrates as a sole source of carbon is one of the important taxonomic criteria. The auxotrophic mutants requiring amino acids are important and useful markers in the study of bacterial genetics. Alteration in chromogenesis usually involves mutation in the direction from pigmentation to nonpigmentation. However, chromogenesis of certain coryneform bacteria is governed by the concentrations of thiamine added in the medium.

### *Drug Resistance Mutation*

Drug resistance mutations raise serious concerns in the control of bacterial plant diseases. The wide distribution of streptomycin-resistant strains in field was recently disclosed with *P. syringae* pv. *papulans* (apple),

*P. syringae* pv. *actinidiae* (kiwi), *E. amylovora* (apple), and *X. campestris* pv. *vesicatoria* (pepper). On the other hand, drug resistance is a useful genetic marker and has been employed as widely as auxotrophs in the study of bacterial genetics.

Resistance to antibiotics may be based on the following mechanisms: (1) enzymatic detoxification of the antibiotic, (2) genetic alteration of an enzyme so that it no longer binds to the antibiotic, (3) alteration of ribosomal RNA by a genetic change restoring inhibition of protein synthesis by antibiotics, (4) alteration of the outer membrane proteins resulting in decreased penetration of antibiotic, (5) diversion to an alternate pathway circumventing blockage by antibiotics (Koch, 1981).

Mutations also occur on susceptibility to cytotoxic heavy metals such as $Hg^{2+}$, $Cd^{2+}$, $Co^{2+}$, $Cu^{2+}$, and $Zn^{2+}$. It has been recently disclosed that copper-resistant mutants are widely spread in plant pathogenic bacteria. Heavy metal resistances are usually inherited by plasmids and derived either from activated efflux or detoxification by proteins equivalent to methallothioneins in animals. When bacterial cells selected for a cytotoxic drug demonstrate resistance to apparently unrelated toxic compounds, this phenomenon is called *multidrug resistance* (MDR). MDR is mediated by the active efflux systems performed by various membrane ATPase pumps.

### *Antigenic Mutation*

Antigenic mutations include those derived from deletion of *O*-antigen from lipopolysaccharide, loss of flagella, or development of different shapes of flagellation. Deletion mutations of *O*-antigen usually accompany colony mutation from smooth to rough. Phase variation in *Salmonella* is the alternate expression of H1- and H2-type flagella antigen, which is regulated by alternate transcription of structural genes. Flagella variation found in some marine *Vibrio* consists of alteration between a polarly flagellated form and peritrichously flagellated one.

### *Mutation in Pathogenicity*

Spontaneous mutation of virulence may occur either in association with other phenotypic mutations affecting colony morphology or the ability to induce hypersensitive reaction (pleiotropic mutation), or independently from these mutations.

Attenuation also occurs in association with induced nutritional mutations. In this case, virulence is usually restored when the requisite nutrients are added to the infection sites. Otherwise, attenuation may result from antagonistic effects of amino acids. Complete loss or decline of virulence derived from the auxotrophic mutations has been demonstrated in many plant pathogenic bacteria including *P. solanacearum, E. carotovora* subsp. *carotovora*, and *P. syringae* pv. *tabaci*.

#### *Phenotype Conversion (Pleiotropic Mutation)*

Pleiotropic mutation is common in plant pathogenic bacteria, particularly in those invading vascular tissues. For example, *P. solanacearum* readily undergoes spontaneous mutation during subculturing. The mutants are characterized by loss of the ability to produce fluidal form colony, disease symptoms, endoglucanase and extracellular polysaccharides, and in reverse by acquisition of the ability to induce hypersensitive reaction and of motility. This conversion is controlled by the gene *phcA* (phenotype conversion) of 2.2 kb fragment of chromosomal DNA.

The pathovars of *X. campestris* that primarily attack xylem also cause similar mutations involving loss of virulence, ability to synthesize extracellular polysaccharides, and phage susceptibility as well as in some cases acquisition of chemotactic trait. The pleiotropic mutation may originate in the failure of function or expression of the genes encoding these phenotypes (such as *phcA*). Such pleiotropic mutations may also be under the control of genes or gene clusters similar to *phcA* of *P. solanacearum*.

## 6.3 Genetic Analysis of Some Phenotypic Characters

### 6.3.1 PIGMENTATION

#### *Fluorescent Pigment of* **Pseudomonas**

*P. syringae* pv. *syringae* produces a yellow-green, water soluble fluorescent pigment, pyoverdin$_{pss}$ in iron limited media (see Chapter 3). By complementation analysis of a genomic library of this bacterium with the pigment minus mutants, it was revealed that at least four genes or gene clusters are involved in the production of the pigment. Another study indicated that the synthesis of pyoverdin and inhibition of fungal growth by *P. fluorescens* were determined by at least five distinct genes or gene clusters.

### *Nondiffusible Yellow Pigment of the "Herbicola" Group of Genus* **Erwinia**

These bacteria produce common nonwater-soluble yellow carotenoid pigments, zeaxanthin-$\beta$-diglucoside, which are considered to increase tolerance to UV and superoxide. The carotenoid biosynthesis genes, *crt* of *E. uredovora* were differentiated into six open reading frames of *crtE, crtX, crtY, crtI, crtB,* and *crtZ* on 6,918 base pairs (bp) fragment of chromosomal DNA. The role of the gene products in carotenoid biosynthesis was elucidated as follows: geranylgeranyl $PP_i$ -CrtB→ prephytoene $PP_i$ -CrtE→ phytoene -CrtI→ lycopene -CrtY→ $\beta$-carotene -CrtZ→ zeaxanthin -CrtX→ zeaxanthin-$\beta$-diglucoside.

*E. herbicola* genes determining pigment production were localized in 12.4 kb chromosomal fragment. The genes are expressed in *Escherichia coli.* Two polypeptides of 37 kDa and 40 kDa are involved in biosynthesis of the pigment. The former is required for yellow pigment production and the latter is required for a pink-yellow pigment, a possible precursor of yellow pigment. The synthesis of these two polypeptides is distinguished by sensitivity to catabolite repression. Although these two bacteria produce the common carotenoid, zeaxanthin-$\beta$-diglucoside, homology has not been detected between their carotenoid biosynthesis genes.

## 6.3.2 ICE NUCLEATION ACTIVITY

Some epiphytic bacteria including plant pathogens can catalyze ice formation at temperatures as warm as $-2°$ to $-3°C$. These bacteria are called *ice nucleation-active* (INA) bacteria. INA bacteria include *P. fluorescens, P. syringae, P. viridiflava, E. ananas, E. herbicola, X. campestris* pv. *translucens,* and an undetermined subspecies of *X. campestris.* The INA genes cloned from genomic libraries of these bacteria have unique but similar sequences. The sequence of ice gene *inaZ* (*P. syringae*) is characterized by a striking degree of repetition with a periodicity of 24 bases that are translated into an octapeptide (Ala-Gly-Tyr-Gly-Ser-Thr-Leu-Thr).

The product of ice gene *inaZ* consists of 122 imperfect repeats of the consensus octapeptide, which makes up 70% of the protein. The octapeptide forms secondary structures by the periodicity of 1 unit (8 residue repetition $\times$ 1), 2 units (8 $\times$ 2) or 6 units (16 $\times$ 3), and forms a tertiary structure which matches the hexagonal structure of ice crystal and plays the role of template (Fig. 6.5).

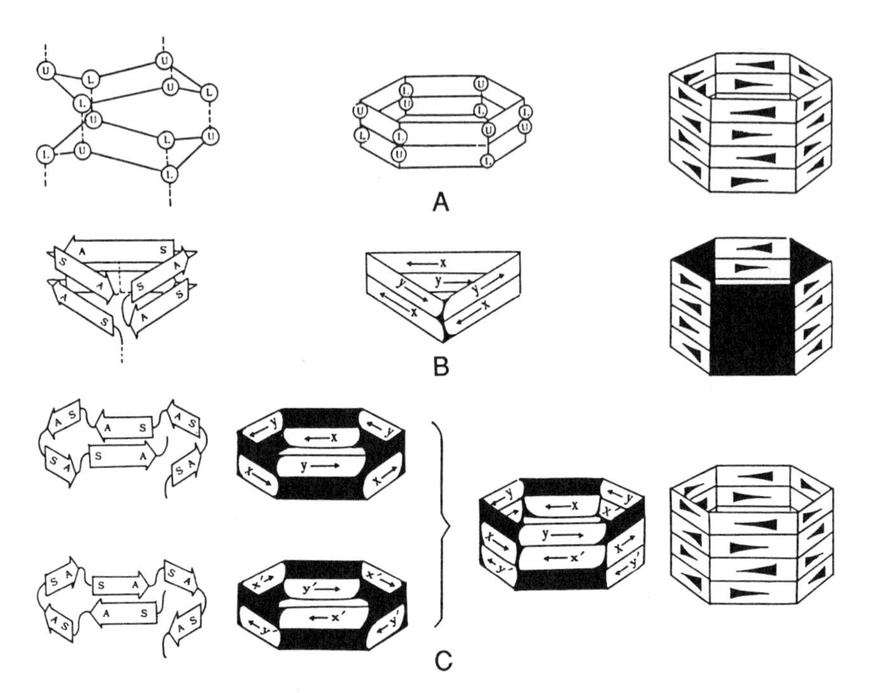


**Fig. 6.5** Schematic models of symmetrical structures of ice nucleation-active proteins and ice crystals. (A) Symmetry of ice; left, spatial arrangement of 12 water molecules; center, symmetry of this structure; right, extended symmetry. Lettered circles are oxygen atoms; dashed lines are O . . . H-O bonds in the *c* crystal axis; U (upper) and L (lower) serve for comparing the drawings at left and center. (B) Triangular model; left, a 48-residue repeat; center and right, symmetry of octapeptides; S, serine; A, alanine; z,y, an alteration of octapeptide types periodicity. (C) Antiparallel double helix model; left, a 48-residue repeat of each of the intertwined chains; center and right, individual, combined, and extended symmetries. [From Warren, G., Corotto, L., and P. Wolber. (1986). Conserved repeats in diverged ice nucleation structural genes from two species of *Pseudomonas. Nucleic Acids Research* **14,** 8047–8060.]

Ice nucleation-active proteins may be differentiated into three types by nucleating activity; class A (or type I) at −5°C or warmer, class B (type II) at −5 to −8°C, and class C (type III) at −10°C, respectively. The class C protein is assumed to be the primary product, and it is structurally modified to class B and to class A proteins with stepwise increase in ice nucleation activity. The class B proteins are considered to be glycoproteins that are modified from class C proteins by binding sugars such as mannose, glucosamine, or others to the amide nitrogen of the three essential asparagine residues in the unique amino-terminal portion of the protein. The class A proteins are lipoglycoproteins that are modified from class B protein by binding phosphatidylinositol to its sugar residues.

The phosphatidylinositol of class A protein plays a role for anchoring the large aggregates of glycoproteins to bacterial cell membrane.

### 6.3.3 COPPER RESISTANCE

Copper ion ($Cu^{2+}$) is an essential micronutrient or trace element of living organisms. It is a component of specific enzymes such as phenolases, laccase, and ascorbic acid oxidase. On the other hand, copper ion shows toxicity at low concentrations, inhibiting bacterial growth. Strains with high copper resistance are known in *Escherichia coli* and *Proteus vulgaris.*

Copper compounds have been widely used for control of bacterial plant diseases. Plasmid-determined copper resistance has been detected in many phytopathogenic bacteria such as *P. cepacia, P. gladioli, P. glumae, P. syringae* pv. *tomato, P. syringae* pv. *actinidiae, X. campestris* pv. *vesicatoria,* as well as some epiphytic saprophytes. The wide distribution of copper resistant bacteria in nature implies the possibility of the active transfer of copper resistance genes (*cop*) between different pathogenic as well as saprophytic bacteria in or on plants.

The resistance level is usually increased by exposure to low concentrations of copper compounds. In *P. syringae* pv. *tomato,* the *cop* genes encoding four proteins of CopA, CopB, CopC, and CopD are located on 4.5 kb fragment of a 35-kb non-self-transmissible plasmid, pPT23D. The products of *cop* operon are found both in periplasm and the outer membrane and seem to prevent entry of toxic copper ions into the cytoplasm.

The copper resistance in *X. campestris* pv. *vesicatoria* is determined by self-transmissible, polymorphic plasmid, pXvCu, of 170–300 kb. One of the plasmids, pXvCu1, also carries an avirulence gene (*avrBsl*), which is responsible for specific hypersensitive response on the pepper cultivar carrying resistance gene *Bsl.*

### 6.3.4 MYCOPLASMALIKE ORGANISMS

Mycoplasmalike organisms (MLO) are found in phloem of the plants showing yellow-dwarf type symptoms as well as salivary glands of vector insects. Genetic analysis of MLO has been difficult because they are not cultivable on artificial media. Recently, however, chromosomal DNA or plasmid DNA were cloned from MLO either in host plants or insect vectors, and probe DNAs were produced, making it possible to elucidate the genetic relatedness of these uncultivable plant pathogenic prokaryotes. It was revealed so far that the genome size of aster yellow MLO and *Oenothera* MLO have been determined to be 1,185 kb and 1,050 kb, respectively, and their 16S rRNA sequences have 99.5% homology.

By dot-blot hybridization and restriction fragment length polymorphism analyses with a cloned clover proliferation MLO-specific DNA probe, it was elucidated that the clover proliferation MLO is most closely related with potato witches'-broom MLO, but distinct from the other 13 MLOs, such as clover phyllody MLO, hydrangea virescence MLO, eastern aster yellows MLO, and western aster yellows MLO.

## 6.4 Some Basic Techniques for Gene Manipulation

### 6.4.1 GENOMIC LIBRARY OR CLONE BANK

The construction of a genomic library or clone bank is the first step in gene manipulation. Genome DNA is extracted from bacterial cells, purified, and partially digested with certain endonucleases to obtain shotgun fragments of about 25 to 50 kb. These fragments are ligated to cosmid vectors such as pVK102 and pLAFR1 that are plasmids with the *cos* region originating from lambda phage that is required for packaging of phage DNA. The resulting recombinant plasmids are packaged into phage heads *in vitro* and transduced into an *E. coli* strain. With other vectors, e.g., pBR322, the *E. coli* strain is transformed with recombinant plasmids. The *E. coli* cells that were transduced or transformed are screened on a medium containing certain antibiotics, depending on the vectors used (Fig. 6.6).

### 6.4.2 VECTORS

Plasmids can be vectors for gene manipulation to deliver foreign DNA molecules to a bacterial cell. The necessary properties of vectors include: (1) ability to replicate in the host bacterial cell, (2) possession of sites for available restriction endonucleases, and (3) possession of genetic markers such as drug resistance and temperature sensitivity genes.

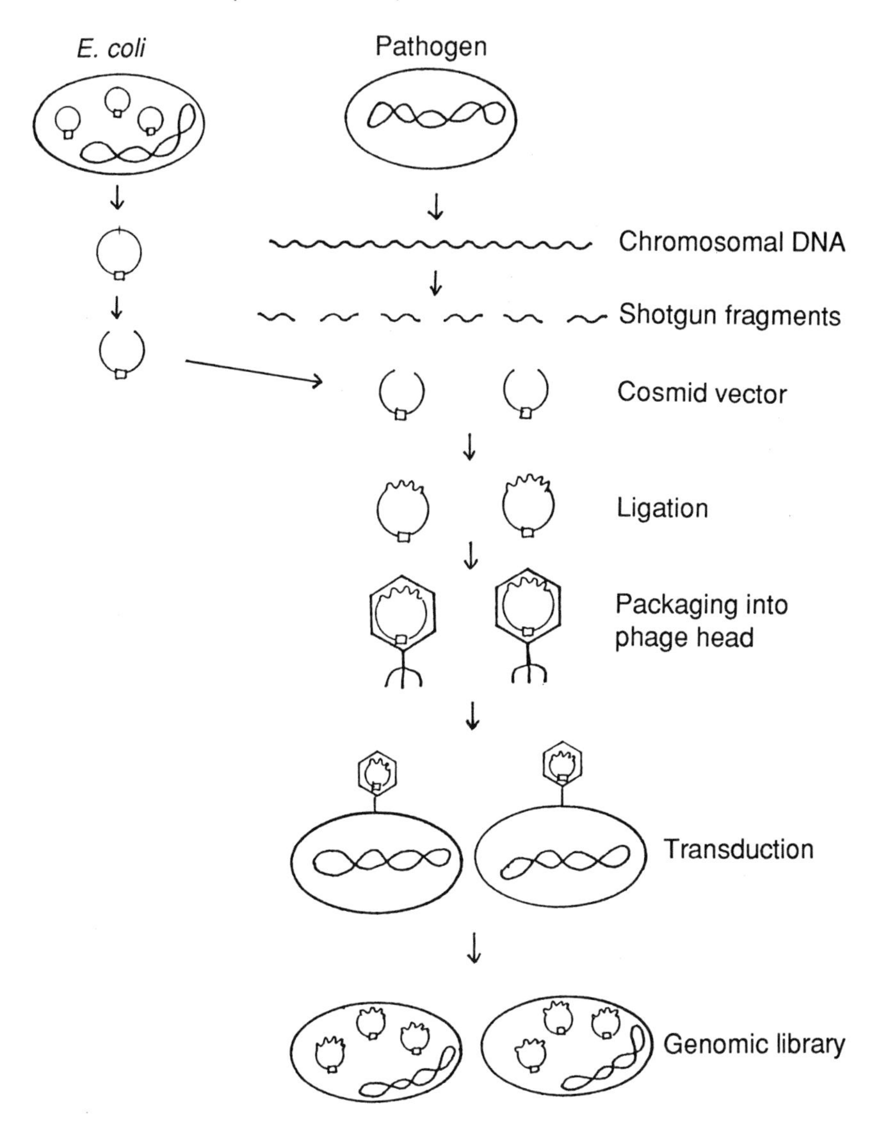

**Fig. 6.6** Schematic illustration of construction of genomic library with a cosmid vector.

### *Suicide Vectors*

Suicide vectors are those that can enter a recipient cell and induce mutation by transposition of a stable mobile element into a site within the genome of the new host, but the vectors themselves fail to replicate so that they soon vanish from the host cell.

### *Chromosome Mobilizing Vectors*

For achieving high chromosome mobilizing efficiency, vectors are often constructed through cointegration between broad host range plasmids such as Inc-P plasmids, the F plasmid or cryptic plasmids. Bacteriophage Mu, transposons, or insertion sequences can be inserted into plasmids to facilitate chromosome mobilization. The interaction between these vectors and chromosome occurs at the sites of their homology.

### *Shuttle Vectors*

These are vectors constructed to express their functions in distantly related hosts of both gram-negative and gram-positive bacteria, or between bacteria and yeasts.

## 6.4.3 MARKER-EXCHANGE MUTAGENESIS

Marker-exchange mutagenesis is a general technique for producing site-directed mutations in particular genes and for conducting a detailed genetic analysis in bacteria for which experimental genetic systems have not yet been established. The procedure consists of the following steps: (1) restriction fragments cloned into *Escherichia coli* are mutagenized either with transposon Tn5 through conjugation with the *E. coli* strain, which has a plasmid carrying Tn5, or with *in vitro* insertion of a DNA fragment with marker genes such as antibiotic resistance, (2) mobilization of the plasmid-carrying mutagenized fragments from *E. coli* to the wild-type parent strain with or without a helper plasmid, and (3) screening on selective media of the mutants generated by replacing the mutated genes with the complementary wild-type parent genes through recombination (Fig. 6.7). Deletion of the DNA sequences in the cloned restriction fragments may also be used for marker-exchange mutagenesis.

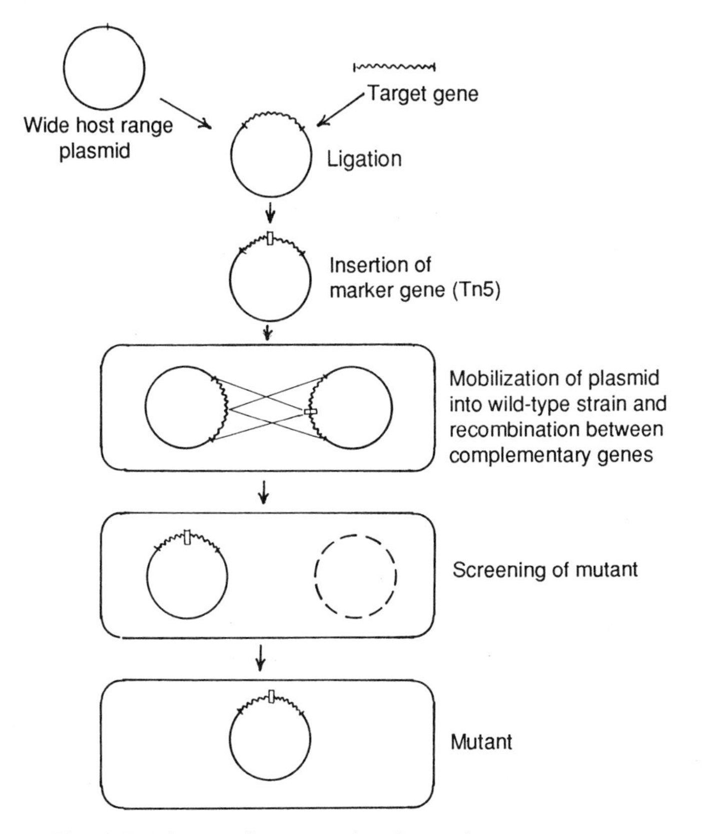


**Fig. 6.7** Schematic illustration of marker-exchange mutagenesis.

## 6.4.4 MOLECULAR HYBRIDIZATION

### *Southern Blot Hybridization*

Southern blot hybridization is a relatively simple method for detecting DNA fragments that are complementary to a probe DNA. After DNA fragments have been separated by gel electrophoresis, they are transferred from the gel to a nitrocellulose or nylon membrane filter after alkaline denaturation. The marker-labeled probe DNA is denatured by heating at 100°C for 4 min and then hybridized. The hybridized DNA fragments are detected by the labeled marker, e.g., radioisotope $^{32}P$. Hybridization performed similarly between RNA and DNA or RNA is called *northern blot hybridization,* and that between two protein molecules *western blot hybridization.*

### *Colony Hybridization*

Colony hybridization is for screening bacterial colonies containing a recombinant plasmid by DNA hybridization at high frequency. The bacterial colonies formed on agar plates are replicated onto a nitrocellulose or nylon membrane filter and lysed with 0.5 *M* NaOH releasing bacterial DNA, hybridized with a radioisotope-labeled probe DNA, and autoradiographed. The positions of the marked colonies are determined on the autoradiograph and the desired colonies are picked from the original plates.

## 6.4.5 POLYMERASE CHAIN REACTION

The polymerase chain reaction (PCR) is a technique for gene amplification. The principle of the method is an oligonucleotide primer-directed synthesis of the target DNA, i.e., the primers determine the direction of replication at the opposite DNA strands at opposite ends. The reaction mixtures of PCR contain four deoxynucleotide triphosphates (dNTPs), either a synthesized or a natural DNA primer, a template DNA, and heat-stable DNA polymerase. The reactions proceed as shown in Fig. 6.8. Each reaction cycle consists of three steps: (1) thermal denaturation of template duplex DNA, (2) annealing of the primer to the template DNA, and (3) elongation of nucleotides by polymerase activity. Within a few hours, the cycles may be repeated 30 to 40 times, accumulating a million or more copies of target DNA. Thus, the PCR technique can quickly produce a great number of copies of a specific region of DNA.

## 6.4.6 ELECTROPORATION

Electroporation is a powerful electromanipulation techniques for gene transfer. In principle, a mixture of bacterial cells and DNA molecules is suspended between electrodes and supplied with high-voltage direct current pulses. A reversible disarrangement is induced in the cell envelope, resulting in the formation of a pore through which the DNA molecule is incorporated into the cell. The pore is then closed by a rearrangement of the cell envelope (Fig. 6.9). Compared to the conventional transformation systems, this alternative technique has several advantages, such as simplicity, speed, and suitability for mass production of transformants.

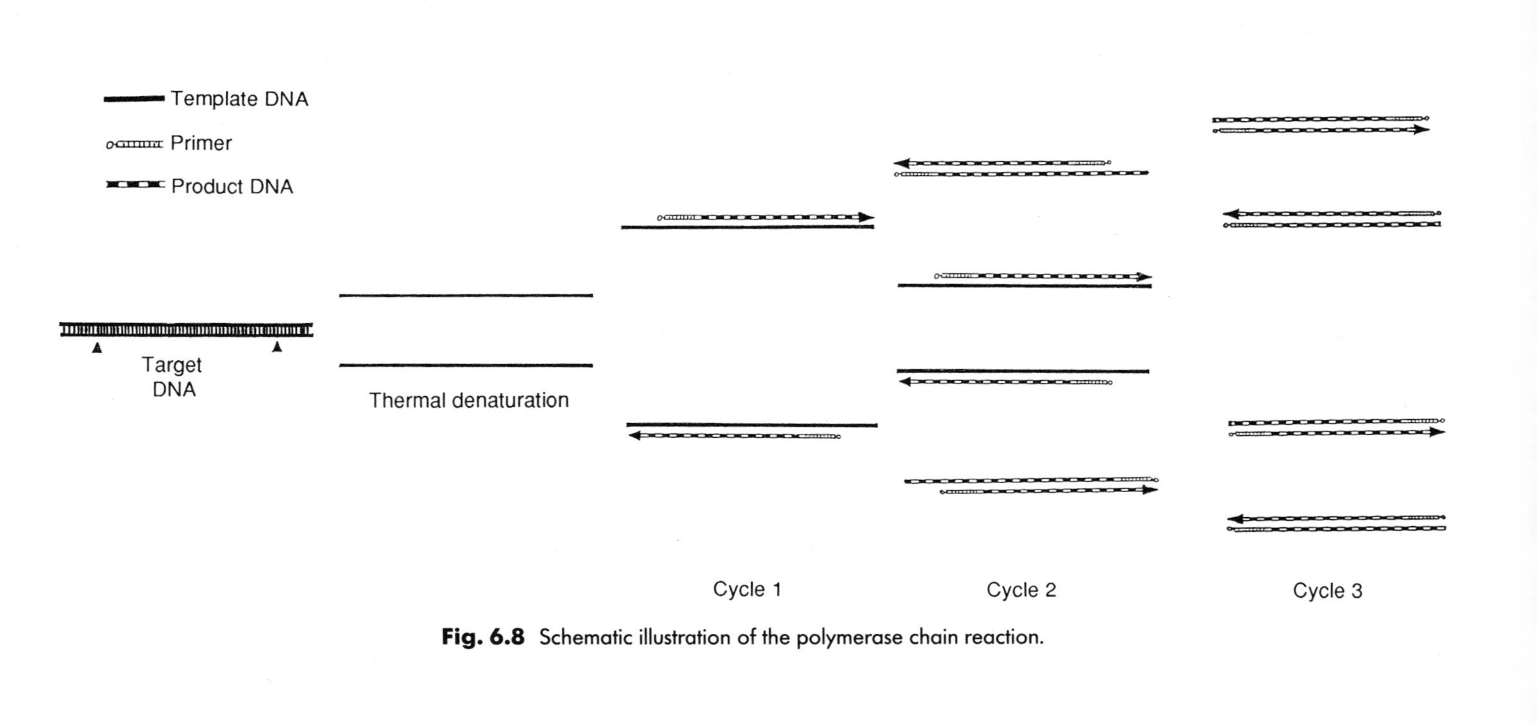


**Fig. 6.8** Schematic illustration of the polymerase chain reaction.

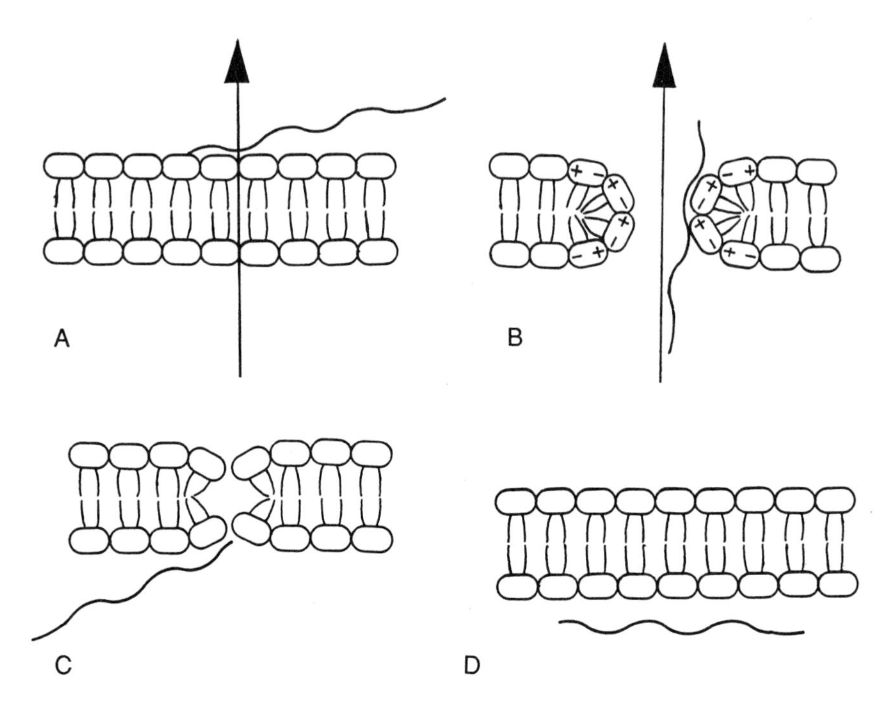


**Fig. 6.9** Introduction of DNA by electroporation. (A) Attachment of DNA fragment on plasma membrane. (B) Pore is formed by electric pulse. (C) and (D) DNA fragment is introduced into cell and pore is repaired.

## 6.4.7 GENETIC ANALYSIS OF PATHOGENESIS

### *Determination of the Genes for Virulence Factors*

In virulence-factor genes that can be expressed in *E. coli,* the genomic library constructed from the wild type of the pathogenic bacteria is screened for virulence factors such as tissue degrading enzymes, phytotoxins, and phytohormones. The cloned fragments, including the virulence genes, are subcloned with another endonuclease to define the smallest DNA fragment or gene cluster that is required for expression of the virulence phenotype.

### *Determination of Genes for Pathogenicity*

The hypersensitive reaction and pathogenicity (*hrp*) genes, avirulence (*avr*) or virulence factor genes that are not expressed in *E. coli* can be determined by *complementation analysis.* In this method, a nonpathogenic mutation (*pat*$^-$) is generated by transposon tagging. The clones from a genomic library constructed from the pathogenic wild-type parent (e.g., *pat*$^+$) are introduced into the nonpathogenic mutant, inducing recombination between the *pat*$^+$ genes in the genomic library and the complementary *pat*$^-$ genes of the mutant recipients. The mutants that restore virulence are screened by inoculation tests. The clone that restores virulence contains the DNA fragments that are essential for development of pathogenicity.

## Further Reading


Black, L. W. (1989). DNA packaging in dsDNA bacteriophages. *Ann. Rev. Microbiol.* **43,** 267–292.

Calos, M. P., and Miller, J. H. (1980). Transposable elements. *Cell* **20,** 579–595.

Cooksey, D. A. (1990). Genetics of bactericide resistance in plant pathogenic bacteria. *Ann. Rev. Phytopathol.* **28,** 201–219.

Gerhardt, P., ed. (1981). "Manual of Methods for General Bacteriology." American Society for Microbiology, Washington, D.C.

Hibi, T. (1989). Electrotransfection of plant protoplasts with viral nucleic acids. *Adv. Virus Res.* **37,** 329–342.

Koch, A. L. (1981). Evolution of antibiotic resistance gene function. *Microbiol. Rev.* **45,** 355–378.

Mills, D. (1985). Transposon mutagenesis and its potential for studying virulence genes in plant pathogens. *Ann. Rev. Phytopathol.* **23,** 297–320.

Panopoulos, N. J., and Peet, R. C. (1985). The molecular genetics of plant pathogenic bacteria and their plasmids. *Ann. Rev. Phytopathol.* **23,** 381–419.

Riley, M., and Anilionis, A. (1978). Evolution of the bacterial genome. *Ann. Rev. Microbiol.* **32,** 519–560.

Silver, S., and Misra, T. K. (1988). Plasmid-mediated heavy metal resistances. *Ann. Rev. Microbiol.* **42,** 717–743.

Smith, C. L., and Condemine, G. (1990). New approaches for physical mapping of small genomes. *J. Bacteriol.* **172,** 1167–1172.

Warren, G., Corotto, L., and Wolber, P. (1986). Conserved repeats in diverged ice nucleation structural genes from two species of *Pseudomonas*. *Nucleic Acids Research* **14,** 8047–8060.

Watson, J. D. (1975). "Molecular Biology of the Gene," 3rd ed. W. A. Benjamin, Inc., Menlo Park, California.

Willetts, N., and Wilkins, B. (1984). Processing of plasmid DNA during bacterial conjugation. *Microbiol. Rev.* **48,** 24–41.

CHAPTER
**SEVEN**

# *Serology*

From a historical point of view, serological studies of plant pathogenic bacteria at the initial stage were taxonomy oriented under the strong influence of the *Salmonella* serology. However, serological taxonomy developed in *Salmonella* was not established in plant bacteriology. The results obtained by agglutination tests with polyclonal antibodies did not necessarily indicate the close relationships with other taxonomic criteria, including host specificity.

The situation has not been altered for long in spite of development of various improved techniques in antigen–antibody interactions, e.g., agar double-diffusion method, fluorescent antibody technique, enzyme-linked immunosorbent assay, and immuno-electronmicroscopy. The innovation of monoclonal antibody technology in the 1970s, however, has brought remarkable improvement in the specificity, sensitivity, and reproducibility in serological analysis of bacterial cell components.

The serological approach with monoclonal antibody technology, combined with improved immunological procedures, is effectively being applied today in various fields of bacterial plant pathology, e.g., identification and diagnosis of bacteria, detection of bacteria from natural habitats, and host–parasite interactions.

## 7.1 Antigen

An antigen is a substance that can stimulate an animal with a functioning immune system to produce certain proteins (antibodies) to react specifically with the substance *in vitro* or *in vivo*.

Substances that are foreign to an animal to be immunized can serve as antigens. Substances such as polysaccharides can react specifically with antibodies *in vitro* but do not stimulate an animal to produce antibodies unless they are combined with a carrier protein. These substances are called *haptens*. Low molecular substances such as coronatine (phytotoxin produced by *P. syringae* pv. *atropurpurea*) may become hapten and induce antibody formation when they have the capacity to bind animal proteins. For immunization of an animal, antigens of weak antigenicity can be mixed with a nonspecific antigen (stimulating agent) to enhance the formation of antibodies against the antigen of interest. Such substances are called *adjuvant*. The well-known complete Freund's adjuvant contains mineral oil, surface active agent (fatty acid), and killed tubercle bacilli.

Some bacterial antigens are degenerated readily by heat treatment at 100°C and lose their agglutinability. Flagella and fimbriae antigens are generally heat-labile, whereas somatic antigens are heat-stable.

For preparation of polyclonal antibodies, bacterial cells to be injected are washed from agar plates and suspended in saline at an approximate concentration of $1 \times 10^9$ cells/ml. The cells may be fixed with glutaraldehyde before injection.

Rabbits are given intravenous injections. The first dose is about 0.1 to 0.5 ml, and further injections are given every 5 to 7 days by doubling the volume. A small amount (about 5 ml) of blood is taken from the animal a week after the last injection to determine the titer of the antibody. When titer is still low, a booster injection is given using the same amount as the last injection and the animal is bled a week after that injection.

Somatic antigens (O-antigens), flagellar antigens (H-antigens) and fimbriae antigens have different antigenicity. The fimbriae have high antigenicity and often mask other antigens on the cell surface. In fimbriae-rich bacteria, therefore, bacterial cells of the fimbriae-absent mutants, or bacterial cells washed with 0.005% HCl at 37°C for 5 min or heated at 100°C for 30 min to destroy both fimbriae and flagella, are used as antigens for both immunization of animals and antigen–antibody interactions.

## 7.2 Antibody

An antibody is the specific protein that is produced in an animal in response to the injection of an antigen. It causes the specific and visible reaction *in vitro* with the antigen. The serum containing antibody is called *antiserum*. Antibodies are immunoglobulins that are divided into several classes with designation abbreviated as IgG, IgM, IgA, IgD, and IgE. Molecular weights of these immunoglobulines range from 160,000 (IgG) to 900,000 (IgM).

Antiserum contains *complement,* that is, a complex enzymatic system of serum proteins composed of at least nine globulin components. The complement potentially has bacteriolytic properties against Gram-negative bacteria. Therefore, antiserum is usually heated at 56°C for 10 min to destroy complement except for the complement-fixation reaction. This procedure is called *inactivation.*

### 7.2.1 POLYCLONAL ANTIBODY

When animals are immunized with antigens with multiple antigenic determinants or epitopes as bacterial cell, antisera contain various antibodies secreted by many antibody-producing $\beta$-lymphocyte (B) cells. Such antibodies are termed *polyclonal antibody* (PCA). PCAs are different in quantity, quality, and affinity of antibodies, even if the same antigen and the same in-bred animals are used for immunization. The spectrum of antibodies varies depending on the individual animal, the method of antigen preparation, and even in different bleedings of the same animal. Furthermore, PCA fails to detect loss of one of the complex antigens because the majority of other antigens still react with the PCA. Thus, in spite of great advances in immunological procedures, the sensitivity, specificity, and reproducibility of PCA are largely dependent on the quality of antisera.

### 7.2.2 MONOCLONAL ANTIBODY

The monoclonal antibody (MCA) technology introduced by Köhler and Milstein (1975) has made a revolutionary advance in serology. MCA has the following advantages over PCA: (1) a large amount of antibody can be produced from a small quantity of antigen; (2) exactly the same antibody can be supplied indefinitely; and (3) pure antibody specific for a single antigenic determinant can be obtained even when complex antigens are used as immunogen. Thus, the majority of disadvantages encountered in PCA could be solved if potent MCA is obtained.

In the practice of the MCA technology, antigen is injected to mice or rats for immunization. After the titer of antibodies reaches an anticipated level, the nucleated $\beta$-lymphocyte cells (B-cells; antibody-producing cells) of the spleen are fused to myeloma cells (malignant cells), producing hybrid cells, or *hybridomas*. The hybridomas carry the properties of parents, i.e., the ability of B-cells to produce specific antibodies and the ability of myeloma cells to divide indefinitely in culture. The hybridomas are separated from unfused myeloma- and B-cells on selective media and further screened for production of antibody to the desired antigen by enzyme-linked immunoassay or radioimmunoassay. Large amounts of the antibody can be produced either by growing the selected hybridoma clones in mass culture or injecting a mouse to form an ascites tumor. The hybridoma lines can be stored by freezing in liquid nitrogen for long periods of time (Fig. 7.1).

## 7.3 Antigen–Antibody Reactions

Since antibodies are divalent and monospecific immunoglobulin molecules, they possess two antibody-reaction sites at both amino ends of the molecule and bind the same antigen. In contrast, bacterial cells have multiple antigenic-determinant sites (*epitopes*) on their surfaces so that they are multivalent and usually multispecific antigens (Fig. 7.2). The relation between the reaction sites of antigen and antibody is highly specific and often expressed as a "lock and key" relationship.

### 7.3.1 AGGLUTINATION REACTION

The particulate antigens (agglutinogens), such as bacteria, are linked together, forming cell clumps by the binding bridges of antibodies (agglutinins) under appropriate salt concentrations and pH. With antigens such as virus particles or protein solutions that show no visible interactions when mixed with antiserum, red blood cells can be used as a carrier for these antigens.

Because bacteria carry multiple antigenic determinants on the surface of the cell envelope, including flagella and fimbriae, some of these antigens may be shared with other bacteria, causing common agglutination between heterologous bacteria. This type of reaction is called *group agglutination* or *cross-reaction* and generally occurs at lower titer than the major agglutination or specific agglutination with the high specificity to homologous bacteria. Because of the possible presence of common antigens, reciprocal agglutination between two bacterial strains at the maximum dilution rate (or titer) does not necessarily mean an identity

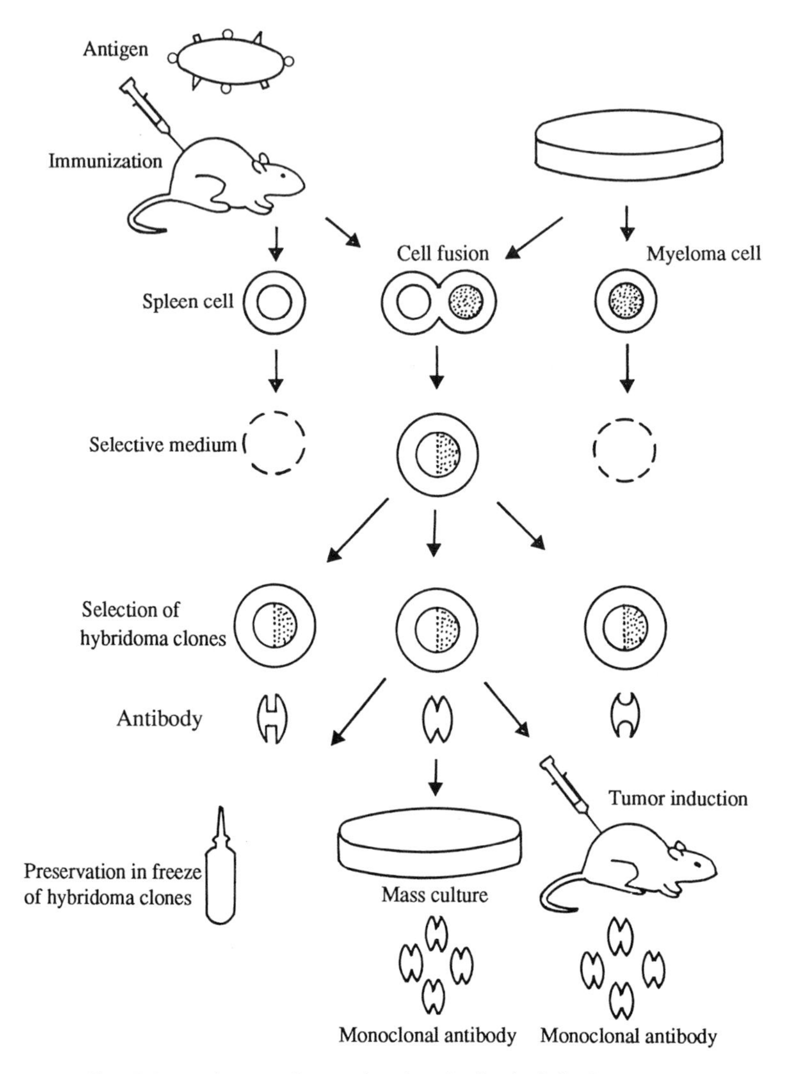


**Fig. 7.1** Production of monoclonal antibodies by hybridoma clones.

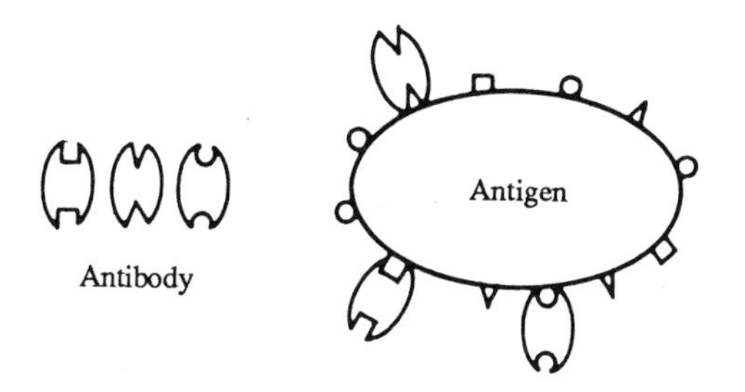


**Fig. 7.2** Multispecific determinants (epitopes) of antigen and monospecific reaction sites of antibody [From Goto, M. (1981). "New Bacterial Plant Pathology" Soft Science Co. Ltd., Tokyo.]

between their antigenic structures. The exact relationship of antigenic structures of two bacterial strains depends on adsorption tests.

### 7.3.2 PRECIPITIN REACTION

In the presence of appropriate electrolytes (0.85% NaCl), a soluble antigen can specifically unite with the homologous antibody forming visible precipitation. This reaction is called the precipitin reaction. The standard procedure to obtain the precipitin reaction is to place a certain amount of antiserum in a series of small test tubes and then gently overlay with twofold dilutions of antigen. Precipitation is first observed in a few tubes in which antigen and antibody are added at the optimal proportions. On both sides of the optimal proportion or equivalence in the series of tubes, there is an antigen excess zone and an antibody excess zone where the degree of precipitation gradually declines. When the proportion of antigen and antibody is extremely unbalanced, precipitation may not occur. This is called a zone phenomenon.

### 7.3.3 AGAR DOUBLE-DIFFUSION TEST

The precipitation reaction may be exploited in the agar double-diffusion tests. Agar containing electrolyte is solidified in a Petri dish, and several wells are cut out of the agar layer in an arrangement illustrated in Fig. 7.3. The antiserum (usually undiluted) is placed in the center well, and the sample antigens and the homologous antigen in the surrounding wells. As both antigen and antiserum diffuse into the agar layer, a zone of the optimal proportion of antibody and antigen for each antigen–

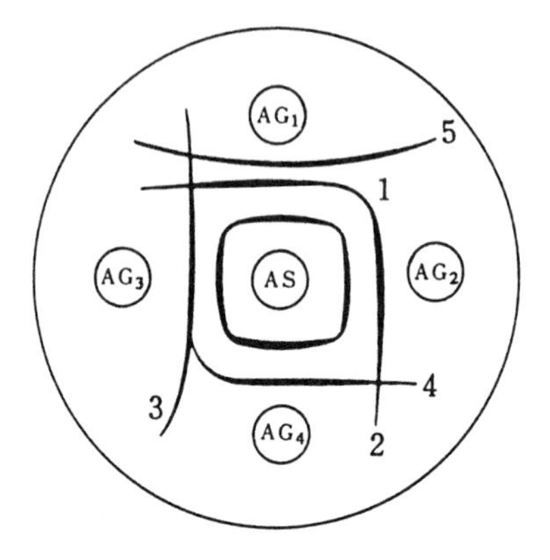


**Fig. 7.3** Schematic illustration of reactions in Ouchterolony double diffusion test. As, antiserum; $AG_1$–$AG_4$, antigens; 1–5, different precipitation bands. [From Goto, M. (1981) "New Bacterial Plant Pathology," Soft Science Co. Ltd., Tokyo.]

antibody combination is established and a white precipitation line is formed. The number and shape of lines indicate the minimal number of antigen–antibody systems in the given samples and their identity. Therefore, if known antigens are included, unknown antigens of the samples may be identified from the precipitation line patterns (Fig. 7.3). This is one of the most reliable procedures for identifying bacteria.

### 7.3.4 IMMUNOELECTROPHORESIS TEST

The agar double-diffusion test is combined with electrophoresis to differentiate proteins that are identical in their behavior in agar-double diffusion tests but differ in their electrophoretic mobilities. The antigen is placed in a well that is cut in the agar layer on a glass plate and set for electrophoresis by exposure to an electric field. After a defined exposure period, the antiserum is placed in lateral troughs and allowed to diffuse in the agar layer. Arc-shaped precipitation lines then develop. The antigen may be identified by the position of the corresponding arc in the gel (Fig. 7.4).

### 7.3.5 COMPLEMENT FIXATION TEST

The complement fixation test consists of the following systems: antigen–antibody reaction, adsorption of complement to the antigen–antibody complex, and demonstration of hemolysis. When the active

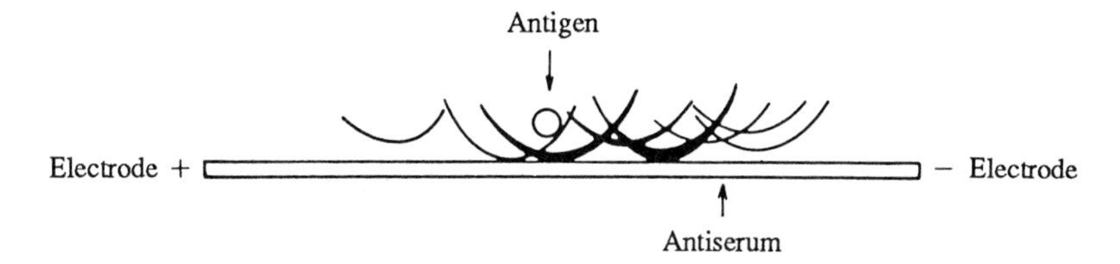


**Fig. 7.4** Schematic illustration of immunoelectrophoretic patterns [From Goto, M. (1981). "New Bacterial Plant Pathology." Soft Science Co. Ltd., Tokyo.]

components of complement are attached on sensitized red cells (or red blood cells binding hemolytic antibody), the red cells burst, turning the reaction mixture red. If the antigen–antibody reaction takes place, there is no free complement left in the antigen–antiserum mixture and no bursting of red cells. In contrast, if the antigen–antibody reaction does not occur, the complement is left free in the mixture and causes the bursting of red cells. This test has not been used often in serological studies of plant pathogenic bacteria.

### 7.3.6 IMMUNOFLUORESCENCE TECHNIQUE OR FLUORESCENT-ANTIBODY TECHNIQUE

Immunoglobulins in an antiserum are separated by precipitation with ammonium sulfate or cold ethanol and conjugated with fluorescein isothiocyanate (FITC). Globulin molecules carrying a low load of fluorescin and highest staining specificity are fractioned for use.

Staining with the labeled antisera can be accomplished either by the direct or indirect method (Fig. 7.5). In the direct method, antisera to be labeled with dye are prepared by immunizing rabbits with bacterial cells, and the labeled antisera are directly applied to antigens. The indirect method is referred to as the antiglobulin method, in which the antisera to be labeled with dye are prepared from a sheep that is immunized against rabbit immunoglobulins. The latter is substituted by the protein A-FITC complex commercially available as a serological reagent.

### 7.3.7 ENZYME-LINKED IMMUNOSORBENT ASSAY

In enzyme-linked immunosorbent assay (ELISA), immunoglobulins separated from antisera are conjugated with enzymes such as alkaline phosphatase. The enzyme-linked antiserum is added in wells of appropriate plastic or glass plates that were previously coated with the antigen.

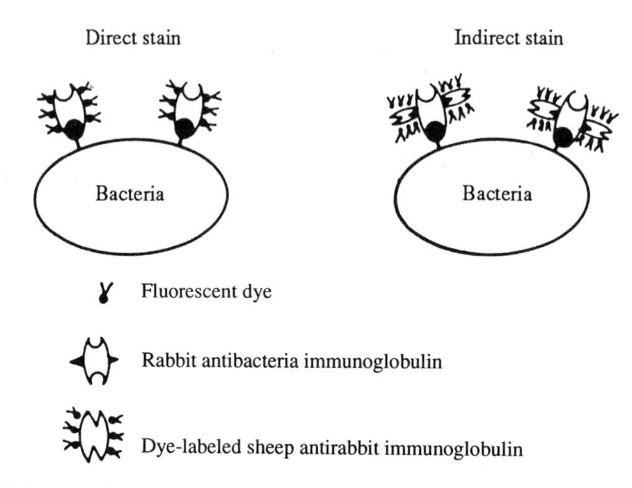


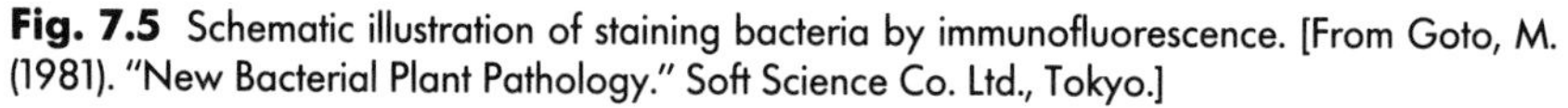

**Fig. 7.5** Schematic illustration of staining bacteria by immunofluorescence. [From Goto, M. (1981). "New Bacterial Plant Pathology." Soft Science Co. Ltd., Tokyo.]

After excess antiserum is washed off with phosphate buffered saline, the substrate of the enzyme (e.g., *p*-nitrophenylphosphate solution) is added to the well. The substrate is hydrolyzed by the enzyme bound to antiserum and releases chromogens (e.g., *p*-nitrophenol), changing the color of the reaction mixture. The color intensity, which is directly proportional to the concentration of antibody, can be measured by a spectrophotometer. Two procedures, direct and indirect, can be used in this method (Fig. 7.6). In the indirect method, protein A-enzyme complex is commercially available as a serological reagent.

## 7.3.8 DOT-BLOT METHOD OR DOT-IMMUNOBINDING ASSAY

The dot-blot method relies on the protein-binding properties of nitrocellulose membrane filters. When applied to plant tissues, bacterial suspensions may be prepared either by macerating a certain area of lesion (e.g., 1 $cm^2$) or by immersing it in 1 ml of distilled water. The bacterial suspension is spotted on the membrane filters, air dried, and fixed. The bacterial cells are reacted with an antiserum prepared with the bacterium under consideration. The bacterial cell–antibody complex is then treated with protein A-enzyme (e.g., alkaline phosphatase) conju-

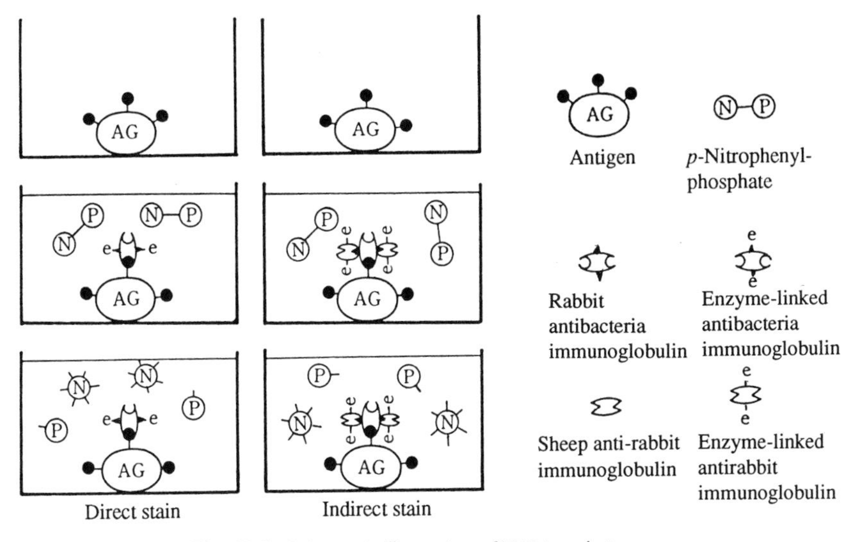

**Fig. 7.6** Schematic illustration of ELISA technique.

gate. The membrane is subsequently exposed to the mixture of a substrate (e.g., naphthol AS-MX phosphate) and a stain (e.g., fast-violet B). The colored precipitates indicate a positive response. The assay is rapid and sensitive, and the lowest detection level is around $10^5$ cells/ml.

## 7.3.9 IMMUNO-GOLD STAINING TECHNIQUE

The immuno-gold staining technique is an immunoelectron microscopic method which is applied for immunocytochemistry of surface antigens as well as for enzyme localization in bacterial cells. In the former, whole bacterial cells can be applied, whereas ultrathin sections are prepared in the latter. In the direct method, immunoglobulins purified from antiserum are labeled with gold particles of 18–20 nm in diameter.

In the indirect method, bacterial cells are first treated with specific antisera and then stained with gold-labeled antirabbit immunoglobulin, which can be substituted with the commercial reagent protein A-gold complex. The bacterial cells thus stained are observed under an electron microscope so that the antigen–antibody reaction can be visualized (Fig. 7.7).

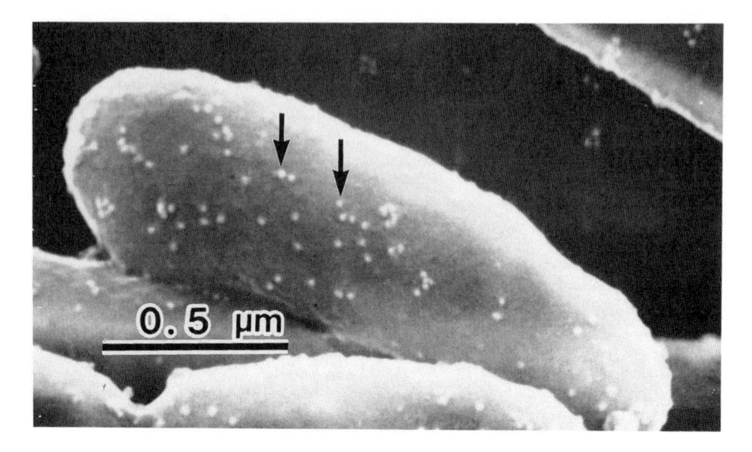


**Fig. 7.7** *Pseudomonas glumae* cell immunostained with gold particles (see arrows) (Courtesy of Dr. I. Matsuda).

## 7.4 Use of Serological Methods in Bacterial Plant Pathology

### 7.4.1 POLYCLONAL ANTIBODY

The advantage of serology lies in the high specificity and speed of the procedures. A large number of papers have been published on the usefulness of the PCA technology in bacterial plant pathology on the basis of correlation between the serological reactions and bacterial taxa, host specificity, or geographical distribution of the strains. With a few exceptions, however, the results have been quite variable, depending either on the bacterial species or the laboratories. Such disagreements may be attributed to the differences in (1) experimental methods, (2) serological heterogeneity of bacteria, (3) the number of bacterial strains used, and (4) host and/or geographical origins of the bacterial strains used.

The most important factor in such disagreement is the insufficient specificity, sensitivity, and reproducibility of PCA. The specificity of PCA may be increased by various treatments, including proper dilution and/or absorption of antisera to eliminate unwanted antibodies. However, these methods do not necessarily guarantee that the specific antiserum thus obtained contain only antibodies for certain components of bacterial cells. This defect can be improved by immunization with the purified components of bacterial cells such as lipopolysaccharides, mucopolysaccharides, flagella, glycoproteins, nucleoproteins, ribosomes, or enzyme proteins (Fig. 7.8).

Although application of serological techniques to the natural habitats may be limited in some cases by the antigenic heterogeneity of bacteria, this approach is quite efficient in the system that is established by artificial infestation with specific strains with known serological properties. The minimum number of bacterial cells to be detected by serological methods varies widely depending on the sensitivity of the techniques used. The average minimum detection levels for fluorescent antibody and ELISA methods range from $10^3$ to $10^4$ cells/ml.

### 7.4.2 MONOCLONAL ANTIBODY

The usefulness of monoclonal antibody (MCA) technology has been confirmed with a number of plant pathogenic prokaryotes since the mid 1980s from various points of view, e.g., identification, diagnosis, epidemiology, immunochemistry, and enzymology.

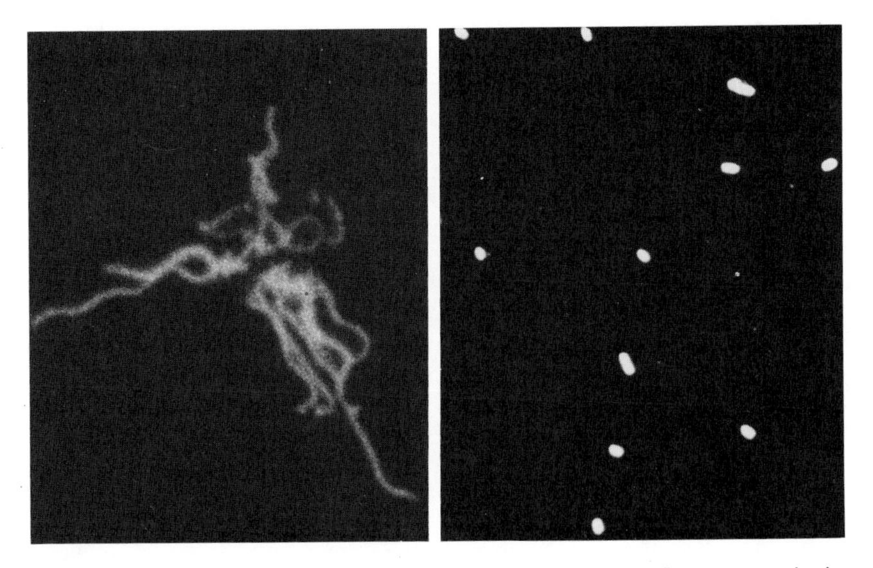

**Fig. 7.8** Bacterial cells and flagella of *Erwinia chrysanthemi* stained by fluorescent antibodies prepared by immunization with purified lipopolysaccharide- (right) and flagella-antigens (left) (Courtesy of Dr. R. Samson).

The MCAs prepared with plant pathogenic prokaryotes are generally divided into several categories on the basis of specificity, i.e., (1) genus specific, (2) species-specific, (3) pathovar-specific, and (3) serovar-specific. Most of these studies have reached the conclusion that the MCA is so specific that it can effectively be used for identification of plant pathogenic prokaryotes at each taxonomic rank, for rapid and correct diagnosis of bacterial plant diseases, and as the reliable marker for epidemiological work. Because the immunological procedures with MCA are same as those with PCA, e.g., ELISA or fluorescent antibody technique, the detection efficiency of plant pathogenic bacteria from natural habitats is around $10^3$ to $10^4$ cells/ml, equivalent to that of the PCA method.

The effectiveness of MCA technology has also been demonstrated in immunochemistry as follows. The MCA prepared against pectate lyase (PL) of *E. carotovora* subsp. *carotovora* could differentiate PL of different origins. It reacted with PL of *E. carotovora* subsp. *atroseptica* and potato strains of *E. chrysanthemi* but not with PL of pseudomonads and xanthomonads. The MCAs have been developed for xanthan, the extracellular polysaccharides of *X. campestris*, and for ferric pseudobactin, the siderophore of *P. putida*. These MCAs specifically recognized the components of these substances, i.e., the acylated side-chains with the pyruvylated terminal mannose and the nonsubstituted trisaccharide with the inner mannose-glucuronic acid in xanthan and the peptide chain in pseudobactin, respectively. These MCAs could detect xanthan at the level of 0.1 $\mu$g and pseudobactin at $5 \times 10^{-12}$ mol.

## Further Reading

Halk, E. L., and De Boer, S. H. (1985). Monoclonal antibodies in plant-disease research. *Ann. Rev. Phytopathol.* **23,** 321–350.

Köhler, G. and Milstein, C. (1975). Continuous cultures of fused cells secreting antibody of predefined specificity. Nature **256,** 495–497.

Oakley, C. L. (1971). Antigen–antibody reactions in microbiology. *In* "Methods in Microbiology" (J. R. Norris and D. W. Ribbons, ed.), Vol. 5A, pp. 173–218. Academic Press, New York.

Pollock, R. R., Teillaud, J.-L., and Scharff, M. D. (1984). Monoclonal antibodies: a powerful tool for selecting and analyzing mutations in antigens and antibodies. *Ann. Rev. Microbiol.* **38,** 389–417.

Schaad, N. W. (1979). Serological identification of plant pathogenic bacteria. *Ann. Rev. Phytopathol.* **17,** 123–147.

Trigalet, A., Samson, R., and Coleno, A. (1978). Problems related to the use of serology in phytobacteriology. *Proc. Int. Conf. Plant Pathathogenic Bacteria, 4th.* Angers, 271–288.

Weiser, R. S., Myrvik, Q. N., and Pearsall, N. N. (1972). "Fundamentals of Immunology for Students of Medicine and Related Sciences." Lea & Febiger, Philadelphia.

CHAPTER
**EIGHT**

# *Pathogenesis and Resistance*

Pathogenesis and disease resistance are closely related to each other because these subjects treat host–parasite interactions from different points of view, i.e., pathogenesis from the side of compatible interactions and resistance from the incompatible ones. Genetic analysis of the ability of plant pathogenic bacteria to induce pathogenic as well as resistant reactions on plants is one of the most rapidly developing fields in bacterial plant pathology.

## 8.1 Pathogenesis

The pathogenicity of plant pathogenic bacteria can be expressed through several infection stages, for example, invasion (or ingress), recognition, multiplication of bacteria, production of virulence factors, and symptom development. These stages, however, often occur in continuity and are difficult to recognize as independent phenomena. Among these stages, there is rather limited information available on bacterial multiplication in the compatible combination because this process has not been given much attention in host–parasite interactions.

### 8.1.1 STAGES OF PATHOGENESIS

The infection stages of the majority of bacterial plant pathogens are entirely different from those of *Agrobacterium tumefaciens,* the causal agent of crown gall. The general concept of pathogenesis is, therefore, discussed with the common bacterial pathogens first, and the pathogenesis of *A. tumefaciens* is separately described later.

### *Invasion*

Invasion of plant pathogenic bacteria through portals of entry such as natural openings and wounds is usually a passive phenomenon. The genetic functions found in the infection process of most fungal pathogens are not necessary. Chemotaxis is sometimes referred to as the active response of bacteria to invasion. However, there is no conclusive evidence that chemotaxis is required for plant pathogenic bacteria to enter into stomata or wounds. Instead, the process of invasion is substantially affected by the mode of dispersal (see Chapter 9).

### *Recognition*

It has been documented that bacterial cells in the incompatible interactions attach to the host cell wall and rapidly and drastically induce structural damage of the plasma membrane, resulting in release of electrolytes and the subsequent death of the host cell. In the process, toxic phenolic compounds are also released and kill the pathogenic bacteria in intercellular space. This process of recognition is referred to as *hypersensitive reaction* (HR). Biological and physiological aspects of HR and its genetic background are described later in Section 8.2.1 on resistance.

Bacterial attachment can also occur in compatible interactions. For example, the specific attachment of *Xanthomonas campestris* pv. *citri* to the wounded surface of susceptible citrus leaf has been reported. This attachment is derived from interactions between bacterial cells and citrus agglutinins. The attachment in the compatible interactions is functionally different from that in the incompatible ones; it induces a release of nutrients or stimulants for bacterial growth through mild degeneration of the host cell membrane.

Most plant pathogenic bacteria exhibit specificity at various degrees with respect to the portals of entry as well as the tissues that they invade. *A. tumefaciens* and other hyperplastic bacteria can infect plants only through wounds but not natural openings. Soft rot bacteria also infect plants mainly through wounds, although they are not as strict as *A. tumefaciens*. Other plant pathogenic bacteria can invade either through wounds or natural openings such as stomata and hydathodes. Bacteria attacking parenchymatous tissues cannot invade xylem and *vice versa*. Nothing has been clarified on the mechanisms of such specificity in plant–bacteria interactions.

However, recent progress in molecular plant pathology suggests that such specificity is controlled by bacterial genes associated with pathogenesis, e.g., *hrp, avr,* or *dsp* (see Section 8.1.3). These genes may be induced by some signal components available at the site of infection as has been elucidated in the induction of virulence genes and nodule development genes of *A. tumefaciens* and *Rhizobium* spp., respectively.

### *Multiplication*

When plant tissues are infected by compatible bacteria, partial degeneration of cell membrane takes place, activating the $K^+$ efflux–$H^+$ influx exchange. The potassium ion released into intercellular space increases the pH of intercellular fluid from 5.5 to 7.0–7.5, and this change further induces efflux of sucrose, amino acids, and inorganic ions without causing structural damage to plasma membrane. As bacterial growth progresses in the intercellular space, water-soaking becomes visible to the naked eye, and creates conditions for accelerated multiplication of bacteria.

### *Production of Virulence Factors*

Some plant pathogenic bacteria growing in the intercellular spaces produce virulence factors such as toxins, enzymes, plant hormones, and extracellular polysaccharides and induce various symptoms of yellowing, soft rot, hyperplasia, necrosis, and wilting. In many bacteria, however, the nature of the virulence factors has not been elucidated.

In general, toxin production alone is not sufficient to cause disease. When a nonpathogenic epiphytic strain of *Pseudomonas syringae* was converted to the "tabtoxin positive" strain by genetic recombination, it still remained nonpathogenic on tobacco. Thus, successful growth of the causal bacteria in plant tissues is essential for the expression of pathogenesis with the exception of crown gall bacteria in which the virulence factors or phytohormones are produced by transformed plant cells.

## 8.1.2 FUNCTIONS OF VIRULENCE FACTORS

### *Toxins*

All toxins produced by plant pathogenic bacteria are non-host-specific and exhibit toxicity to various plants and microorganisms at low concentrations. A direct relationship is found between a particular toxin and a characteristic symptom. The bacterial phytotoxins can be differentiated into two groups by structural characteristics and modes of action.

The first group includes the organic substances of low molecular weights that adversely affect the normal metabolism of host plants. This type of toxin includes tabtoxin, phaseolotoxin, tagetitoxin, coronatine, rhizobitoxin, fervenulin, toxoflavin, 3(methyl) propionic acid, and albicidin (Fig. 8.1).

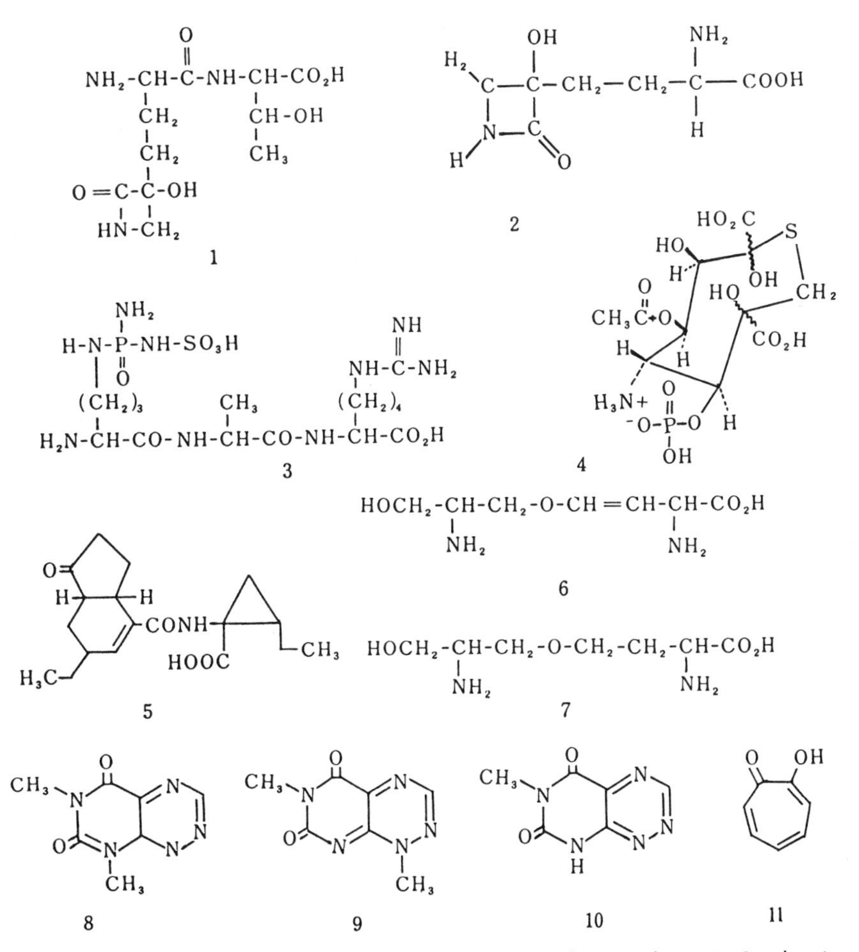


**Fig. 8.1** Structures of phytotoxins produced by plant pathogenic bacteria. 1, tabtoxin; 2, tabtoxin-$\beta$-lactam; 3, phaseolotoxin; 4, tagetitoxin; 5, coronatine; 6, rhizobitoxine; 7, dihydrorhizobitoxine; 8, fervenulin; 9, toxoflavin; 10, reumycin; 11, tropolone.

The second group that has been recently elucidated includes the lipodepsipeptide compounds with usually larger molecular weights than the former. These toxins affect the functions of the host cell membrane through the formation of ion channels causing leakage of electrolytes. This type of toxin includes syringomycins, syringotoxins, syringostatins, syringopeptins, and tolaasin (Fig. 8.2 and 8.3).

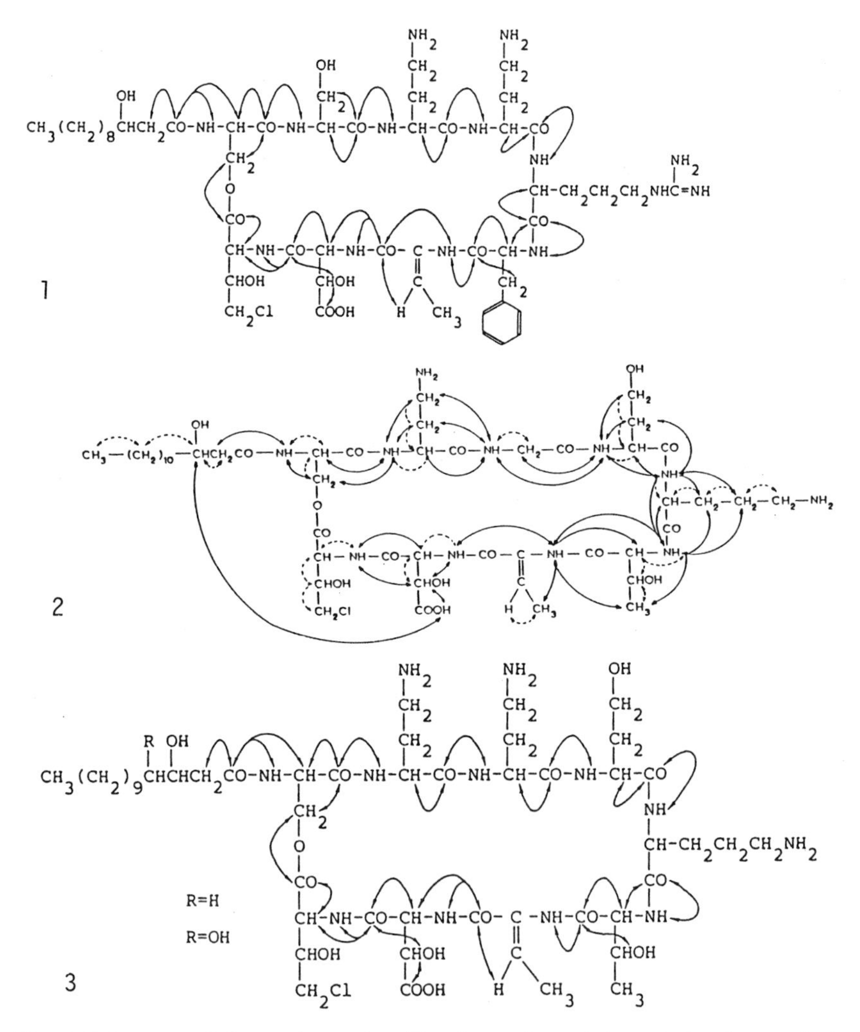


**Fig. 8.2** Structure of phytotoxins produced by plant pathogenic bacteria 1, syringomycin; 2, syringotoxin; 3, syringostatins A and B. [From Fukuchi *et al.*, 1990, Ballio *et al.*, 1990, Isogai *et al.*, 1990].

1

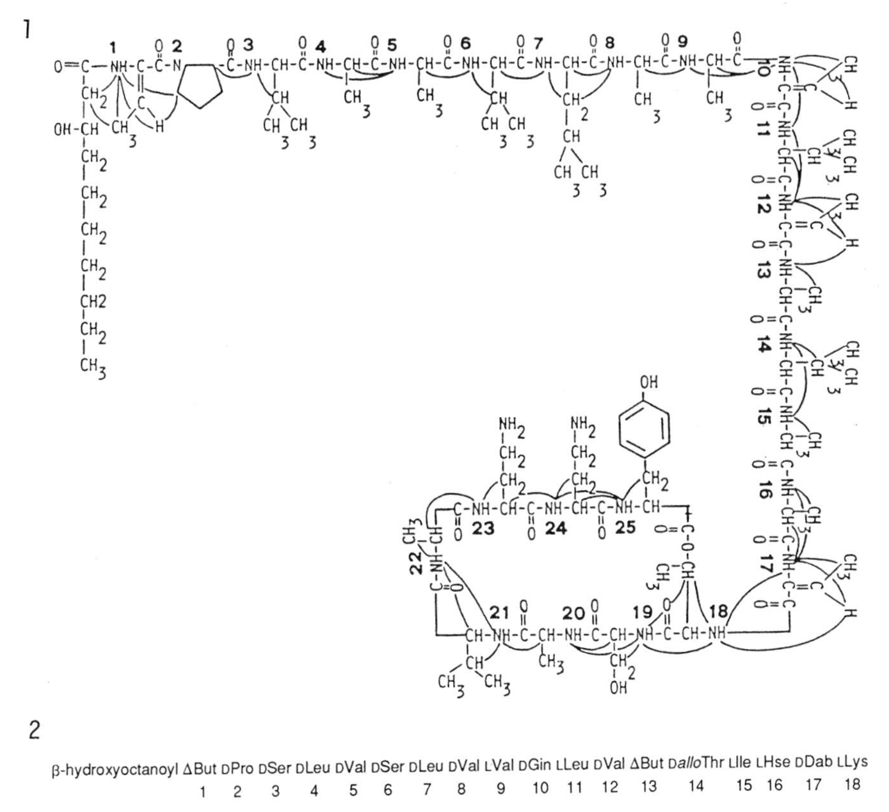

2

β-hydroxyoctanoyl ΔBut DPro DSer DLeu DVal DSer DLeu DVal LVal DGln LLeu DVal ΔBut D*allo*Thr LIle LHse DDab LLys

1 2 3 4 5 6 7 8 9 10 11 12 13 14 15 16 17 18

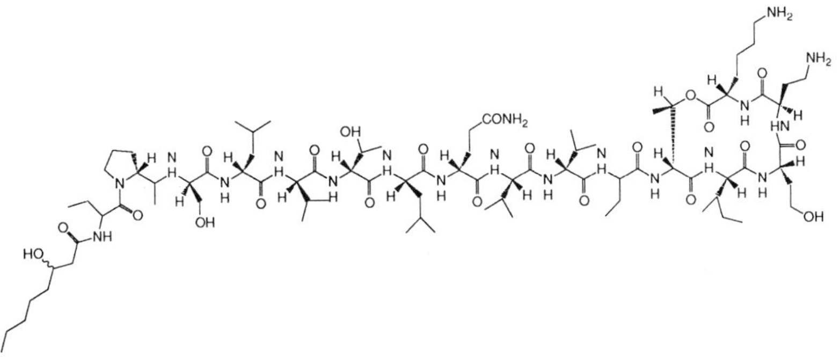

**Fig. 8.3** Structure of phytotoxins produced by plant pathogenic bacteria (3). 1, syringopeptin ($Sp_{25}$-A); 2, tolassin. [From Ballio *et al.,* (1991); Hutkins *et al.,* (1991)].

*Tabtoxin*

Tabtoxin is a chlorosis-inducing dipeptide produced by *P. syringae* pv. *tabaci, P. syringae* pv. *coronafaciens,* and some other pathovars of *P. syringae.* Tabtoxin consists of a new amino acid, tabtoxinine and L-threonine (Fig. 8.1). The toxin containing L-serine instead of L-threonine is called (2-serine)-tabtoxin.

Tabtoxin does not induce chlorosis but inhibits ribulose-1,5-bisphosphate carboxylase activity, which is one of the key enzymes in photosynthesis. The active moiety is tabtoxinine-β-lactam (TβL), which is released on hydrolysis by peptidases of bacteria and plant origin in the presence of zinc (Fig. 8.1). TβL is taken by plant cells through the amino acid transport system and irreversibly inhibits glutamine synthetase under light. Consequently, assimilation of ammonia evolved from photorespiration is inhibited, causing its rapid buildup in the cell. The accumulation of ammonia is responsible for the induction of chlorosis.

*P. syringae* pv. *tabaci* has also glutamine synthetase, a physiological target of tabtoxin. However, the bacterium can grow without being affected by the toxin because the bacterium detoxicates by cleaving TβL with β-lactamase.

*Phaseolotoxin*

Phaseolotoxin is the chlorosis-inducing tripeptide toxin produced by *P. syringae* pv. *phaseolicola,* the causal bacterium of halo blight of beans (Fig. 8.1). Phaseolotoxin is transferred from lesions to lamina of trifoliate leaves through phloem and produces chlorosis. In addition to beans, this toxin produces chlorosis on the leaves of various plants.

Phaseolotoxin causes a reversible inhibition of ornithine carbamoyltransferase (OCTase, E.C.2.1.3.3), causing accumulation of ornithine. In diseased plant tissues, phaseolotoxin removes the terminal amino acids L-alanine and L-homoarginine by a peptidase, producing a substance termed *octicidin,* which is 20 times more toxic than phaseolotoxin.

Toxin-producing strains of the pathogen are also sensitive to phaseolotoxin. They acquire tolerance to the toxin by producing the toxin-insensitive OCTase, which has a markedly different kinetic. It is not clear, however, whether or not this OCTase with the distinct kinetic is the different gene product or derived from structural alteration of the protein after translation.

*Tagetitoxin*

Tagetitoxin is produced by *P. syringae* pv. *tagetis,* which is the pathogen causing a distinctive apical chlorosis in common ragweed, Jerusalem artichoke, zinnia, marigold, and sunflower (Fig. 8.1). Marigold or zinnia plants develop an apical chlorosis 2-3 days after treatment with as little as

20 ng of tagetitoxin. In chloroplast treated with the toxin, ribose-1,5-bisphosphate carboxylase activity is markedly reduced, and grana and stroma lamella become disorganized. There is no detectable amount of 70S ribosome. The chloroplasts begin to degenerate within 1-2 days after toxin treatment and collapse 4 days after treatment. It seems that the toxin directly inhibits RNA synthesis by affecting RNA polymerase.

*Coronatine*

This chlorosis-inducing toxin is produced by *P. syringae* pv. *atropurpurea,* the causal agent of chocolate spot disease of Italian ryegrass (Fig. 8.1). Production of the toxin has also been revealed with other bacteria such as *P. syringae* pv. *glycinea. P. syringae* pv. *maculicola, P. syringae* pv. *morsprunorum, P. syringae* pv. *tomato,* and *Xanthomonas campestris* pv. *phormiicola.* Coronatine induces chlorosis on the leaves of many plants as well as hypertrophy of potato tuber tissues.

The mechanisms of chlorosis induction is assumed to be a result of reduced chlorophyll synthesis due to the inhibition of δ-amino levulinic acid synthesis or due to ethylene evolved from cyclopropane moiety of the toxin. The hypertrophic effect of coronatine is similar to that of indoleacetic acid in physiological effect but differs in tissue specificity.

*Rhizobitoxine*

Rhizobitoxine is produced by *Bradyrhizoblum iaponicum,* soybean nodule bacteria, and *P. andropogonis,* the causal agent of bacterial stripe of sorghum (Fig. 8.1). The saturated analogue of the toxin is called dihydrorhizobitoxine. Rhizobitoxine irreversibly inactivates β-cystathionase, the enzyme that catalyzes the conversion of cystathionine to homocysteine, inhibiting the incorporation of $SO_4^{2-}$ as well as the conversion of methionine into ethylene in plants.

*Fervenulin and Toxoflavin*

Fervenulin and toxoflavin are produced by *P. glumae,* the causal agent of bacterial grain rot of rice (Fig. 8.1). Toxoflavin is unstable and readily changes to reumycin by elimination of N(1)-Me. These toxins show yellowish-green fluorescent color under near-ultraviolet light. These toxins produce chlorosis of leaves as well as growth inhibition of leaves and roots of rice seedlings at concentrations of about 10 μg/ml or more. Toxoflavin is an electron transfer which produces hydrogen peroxide in the presence of oxygen. However, the exact mechanisms of symptom development due to these toxins have not been elucidated.

*Tropolone*

This toxin is a colorless, needle-like crystal produced by *P. plantarii,* the causal bacterium of bacterial seedling blight (Fig. 8.1). This compound induces iron deficiency by chelating ferrous ions with the formation of water-insoluble, red crystals (Fig. 4.2). It causes chlorosis, root growth inhibition, and wilting of rice seedlings at concentrations of 3 to 25 μg/ml.

*3(Methylthio) Propionic Acid*

This compound ($CH_3S(CH_2)_2COOH$) is produced by *X. campestris* pv. *manihotis,* the causal agent of bacterial blight of cassava, by the following pathway: methionine → methylthiopropionic acid → 2-keto-4-methylthiobutyric acid → methylthiopropionic acid. This compound is toxic to cassava, tobacco, and tomato plants at a concentration of 9.9 m*M*. The mechanisms are, however, unknown.

*Albicidin*

Albicidin is produced by *X. albilineans,* the causal agent of leaf scald disease of sugarcane. The chemical structure is unknown except that it has several benzene rings and about 30 carbon atoms. Albicidin preferentially inhibits plastid DNA replication and blocks chloroplast differentiation.

*Syringomycins, Syringotoxins, Syringostatins, and Syringopeptins*

Syringomycin was first reported as the necrosis-inducing toxin produced by most strains of *P. syringae* pv. *syringae,* irrespective of its host plants. Likewise, the toxin produced by the citrus strain of the bacterium was named syringotoxin. Recent studies revealed, however, that the phytotoxins produced by *P. syringae* pv. *syringae* are a group of unique lipodepsipeptides.

Syringomycins are cyclic compounds consisting of a peptide chain and fatty acid residues. Syringomycin produced by a sugarcane strain consists of each of two residues of serine and diaminobutanoic acid, and each of one residue of arginine, phenylalanine, dehydrothreonine, beta-hydroxyaspartic acid, gamma-chlorothreonine as amino acids, and 3-hydroxydodecanoic acid as an acidic residue (Fig. 8.2).

Syringotoxin produced by a citrus strain of *P. syringae* pv. *syringae* consists of one each of a residue of serine, 2,4-diaminobutyric acid, glycine, homoserine, ornithine, allothreonine, 2,3-dehydro-2-aminobutyric acid, 3-hydroxyaspartic acid, and 4-chlorothreonine, with the terminal carboxy group closing a macrocyclic ring on the OH group of the N-terminal serine, which in turn is *N*-acetylated by 3-hydroxytetradecanoic acid (Fig. 8.2).

Syringostatin produced by a lilac strain of *P. syringae* pv. *syringae* consists of one each of a residue of serine, threonine, ornithine, homoserine, beta-hydroxyaspartic acid, 2-amino-2-butenoic acid, and 2-amino-4-chloro-3-hydroxybutanoic acid, and two residues of 2,4-diamino butanoic acid as amino acids and a long chain of 3-hydroxytetradecanoic acid (syringostatin A) or 3,4-dihydroxytetradecanoic acid (syringostatin B) (Fig. 8.2).

Syringopeptin ($Sp_{25}$-A) produced by three strains of *P. syringae* pv. *syringae* consists of one each of a residue of proline, leucine, allothreonine, serine, 2,4-diaminobutyric acid, tyrosine, four residues of 2,3-dehydro-2-aminobutyric acid, five residues of valine, and nine residues of alanine, with tyrosine and allothreonine forming the macrolactone ring, and smaller amounts of the 3-hydroxydodecanoyl homologue (Fig. 8.3).

These toxins are responsible for the antimicrobial and the phytotoxic activity of *P. syringae* pv. *syringae*. Syringomycins are unstable in an alkaline environment of pH 10, whereas syringopeptin ($Sp_{25}$-A) is stable. The former substances are 30 times more active than the latter in antimicrobial assay but at least 40 times less active in electrolyte leakage assay with carrot tissues.

In these phytotoxins, the mode of action is related to destruction of plasma-membrane-associated functions such as $K^+$ efflux, $Ca^{2+}$ influx, $H^+$-ATPase activity, membrane potential, and protein phosphorylation. Among these functions, $Ca^{2+}$ transport associated with elevation of membrane potential is supposed to be most important in terms of ion channel formation in the lipid bilayer of host cell membrane.

*Tolaasin*

This is another lipodepsipeptide toxin produced by *P. tolaasii,* the causal agent of brown blotch disease of mushrooms. It consists of four residues of valine, three residues of leucine, two residues of serine, each of one residue of proline, glutamine, threonine, isoleucine, homoserine, diaminobutanoic acid, and lysine, with a lactone ring composed of five residues (Fig. 8.3). As in syringomycins, tolaasin affects the function of the host cell membrane by formation of a membrane-potential dependent and cation-selective ion channel in the lipid bilayer.

***Enzymes***

Enzymes associated with pathogenesis include pectinase, cellulase, protease, esterase, and cutinase. The role of pectinase on pathogenesis is described later in this Chapter. The significance of other enzymes in pathogenesis has not been elucidated in detail.

*P. syringae* pv. *lachrymans* and *P. syringae* pv. *tomato* produce protease in diseased tissue. It is assumed that the enzyme degrades plant proteins to amino acids, which finally results in release of ammonia, causing damage to plant cells. In contrast, studies with *E. carotovora, E. chrysanthemi,* and *X. campestris* pathovars have revealed that these bacteria produce metalloproteases and that protease-deficient mutants showed reduced bacterial multiplication in host plants and disease severity. In these cases, the role of protease may simply be explained by nutrition of bacteria.

### *Plant Hormones*

#### *Auxin*

The bacteria inciting hyperplasia or gall formation on plants include *A. tumefaciens. Erwinia milletiae, P. syringae* pv. *savastanoi, P. syringae* pv. *myricae*, etc. Except with *A. tumefaciens,* outgrowth is induced through continuous stimulation by auxin such as indoleacetic acid produced by the causal bacteria in intercellular spaces. Pathogenic bacteria produce indoleacetic acid through the indole-3-acetamide pathway and/or the indole-3-pyruvate pathways (Fig. 8.7). Auxin exhibits diverse physiological functions: (1) enhancement of the elastic extensibility of the plant cell wall by exchange of $Ca^{2+}$ or $Mg^{2+}$ bound to wall pectins to H or $CH_3$ groups, (2) increase in cell permeability, (3) control of pectin methylesterase activity, (4) formation of complexes with phenolic substances, (5) stimulation of respiration, and (6) promotion of the biosynthesis of RNA and proteins such as cell wall protein extensin.

#### *Cytokinins*

Cytokinins are derivatives of adenine and function in cell division and cytodifferentiation in cooperation with indoleacetic acid and exhibit an important role in the development of hyperplastic symptoms.

Cytokinins are the major virulence factor of fasciation (leafy gall) of peas, sweet peas, chrysanthemums, and other plants caused by *Rhodococcus fascians.* The disease is characterized by a loss of apical dominance, leading to the development of many short, thick, and fasciated shoots with deformed leaves arising from the lower portions of stems. Cytokinins have another function, which is to inhibit the degradation of nucleic acids and proteins and thus delay the onset of senescence.

#### *Ethylene*

Ethylene production *in vitro* has been proved with a limited number of plant pathogenic bacteria such as *P. solanacearum, P. syringae* pv. *phaseolicola* isolated from *Pueraria lobata,* and *Erwinia rhapontici.* However, ethylene production in diseased plant tissues is a common phenomenon, and most of the ethylene thus produced is generated by infected tissues of host plants. In citrus canker caused by *X. campestris* pv. *citri* and bacterial leaf spot of tomato caused by *X. campestris* pv. *vesicatoria,* defoliation starts when ethylene formation reaches a peak.

### *Extracellular Polysaccharides*

In *P. syringae* pathovars, a high correlation is observed between production of alginate and the appearance of water-soaked lesions. In contrast, the bacterial pathogens that do not induce water-soaked lesions produce only trace amounts of bacterial extracellular polysaccharides (EPS). The EPS thus contribute to pathogenesis by inducing water congestion, which produces favorable conditions in plant tissues for the multiplication and translocation of phytopathogenic bacteria.

Many plant pathogenic bacteria induce wilt syndrome. On the mechanism of wilting, two principal hypotheses have been proposed on the basis of toxins and mechanical blocking. In the former, it is postulated that low molecular toxins produced by bacteria irreversibly damage cytoplasmic membranes, inducing loss of turgor of host cells. The latter postulates that mechanical blocking of xylem with a mixture of bacterial cells and EPS dysfunctions water-conducting system. The mechanical blocking may also be interpreted by embolism that is induced by air microbubbles and/or picomole quantities of EPS which cannot pass through the minute pit membrane pores.

None of the wilting toxins reported in the past, however, has been reconfirmed by subsequent studies. Instead, physical blocking of xylem vessels with bacterial cells and EPS has been widely accepted as the mechanism of wilting. It is based on the following observations: (1) the efficiency of water conduction is greatly reduced in infected plants, (2) only virulent strains of the vascular pathogens produce EPS, (3) EPS are produced in xylem of host plant, (4) partially purified EPS produced by virulent strains also induce wilting of cuttings at low concentrations, (5) fluctuation patterns of photosynthesis and transpiration of the diseased plants are identical to those in plants with artificially reduced water supply.

Vascular pathogens entered into vessels appear to extract nutrients from xylem-parenchyma cells, multiply, and produce EPS. The bacteria spread to adjacent vascular bundles by penetrating the vessel wall or pit membrane either by enzymatic action or by pressure of bacterial masses. In vascular diseases, water transport into the affected areas is entirely blocked before symptom appearance (see Fig. 10.3). The parenchyma cells in the affected areas can stay intact for some time by the complementary water supply through neighboring healthy vessels.

Wilting and/or decline symptoms induced by the xylem-limited bacterium, *Xylella fastidiosa,* also results from mild but chronic water deficiency of the host plant. This is caused by blocking of xylem vessels with bacterial cells or pectic substances. Leaf scorch symptom develops when plants suffer acute water deficiency due to accelerated transpiration under sunlight.

In *P. solanacearum,* however, some contradictions have arisen with the role of EPS on wilt induction because [$EPS^-$ : virulent] or [$EPS^+$ : avirulent] strains were isolated from diseased plants or produced by genetic engineering. These findings suggest the involvement of some other virulence factors. Because active multiplication of the bacterium in vessel is essential for pathogenesis, however, there is a possibility that the genetic alterations associated with growth capacity that are independent from EPS production are involved in avirulence mutants of this bacterium.

## 8.1.3 PATHOGENICITY GENES AND THEIR EXPRESSION

The molecular aspects of host–parasite interactions have been greatly elucidated in the past decade, with a few selected bacterial plant pathogens as the model.

### *Pathogenesis of* **Agrobacterium tumefaciens** *and* **A. rhizogenes**

Symptoms of crown gall and hairy root are distinguished by predominant development of autonomous outgrowths in the former and hair-like roots in the latter (see Fig. 13.16). No essential differences are found, however, between the two diseases in the infection stages as well as the mechanism of transformation of plant cells.

*A. tumefaciens* is a typical wound parasite (Fig. 8.4). The host range is very wide and includes dicotyledons of more than 643 species from 331 genera of 93 families. Monocotyledons are generally much less susceptible, although transformation to tumor cells was confirmed with several herbs such as corn, asparagus, gladiolus, and pinaceous gymnosperms such as *Pinus, Tsuga, Pseudotsuga,* and *Abies.* Autonomous growth has been observed in tumor cells of gladiolus and some conifer species. These facts imply that the resistance of monocots to crown gall disease is a result of some of the following reasons: (1) bacterial attachment to the surface of host plant cell is insufficient, (2) transfer of T-DNA is incomplete because wound healing reactions do not accompany the hypertrophy

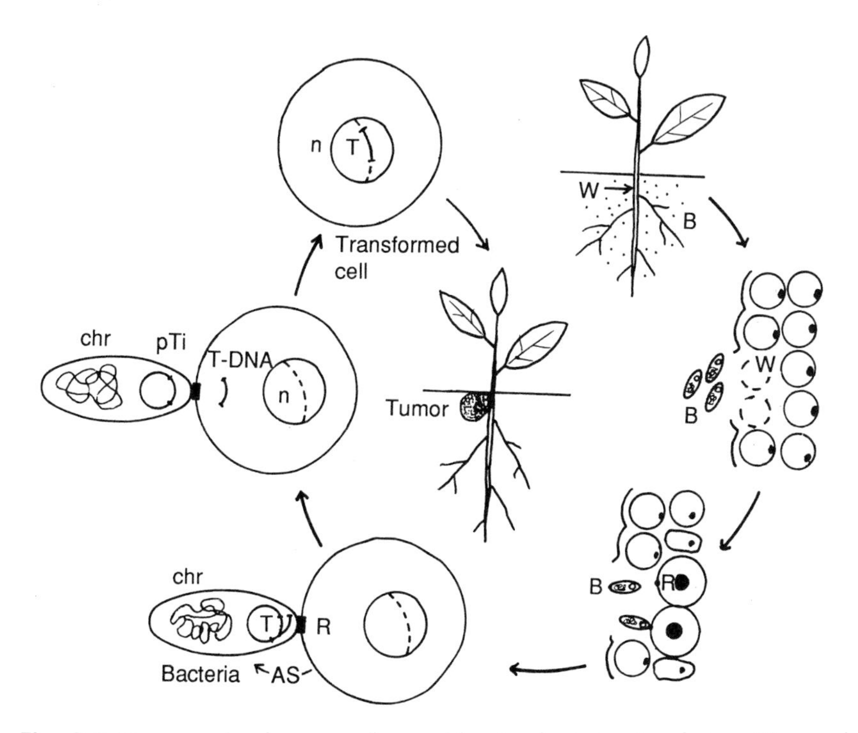


**Fig. 8.4** Disease cycle of crown gall caused by *Agrobacterium tumefaciens.* W, wound; R, receptor; B, pathogenic bacteria; chr, chromosome; pTi, Ti plasmid; T, T-DNA; n, plant genome; AS, acetosyringone.

and/or hyperplasia (3) T-DNA cannot be incorporated in plant genome, and (4) host plant cells do not respond to plant hormones that are produced by the transformed tumor cells.

*Tumor Inducing Plasmids*

The virulent strains of *A. tumefaciens* carry large plasmids termed tumor inducing (Ti) plasmids and *A. rhizogenes* Ri (for root inducing) plasmids. Ti plasmids have molecular size of 150-250 kb and are divided into three principal types—octopine-type, nopaline-type, and agropine-type—based on the opines, unique amino acid derivatives, that are produced by transformed tumor cells.

Ti plasmids carry the genetic determinants associated with pathogenicity, i.e., the transferred DNA (T-DNA) and the virulence region (*vir*) (Fig. 8.5). In addition to these pathogenicity genes, Ti plasmids also carry the genes for biosynthesis and utilization of opines, replication and conjugal transfer of the plasmids, plasmid incompatibility, exclusion of a temperate phage Ap1, and susceptibility to bacteriocin agrocin 84.

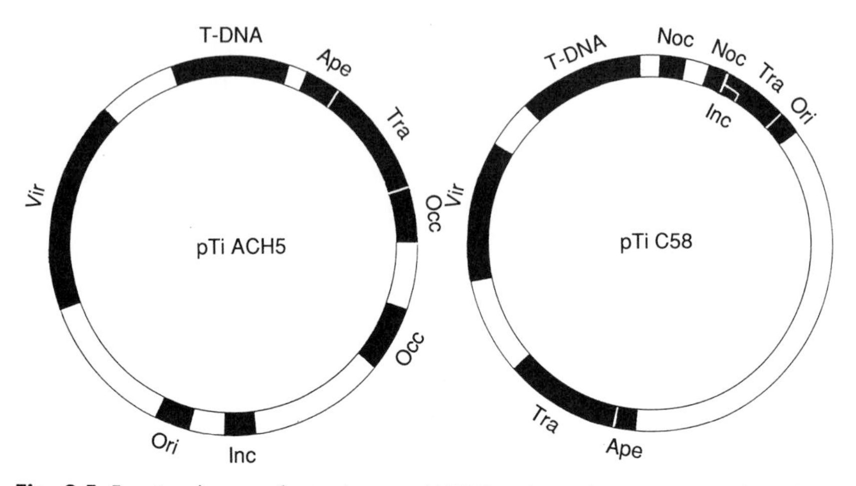


**Fig. 8.5** Functional maps of octopine type (ACH5) and nopaline type (C58) plasmids. Vir, virulence region; T-DNA, transferred DNA; Ape, exclusion of phage Apl; Tra, transfer function; Occ and Noc, catabolism of octopine and nopaline; Ori, origin of replication; Inc, incompatibility.

*Bacterial Attachment to Host Cell*

The first step of infection is the attachment of bacterial cells to the host cell surface. Attachment-defective mutants fail to infect host plants. The attachment of bacterial cells is determined by three chromosomal genes of *chvA* (chromosomal virulence), *chvB*, and *exoC* (exopolysaccharide) (=*pscA*). The loci *chvA* and *chvB* determine the biosynthesis of β-1, 2-D-glucan and its secretion, respectively, and the *exoC* genes the production of a heteropolysaccharide succinoglycan.

The major bacterial cell component for attachment is considered to be β-1, 2-D-glucan, which is assumed to bind to pectic substances or glycoproteins on plant cell wall. This step in *A. tumefaciens* is similar to the attachment of root nodule bacteria (*Rhizobium meliloti*) onto root hair cells of leguminous plants. The loci *chvA, chvB*, and *exoC* of *A. tumefaciens* are equivalent to *ndvA* (nodule development), *nodB*, and *exoC* of *R. meliloti* in their structures and functions.

In addition, the chromosomal genes *chvD, chvE* and *picA* of *A. tumefaciens* are involved in pathogenesis; *chvD* by activation of bacterial growth in acidic conditions of plant tissues, *chvE* by chemotaxis and uptake of sugars, and *picA* (plant inducible chromosomal) by alteration of surface properties of bacterial cells resulting in increased aggregation, respectively.

*Structure and Function of Virulence Region*

The *vir* region (35 Kb) is highly conserved both in Ti and Ri plasmids and absolutely required to incite tumor or hairy roots. The major functions of the *vir* region are processing of T-DNA and its transfer to plant cells. The *vir* region of the nopaline-type plasmids encodes six transcriptional loci: *virA, virG, virB, virC, virD,* and *virE,* whereas the *vir* of the octopine-type has *virF* on the right of *virE*. In the Ti plasmids of *A. vitis* (a narrow host range strain or grapevine strain), the *vir* region is short in comparison to those in the wide host range strains.

The *virA* (2 Kb) and *virG* (1 Kb) genes compose one cistron and are required for the regulation of all other *vir* genes. They are constitutively expressed and encode two proteins, VirA and VirG, which are members of the two-components positive regulatory system. The *virD* (45 Kbp) has the function to process T-DNA by topoisomerase and endonucleases, which are the products of the locus. This locus is further divided into $virD_1$ and $virD_2$. The protein $VirD_2$ binds to the 5′ ends of T-intermediate DNA and facilitates its transfer to plant cells as a "pilot" protein.

The *virE* locus (2 Kb) is also divided into $virE_1$ and $virE_2$. The VirE protein binds to the processed single-strand T-DNA and protects it from endonucleases of bacteria and plant origin. The *virB* locus (9.5 Kb) encodes 11 proteins ($VirB_1$–$VirB_{11}$), three of which function in the formation of transfer system of T-DNA from bacterial cell to plant cell. The *virC* locus (2 Kb) encodes two proteins $VirC_1$ and $VirC_2$. The $VirC_1$ binds to the overdrive of T-DNA and enhances its processing. The $VirC_2$ is required for T-DNA transfer to maize.

*Induction of* vir *Region*

The pathogenic bacteria attached to host cell surface respond to phenolic compounds such as acetosyringone (AS)(Fig. 8.6). These compounds are synthesized in and secreted from wounded plant cells and initiate the T-DNA processing by inducing the *vir* region. In this induction, VirA protein (70 kDa) localized to the inner membrane of the

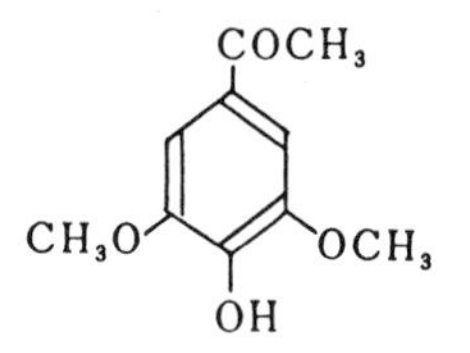


**Fig. 8.6** Structure of acetosyringone.

bacterium acts as a sensor for detecting the presence of AS and activates a regulatory protein VirG (30 kDa) by transmitting this signal through autophosphorylation. The phosphorylated and conformationally changed VirG protein binds to the promoter upstream of *vir* regions and activates the transcription of all *vir* genes through positive regulation.

Other compounds that can serve as the *vir* inducer include pyranose sugars and uronides, some of which are chemoattractants of the bacterium. The opines stimulate the induced *vir* genes and enhance the efficiency of transformation of plant cells. The induction of *vir* genes by acetosyringone is analogous to that of the *nod* genes in *Rhizobium* spp. by flavonoid and its derivatives, the components from unwounded healthy plants.

There are other *vir* regulatory systems which are expressed by three chromosomal genes *ros* (rough surface of colony) and *chvD* in the absence of plant inducers. The *ros* locus encodes a negative regulator of *virC* and *virD*. The *chvD* activates *virG* in phosphate starvation. The host ranges are also affected by these chromosomal genes in addition to *vir* region.

*Structure and Function of T-DNA*

The T-DNA of nopaline-type plasmid (pTiC58) consists of a single fragment (22 Kb) on which the information for the biosynthesis of phytohormones is localized (Fig. 8.7). The octopine-type plasmid (pTiA6) is composed of three segments of $T_L$-DNA (13 Kb), $T_R$-DNA (7.8 Kb), and $T_C$-DNA (central region of T-DNA, 1.5 Kb). The loci for phytohormones are localized on $T_L$-DNA.

The limited host range plasmid of *A. vitis* (grapevine strain) has a 25 Kb $T_C$-DNA which is not integrated into plant genome in transformation. In addition, genetic analysis has revealed that *tmr* genes of the limited host range plasmid is not active. The structural variation of T-DNA in Ti plasmid of *A. vitis* is presumably associated with the transpositional activity of insersion sequence IS868.

The Ri plasmid of *A. rhizogenes* (pRiA4b) has a distinct structure. This plasmid is also composed of two widely separated $T_R$-DNA and $T_L$-DNA regions. The $T_R$-DNA carries only auxin locus *tms*, whereas the $T_L$-DNA carries cT-DNA (or cellular T-DNA), which shares homology with the genome of uninfected carrot and *Nicotiana glauca*.

All T-DNA regions have in both ends the border sequence (BS) or incomplete 25-bp direct repeats. The right BS is more important than the left BS because the processing of T-DNA has polarity starting from the right BS. In the octopine-type T-DNA, a 24-bp flanking sequence called overdrive is localized on the right of the right border and stimulates excision of T-strand, enhancing T-DNA transmission to plant cell.

The T-DNA carries three loci for the biosynthesis of auxin, cytoki-

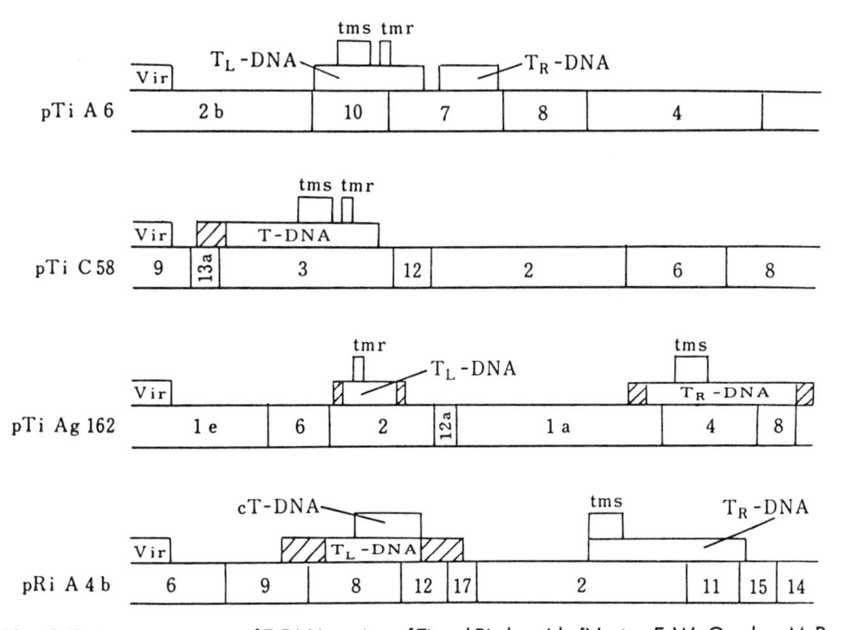


**Fig. 8.7** Restriction maps of T-DNA region of Ti and Ri plasmids. [Nester, E. W., Gordon, M. P., Amasino, R. M., and Yanofsky, M. F. (1984). Crown gall: A molecular and physiological analysis. *Ann. Rev. Plant Physiol.* **35,** 387–413. Reproduced, with permission, from the Annual Review of Plant Physiology].

nin, and opines. The auxin loci (*tms*, or tumor morphology shoot) consists of two genes of *tms1* or tryptophan monooxygenase gene (*iaaM*) and *tms2* or indoleacetoamide hydrolase gene (*iaaH*). The genes *iaaM* and *iaaH* have the high level of homology to those of *P. syringae* pv. *savastanoi*. The *tmr* (for tumor morphology root) encodes the isopentenyltransferase, which catalyzes the biosynthesis of the cytokinin iso-pentenyladenosine 5′-monophosphate from dimethylallyl-pyrophosphate and 5′-AMP.

*Processing of T-DNA and Its Transfer to Plant Cell*

Processing of T-DNA begins with a site-specific nick in the bottom strand at each of 25 bp border sequences by the endonucleases encoded on $virD_1$ and $virD_2$. These enzymes nick the bottom strand of T-DNA at the 5′ end of the right BS and cleave a single-stranded T-DNA or T-strand from right to left (5′→3′), and release it at the 3′ end.

The T-strands capped at the 5′ end by the $VirD_2$ protein are coated along its length with a single-strand-binding protein encoded on *virE*. These proteins protect T-strands from endonucleases of bacteria or plant origin during the infection process. The T-strands are produced in the ratio of 0.1-1 per cell on average.

The transfer of T-strands to plant cells is considered to occur by a mechanism analogous to bacterial conjugation (see Fig. 6.1). It seems that DNA replication in the bacterium produces a replacement strand, and DNA replication in plant cell converts the single-stranded intermediate into duplex DNA. The genes associated in the movement of the T-strands to plant cells differ from those involved in the conjugal transfer of Ti plasmids to other bacterial cells.

*T-DNA Integration Into Plant Genome*

Three copies of T-DNA on average can be detected from the nuclear genome of the transformed plant cell. The number of copies may sometimes account 20-50 per plant genome. The insertion can randomly take place throughout the genome. There are no specific sequences in the plant genome at the site of T-DNA insertion except that the adenine- and thimine-rich sequences are often found at the junctions of T-DNA inserts. The 158-bp direct repeats are detected in both the right and left T-DNA border/plant DNA junctions. It is suggested that these direct repeats result from repair of the staggered nick caused by torsional strain in the target plant DNA.

The T-strand transferred from bacteria to plant cell as a protein-DNA complex is integrated into plant DNA at the 5′ end within 10-40 bp of the right border. The left junction can, however, be spread over 30 to 2000 bp, including the left border sequence.

*Expression of T-DNA in Plant Cells*

All T-DNA integrated into the plant genome is not necessarily expressed because the expression of T-DNA is affected by neighboring sequences. The transformed plant cells acquire the biosynthetic capacity of auxin and cytokinin and come to perform autonomous growth on the artificial media in the absence of phytohormones. Morphological development of the transformed tumor cells into unorganized galls, partially organized teratomas, or hairy roots depends on the intensity of the functions of *tms* and *tmr* as well as the plant-specific sensitivity to these phytohormones.

The plant tumor cells produce the specific amino acid derivatives or opines. The opines are utilized by the pathogens as nutrients. In addition, opines can specifically induce the conjugative transfer of Ti plasmid in the presence of an unknown diffusible conjugative factor, between different strains of *A. tumefaciens* as well as between *A. tumefaciens* and other gram-negative bacteria such as *R. trifolii* and *E. coli.*

### *Pathogenesis of* **Erwinia** *Soft Rot Bacteria*

#### *Recognition and Multiplication*

Bacterial soft rot of vegetables and fruits is induced by many bacteria belonging to *Erwinia, Pseudomonas, Xanthomonas, Bacillus,* and *Clostridium.* The most important bacteria are those referred to collectively as *E. carotovora* and *E. chrysanthemi.* These bacteria usually attack a wide range of dicotyledons and a limited number of monocotyledons and woody plants. There is the possibility that the active defense responses are involved in the insusceptibility of monocots and woody plants in association with the signal compounds derived from plant cell wall fragments, although nothing has been elucidated in the mechanisms of recognition or active defense reaction in soft rot diseases.

#### *Pectic Enzymes and Pathogenicity*

The activity of pectic enzymes is generally analyzed with sliced or excised plant tissues. The pathogenicity determined with sliced tissues, particularly of storage organs such as tubers, roots, and succulent leaves, often cannot be reproduced on the actively growing, intact portion of the same plants because of active defense reactions.

*Pathogenesis and Pectinases* *Erwinia* soft rot bacteria have the capacity to produce various kinds of pectinases such as pectate lyase or pectic acid transeliminase (endo-PL; E.C.4.2.2.2), pectin lyase or pectin transeliminase (endo-PNL; E.C.4.2.2.10), pectin methylesterase (PME; E.C.-3.1.1.11), and polygalacturonase (endo-PG; E.C.3.2.1.15, exo-PG; E.C.-3.2.1.67).

Genetic analysis has clarified that endo-PL is the major virulence factor among the pectic enzymes mentioned above. However, endo-PG of *E. carotovora* can also induce maceration of potato tissues and PME of *E. chrysanthemi* seems to be involved in pathogenesis because the $PME^-$ mutant was noninvasive when inoculated on leaves of *Saintpaulia.* In *E. chrysanthemi* (strain EC16), exo-PG contributes to bacterial utilization of polygalacturonate and induction of pectate lyase.

*Induction of Pectinases* The inducers of endo-PL and oligogalacturonate lyase (OGL; E.C.4.2.2.6) are the metabolic intermediates 2-keto-3-deoxygluconate (KDG), 5-keto-4-deoxy-uronate (DKI), and 2,5-diketo-3-deoxygluconate (DKII). In contrast, endo-PNL is induced by DNA-injuring agents such as UV, mitomycin C, and nalidixic acid as well as DNA-injuring substances derived from wounded plant tissues. Thus, endo-PNL is induced by the specific mechanism which is equivalent to *recA*-dependent SOS reaction (see Chapter 4). The DNA sequences homologous to endo-PNL are detected in *E. carotovora* subsp. *carotovora, E. carotovora* subsp. *atroseptica,* and *E. rhapontici* but not *E. chrysanthemi.*

*Secretion of Pectinases*

Secretion of pectinases is controlled by a protein export system that is constitutive. The endo-PL of *E. chrysanthemi* is dependent on two constitutive enzymes encoded on two loci *out1* and *out2*. The locus *out1* controls the export of endo-PL from the cytoplasm to the periplasm which is synchronized with the translation of mRNA. The complete enzyme protein is excised from the leader peptide, and released outside cytoplasmic membrane (see Fig. 4.11). The locus *out2* controls the transfer of endo-PL which was accumulated in the periplasm across the outer membrane. No structural alteration occurs on the enzyme molecules in this step. The export-defective mutants ($out^-$) also show pleiotropic defects in secretion of cellulase and protease, motility, and virulence. The *out* genes scatter on the chromosome in *E. chrysanthemi,* whereas they form a cluster in *E. carotovora.*

*Isozymes of Endo-PL*

The endo-PL of *E. chrysanthemi* consists of five isozymes designated PLb, PLc, PLa, PLd, and PLe that are encoded as the independent transcriptional units on the loci *pelB, pelC, pelA, pelD,* and *pelE,* respectively. The isozyme PLa is acidic, PLb and PLc are neutral, and PLd and PLe are alkaline. The endo-PL isozymes are different in their isoelectric point (pI), but are common in the properties such as functions, optimum pH in alkaline range, requirement for a divalent cation ($Ca^{2+}$), and the capacity to split pectic acid by $\beta$-elimination.

The endo-PL of *E. carotovora* is also differentiated into five isozymes of $PL_1$ to $PL_5$ with different isoelectric points, and that of *E. carotovora* subsp. *atroseptica* into six isozymes. The endo-PL isozymes of *E. carotovora* subsp. *carotovora* are also distinct in that the loci *peh1* encoding endo-PG and *pel3* form a cluster. There is the 72% DNA homology between the locus *pelB* of *E. chrysanthemi* and *pel2* of *E. carotovora.*

The loci *pelB* and *pelC* of *E. chrysanthemi* form a cluster nearby the *ilv* locus and have 83–93% homology in DNA sequences. The loci *pelA, pelD,* and *pelE* also form a cluster and are arranged in the order of *pelA-pelD-pelE,* forming three independent transcriptional units.

*The Capacity of Enzymes to Macerate Plant Tissues*

The isozyme PLe has a major role in the pathogenesis of *E. chrysanthemi* because the amount of PLe protein reaches to 50 to 60% of total isozyme proteins produced in plant tissue. In contrast, the maceration capacity of PLa is roughly equivalent to 1/1,000 of PLe and cannot macerate potato tissues by itself. This is considered to be the result of its low isoelectric point, which prevents the binding of PLa protein onto the negatively charged plant cell walls.

To fully express its pathogenicity to *Saintpaulia* leaf tissues, on the other hand, *E. chrysanthemi* requires the presence of PLa in addition to PLe but not PLb and PLc. The synergistic effect seems to be present between the different isozymes. Likewise, endo-PNL, which preferentially cleaves the pectic substances with high methoxyl content, enhances tissue maceration under coexistence with endo-PL.

The pathway of catabolism of pectic substances in *E. chrysanthemi* was shown in Fig. 4.5. There is a report that the $OGL^-$ and $Exu^-$ mutants of the bacterium still macerate potato tissues. This fact suggests the possibility that the bacteria have another set of pectic enzymes that can be induced only in plant tissues by substances other than galacturonate-degradation products or KDG. If this is the case, the role of pectic enzymes in pathogenesis should also be reviewed carefully in other phytopathogenic bacteria that lack the *in vitro* activity of KDG-inducible pectic enzymes.

### *Genetic Analysis of Pathogenicity Genes*

*Pseudomonas syringae* is a fluorescent pseudomonad, originally described as the lilac blight pathogen but subsequently found to have a wide host range. Many nomenspecies that are similar in physiological and biochemical properties but distinct in pathological traits were lumped to *P. syringae* in 1980 and differentiated at the infrasubspecific rank of pathovars. Therefore, *P. syringae* includes various pathogenic bacteria that cause leaf spot, chlorosis, blight, canker, and galls.

The virulence factors of *P. syringae* have been elucidated in a limited number of pathovars inducing chlorosis, necrosis, and galls. Genetic studies in this group may be placed into three categories, i.e., (1) analysis of genes for virulence factors, (2) analysis of genes for pathogenicity, and (3) analysis of race-specific genes. The last two categories deal with the fundamentally same function, i.e., host specificity at the species level or at the cultivar level.

#### *Genetic Analysis of Biosynthesis of Toxins and Phytohormones*

The genes governing the biosynthesis of indoleacetic acid (*P. syringae* pv. *savastanoi*), coronatine (*P. syringae* pv. *atropurpurea*), phaseolotoxin (*P. syringae* pv. *phaseolicola*), tabtoxin (*P. syringae* pv. *tabaci*), and syringomycins and syringostatins (*P. syringae* pv. *syringae*) have been determined either on the chromosome or plasmids. The genes for these virulence factors are not expressed in *E. coli* except those for the biosynthesis of indoleacetic acid in *P. syringae* pv. *savastanoi.* The virulence factor deficient mutants of these bacteria retain the capacity to grow in the intercellular space and produce small lesions with a water-soaked appearance.

The coronatine biosynthesis genes are located on an at least 30 Kb fragment that is strongly conserved in large plasmids (90 to 105 Kb) that reside in *P. syringae* pathovars-*atropurpurea, glycinea, tomato* and *morsprunorum.* For syringomycin biosynthesis genes, two loci of *syrA* (2.3–2.8 Kb) and *syrB* (2.4–3.3 Kb) have been identified on chromosome of *P. syringae* pv. *syringae.* The latter is assumed to be regulator genes that can be induced by the presence of plant signal components such as arbutin and D-fructose. These loci encode the SR4 (350 Kda) and the SR5 (130 Kda) proteins, respectively, and which seem to be the members of five toxin-producing enzymes.

*P. syringae* pv. *savastanoi* causes knot or gall formation on olive and oleander. The oleander strains show pathogenicity both on oleander and olive, whereas the olive strains only on olive. The biosynthetic pathway of indoleacetic acid (IAA) in this bacterium involves tryptophan 2-monooxygenase for conversion of L-tryptophan to indoleacetoamide and indoleacetoamide hydrolase for conversion of indoleacetoamide to IAA (Fig. 8.8). The former enzyme, a 62 kDa protein requiring flavin-adenine dinucleotide as a coenzyme, is encoded on the gene *iaaM* and the latter, a 47 kDa protein, on the gene *iaaH* either on the chromosome or a plasmid (pIAA), depending on the strain. The function of the *iaaM* gene is under the feedback regulation of IAA. The oleander strains produce cytokinins (trans-zeatin) in addition to IAA. The cytokinin genes (*ptz*) were found

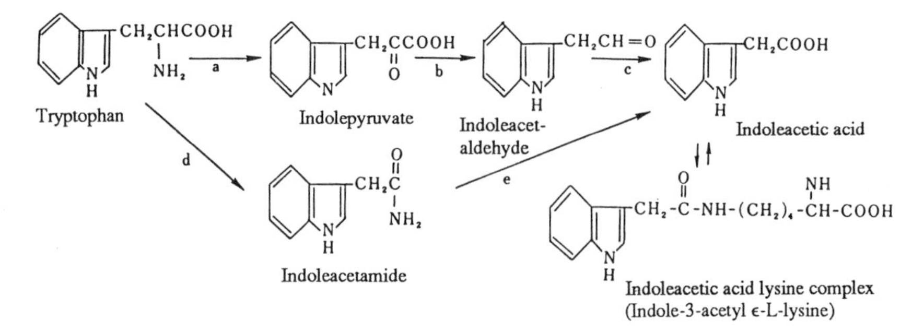


**Fig. 8.8** Biosynthetic pathways of indoleacetic acid. a, tryptophan aminotransferase; b, indolepyruvate decarboxylase; c, indoleacetaldehyde dehydrogenase or indoleacetoaldehyde oxidase; d, tryptophan monooxidase (oxidative decarboxylase); e, indoleacetamide hydrolase.

on a 42 Kbp plasmid and cloned in *E. coli,* verifying the role of cytokinins as another virulence factor of the bacterium.

*P. syringae* pv. *savastanoi* is the first plant pathogenic bacteria in which the virulence genes were determined by cloning. When the wild type strains with pIAA are mutated to resistance to 5-methyltryptophan, the plasmid was eliminated (curing) and at the same time both the virulence and the capacity to synthesize IAA were lost, although the ability to produce cytokinins was retained. The plasmid DNA fragment including the locus *iaaM* was integrated into a vector and transferred into *E. coli* by transformation. The cloned gene was expressed in *E. coli,* producing IAA in culture media. When the *iaaM* plasmid thus cloned was transferred by conjugation to a plasmidless strain of *P. syringae* pv. *savastanoi,* the bacterium was restored virulence and the capacity to synthesize IAA.

The nucleotide sequences of *iaaM* and *iaaH* were determined, revealing that the *iaaM* protein consisted of 557 amino acids and the *iaaH* protein of 455. There was 50% homology between *iaaM* and *tms1* of T-DNA of *A. tumefaciens* and 27% homology between *iaaH* and *tms2* of T-DNA, respectively.

Genetic analysis of *P. syringae* pv. *savastanoi* also showed a wide distribution of the insertion sequences IS51 and IS52, both on pIAA and chromosome, suggesting the origin of naturally occurring avirulent strains of *P. syringae* pv. *savastanoi.* The insertion of these IS elements in the *iaaM* locus (2.75 Kb) induces loss of virulence, defective production of IAA, and inactivation of both tryptophan 2-monooxygenase and indoleacetoamide hydrogenase. No homology is found between the two IS elements. About 70% homology was detected between IS51 of the bacterium and IS868 of the octopine-type Ti plasmid of *A. tumefaciens,* showing that these IS elements are highly conserved in the entirely different bacteria.

### *Genetic Analysis of Biosynthesis of Extracellular Polysaccharides*

Extracellular polysaccharides (EPS) have important roles in pathogenesis and in ecological behaviors such as survival and dissemination of plant pathogenic bacteria. However, genetic analysis of EPS biosynthesis in plant pathogenic bacteria is still in it's initial stage.

*Extracellular Polysaccharides of* Erwinia stewartii, *the Causal Agent of Stewart's Disease of Corn* In this bacterium, EPS is the primary virulence determinant that occludes xylem vessels and induces water soaking on diseased leaves. The mutation in EPS production occurs pleiotropically with the ability to induce wilting and water soaking as well as colony morphology. The EPS is an acidic heteropolysaccharide composed of glucose, galactose, and glucuronic acid. The EPS biosynthesis genes (*cps*) have been identified on a 27 Kbp region of chromosome. This region was distinguished as five clusters (*cpsA* to *cpsE*), of which *cps* to *cpsD* were contiguous on a 10 Kbp fragment. Complementation analysis showed that all of the mutants in these clusters failed to induce wilting but varied in the ability to elicit water soaking as well multiplication in plant tissue; some mutants produced only restricted necrotic lesions with normal bacterial growth in tissue, but others showed complete avirulence without appreciable growth. Although the *cps* genes are poorly expressed in *Escherichia coli,* the regulatory function of *E. stewartii* shares a number of similarities with the positive regulator for capsular polysaccharide synthesis genes (*rcsA*) of *E. coli* K12.

*Extracellular Polysaccharide (Xanthan) of* Xanthomonas campestris *X. campestris* produces xanthan (see Chapter 2) through following steps: (1) formation of a pentasaccharide subunit from the precursors UDP-glucose, GDP-mannose, UDP-glucuronic acid, acetyl coenzyme A, and phosphoenolpyruvate; and (2) subsequent polymerization of polysaccharide units. The complementation analysis showed that the xanthan biosynthesis genes (*xps*) are identified on a 13.5 Kb region of chromosome and distinguished as at least five clusters. A gene affecting pyruvylation of xanthan gum is also linked to these clusters.

*Genetic Analysis of Hypersensitive Reaction and Pathogenicity Genes* (hrp)

The genes of plant pathogenic bacteria that elicit the race-nonspecific hypersensitive response on incompatible hosts and the pathogenic response on compatible hosts are clustered usually on the chromosome loci designated *hrp.* In *P. syringae* pv. *phaseolicola, hrpM* is located elsewhere in the genome in addition to the *hrp* cluster. Mutation in the hypersensitive reaction is usually pleiotropic and causes concurrent mutations in pathogenicity, ability to colonize plants, colony form, or motility.

The *hrp* genes have been cloned in many plant pathogenic bacteria such as *E. amylovora, P. solanacearum,* and several pathovars of *P. syringae* and *X. campestris.* The DNA sequences that are homologous to those of *hrp* genes are highly conserved in many different pathovars of the same species or different species of the same genera.

The *hrp* cluster of *P. syringae* pv. *phaseolicola* covers a 17.5 kb fragment and is organized into the seven distinct complementation groups of *hrpL, hrpAB, hrpC, hrpD, hrpE, hrpF, and hrpSR.* The expression of *hrp* genes is influenced by nutritional conditions. It is actively transcribed *in planta* and in defined minimal medium but only weakly in complex media. Mutations in all loci except *hrpC* induce remarkable reduction in the capacity of bacteria to multiply in bean leaves.

The *hrp* cluster of *P. syringae* pv. *syringae* is unique in that it consists of 11 complementation groups on a 36 kb fragment and its cosmid clone enables *P. fluorescens* and *E. coli* to produce a full or a weak HR on tobacco. It has also been reported that *E. coli* contains genes which complement some of the *E. amylovora hrp* mutants. The *hrp* of *X. campestris* pv. *vesicatoria* also consists of six loci designated *hrpA* through *hrpF* on a 25 kb fragment and that of *P. syringae* pv. *syringae* strain 61 of 13 loci on a 25 kb fragment.

Thus, the *hrp* clusters generally consist of many loci, and each locus is likely to have a distinct function. The elicitor proteins seem to be encoded on the core portion of the *hrp* cluster. A portion of the *hrp* cluster may function as an environmental sensor and positively regulate some other *hrp* genes during plant-bacteria interactions as in the case of *vir* region of *A. tumefaciens*. Other portions of the *hrp* cluster may encode the membrane proteins that work in secretion of elicitor. It is assumed that functions of *hrp* genes relate with basic metabolic traits such as utilization of carbon source. Although the precise functions of *hrp* genes in plant-bacteria interactions have not been elucidated, a hypothesis has been proposed that *hrp* genes may be involved in the suppression of plant defense responses in the compatible host plant that have been well documented in fungal diseases.

The interrelationship between *hrp* genes and avirulence (*avr*) genes, which will be described next, in terms of functions in plant–bacteria interactions is a problem to be elucidated. *X. campestris* pv. *phaseoli* was converted to induce the cultivar-specific HR when a cosmid clone from *X. campestris* pv. *vesicatoria* was mobilized into the former. This gene, referred to as the nonhost avirulence gene *avrRxv*, also inhibited disease development by several other pathovars of *X. campestris* on their own hosts. Similarly, the avirulence gene *avrD* of *P. syringae* pv. *tomato* also gave a unique cultivar-specific HR pattern to soybean when it was introduced into *P. syringae* pv. *glycinea*. On the basis of these findings, it is suggested that the expression of avirulence genes underlies nonhost resistance; or nonhost resistance is under a similar control as that in cultivar–race interactions. In *X. campestris* pv. *vesicatoria*, for instance, it was reported that *hrp* gene is required for the *in planta* expression of *avrBs3* but not for its *in vitro* expression, implying direct interaction between these two gene products in the elicitation of HR response.

*Genetic Analysis of Race-Specific Genes*

*P. syringae* pv. *glycinea* is the causal agent of bacterial blight of soybean (*Glycine max*). This pathovar is further differentiated into several races on the basis of unique patterns of pathogenicity on soybean cultivars (Table 8.2). The cultivar-specific pathogenicity can be interpreted according to the gene-for-gene hypothesis. In this hypothesis, recognition of incompatibility or race-specific resistance occurs only when the interactions take place between the dominant R gene for resistance of the host plant and the dominant A gene conditioning the avirulence of the bacterium. Resistance is phenotypically expressed as HR. The avirulence gene is referred to as the gene of potentially virulent bacteria which are unable to cause disease in specific cultivars of soybean. Flor's gene-for-gene hypothesis has been well documented in fungal diseases, but the system was also confirmed recently in many plant pathogenic bacteria such as *P. syringae* pv. *glycinea, P. syringae* pv. *tomato, X. campestris* pv. *malvacearum, X. campestris* pv. *oryzae,* and *X. campestris* pv. *vesicatoria.*

In the pioneer work of Staskawicz *et al.* (1984) with *P. syringae* pv. *glycinea,* the avirulence gene A cloned from race 6 (*avrA*) is dominant and confers incompatibility to race 6 on cultivars "Harosoy," "Acme," and "Peking," whereas race 5 does not carry the avirulence gene. The *avrA* gene was identified on a 27.2 kb chromosomal DNA fragment in a genomic library of a race 6 strain which was constructed in the mobilizable cosmid vector pLAFR1 and cloned in *E. coli.* A single clone (pPG6L3) out of 680 random cosmid clones mobilized from *E. coli* to the race 5 strain showed alteration in the race specificity of race 5 from virulent to avirulent on cultivar "Harosoy" and "Peking." This change was associated with the appearance of HR, reduction of bacterial growth, and accumulation of the phytoalexin glyceollin.

The clone was also mobilized from *E. coli* to race 1 and race 4, changing the race specificity on cultivars "Acme" and "Chippewa" from virulent to avirulent, respectively, and conferring on these transconjugants the same incompatibility patterns as the wild-type race 6 strains. The original incompatibility pattern of the race 5 strain was restored when transposon Tn5 was inserted into the fragment governing the race 6 specificity in these transconjugants. The *avrB, avrC,* and *avrD* which interact with soybean resistance genes *Rpg1, Rpg3,* and *Rpg4* were subsequently identified, and the nucleotide sequences were determined with some of these genes.

In *X. campestris* pv. *vesicatoria,* the causal agent of bacterial leaf spot of pepper, the *avrBs1* is localized on a transmissible copper resistance plasmid, XvCu1, and induces HR on the resistant cultivar ECW1OR, which carries the dominant resistance gene *Bs1*. The *avrBs1* gene is inactivated by insertion of IS476 (1.2 kb) expressing virulence to the cultivar ECW1OR. This mutant may be isolated in nature as the spontaneous mutant with the altered host range.

Most avirulence genes are chromosomal except those in *X. campestris* pv. *vesicatoria* and *X. campestris* pv. *malvacearum*. Plant pathogenic bacteria are usually equipped with several or more avirulence genes, but they do not closely link. Some avirulence genes of a cluster seem to confer the pleiotropic phenotypes by inducing HR on resistant cultivars and conditioning virulence on susceptible cultivars. It has been suggested that many bacteria with different taxonomic affiliation such as races, pathovars, species, or even genus retain the loci which have the homology of nucleotide sequence with the *avr* genes. These loci are likely to be repressed in these bacteria, resulting in the negative expression of HR.

The biochemical basis of gene-for-gene incompatibility has been interpreted by three hypothesis, the elicitor-receptor model, the dimer model, and the ion-channel model. The elicitor-receptor model may be associated with generalized defense mechanisms, including the pathogen/nonhost interaction. The other two models are more likely to be involved in the unique incompatibility of cultivars. In the dimer model, the R gene products (protein) of host plants may interact with the *avr* gene products of bacteria, forming a heterodimer complex that induces the inhibition of bacterial growth, accumulation of phytoalexin, and development of HR. The ion-channel model presumes opening of an ion channel in the cell membrane inducing efflux of electrolytes followed by subsequent host-cell death and/or elicitation of active defense.

*Disease-Specific Genes*

There is only limited information available on the structures and functions of the disease-specific (*dsp*) genes. Mutation of the *dsp* genes causes loss of pathogenicity without missing the HR-inducing ability. In *P. solanacearum,* three *dsp* regions scatter in the genome. One of these genes covers about 15 kb and contains genes that simultaneously regulate pathogenicity to tomato and the growth of this bacterium under acidic conditions of host plants. The *dsp* genes are assumed not to be directly involved in pathogenesis but have a role in adaptation of *P. solanacearum* to the new environments of plant tissues through control of some general metabolic pathways.

# 8.2 Resistance

Disease resistance includes active or dynamic resistance and preformed or static resistance on the basis of the mechanisms involved. The former is the postinfectional defense reaction associated with morphological and biochemical changes. The latter is the preinfectional resistance based on inadequate or unfavorable structural barriers and/or chemical components.

## 8.2.1 ACTIVE RESISTANCE

### *Hypersensitive Reaction and Phytoalexins*

*Hypersensitive Reaction*

When living bacteria are injected into mesophyll of leaves of incompatible plants at a concentration of around $10^8$ cells/ml, a rapid and confluent necrosis develops within 8 to 12 hr after inoculation. The bacteria trapped in the necrotic lesions are generally killed rapidly. This swift response is called hypersensitive reaction (HR). The HR may occur in incompatible combinations between nonhost plants and virulent strains of plant pathogenic bacteria, resistant cultivars and virulent strains, and susceptible cultivars and avirulent mutants. The genetic background of HR was described above in relation to pathogenesis.

Investigations on avirulence genes (*avr*) of several bacterial plant pathogens indicated that the HR between races of pathogenic bacteria and resistant cultivars can be explained on the basis of Flor's gene-for-gene hypothesis. In this type of HR, resistance genes are directly involved in interactions with *avr* genes. On the other hand, the HR in incompatible combinations between nonhost plants and species, subspecies, and pathovars of plant pathogenic bacteria is induced by hypersensitive reaction and pathogenicity genes (*hrp*). In this case, however, participation in HR of the resistant genes of nonhost plants has not been clarified.

The typical HR is always induced by living bacteria. The process consists of three stages: an induction time of 3–4 hr, a latent period of 4–5 hr, and a tissue collapse period of 1–2 hr. The individual components of the bacterial cell envelope such as outer membrane protein, lipopolysacharides, lipoprotein-peptidoglycan complex have no function as the specific HR elicitor. The novel elicitor of HR is likely produced in affected plant cells: in HR induced by the avirulence gene *avrD* of *P. syringae* pv. *tomato,* the low molecular weight compound produced by 34 kDa protein was suggested as the possible elicitor.

The HR induced by bacteria is thus characterized by the extremely rapid response. It develops before any detectable increases occur in oxidase activities. Therefore, the necrotic lesions produced by HR usually become white to light brown and dry up into thin film. It is distinctly different from dark brown lesions of HR induced by viruses and fungi.

In general, many phytopathogenic bacteria that produce local necrotic lesions can induce HR in incompatible hosts. Soft rot bacteria of genus *Erwinia* and *Pseudomonas,* the tumor-inducing bacteria of *Agrobacterium,* and some pathovars of *Xanthomonas* usually do not induce HR. In most cases, however, these HR negative bacteria are significantly suppressed in their growth in incompatible plant tissues.

The mechanisms of HR induction are not necessarily the same in detail, depending on the combination of plants and bacteria. In the initial stage of HR, bacterial cells attach to the plant cell wall with their fimbriae and are immobilized. These bacterial cells are usually enveloped by a membranous, cuticle-like substance. However, in the HR of soybean induced by *X. campestris* pathovars, immobilization or envelopment is a secondary reaction. In this case, HR can be induced without immobilization when bacteria produce extracellular polysaccharides or multiply in a limited degree in intercellular space.

When HR is induced, adenosine triphosphatase-dependent efflux of $K^+$ and influx of $H^+$ take place leading to the intracellular acidification and intercellular alkalization in association with rapid stimulation of lipoxygenase activity resulting in the increased production of superoxide ($O_2^-$) and hydrogen peroxide ($H_2O_2$). These chemicals injure the lipid bilayer of the cell membrane through peroxidation, causing decrease of its diffusion potential and leakage of electrolytes. The membrane undergoes substantial damage in its structure, leading to the final step of cell collapse.

Hypersensitive reaction may play an important role in plant infection under natural conditions. For example, disease severity incited by the cherry strain of *P. syringae* pv. *morsprunorum* on cherry plants is markedly decreased if the plum strain is mixed with the cherry strain in the inoculum. The decrease of infection has been explained on the basis that an HR is induced by the plum strain in the incompatible cherry host. A similar phenomenon has been reported with many plant pathogenic bacteria.

*Synthesis of Phytoalexin*

Induction of HR is followed by the production of phytoalexin, which is responsible for the rapid decrease of the bacterial population. In soybean leaves, however, the phytoalexin, glyceollin, accumulates in mesophyll inoculated with xanthomonads either of incompatible or compatible combination, implying that phytoalexin is not directly involved in resistance. In general, illumination is required for the development of HR and synthesis of phytoalexin. Bacterial growth is, therefore, not suppressed in the dark.

Some of the phytoalexins produced by infection of plant pathogenic bacteria are shown in Table 8.1. Kievitone and phaseollin are characteristic in the selective inhibition of the growth of gram-negative bacteria. A large amount of rishitin and phytuberin accumulates in potato tubers after inoculation with *Erwinia* soft rot bacteria and incubation in air, but not in low oxygen. This may explain the reason why potato tubers suffer soft rot severely in anaerobic storage conditions.

For biosynthesis of phytoalexins, the phenylpropanoid pathway plays an important role. Phenolic compounds such as phytoalexin and lignins generally synthesized through the phenylpropanoid pathway: phenylalanine is deaminated to *trans*-cinnamic acid by phenylalanine ammonia-lyase (PAL) and tyrosine to *p*-coumaric acid by tyrosine ammonia-lyase (TAL), and further converted to various phenolic compounds (Fig. 8.9).

### *Induced Resistance*

When heat-killed cells of *P. syringae* pv. *tabaci* are infiltrated into one-half of a tobacco leaf and water into the other half and the plant is kept under illumination for 24 hr, growth of the bacterium that is challenge inoculated as well as subsequent symptom development are com-

**Table 8.1**
Phytoalexins Produced by Plants Infected by Plant Pathogenic Bacteria

| Plant | Bacteria | Phytoalexin |
|---|---|---|
| Cotton | *X. campestris* pv. *malvacearum* | Dihydroxycadalene<br>Lacinilene |
| Kidney bean | *P. syringae* pv. *phaseolicola* | Phaseollin, phaseollinisoflavan<br>Coumestrol, kievitone |
| Lettuce | *P. cichorii* | Costunolide, lettucenine A |
| Loquat | *P. syringae* pv. *eriobotryae* | Aucuparin |
| Potato | *E. carotovora* subsp. *carotovora* | Rishitin |
| | *E. carotovora* subsp. *atroseptica* | Phytuberin |
| Soybean | *P. syringae* pv. *glycinea* | Glyceollin I, II, III,<br>Hydroxyphaseollin |

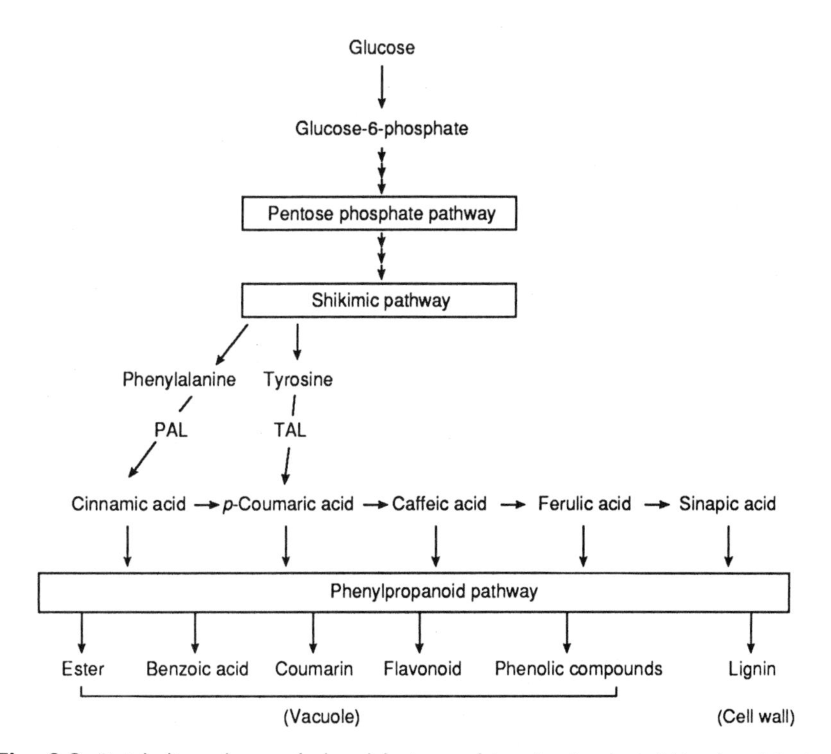


**Fig. 8.9** Metabolic pathway of phenylalanine and tyrosine in plant. PAL, phenylalanine ammonia lyase; TAL, tyrosine ammonia lyase.

pletely prevented in the former but not in the latter. The protective effect is light dependent but not specific to bacteria because the same effects can be induced by prior infiltration of heat-killed incompatible pathogenic bacteria or living saprophytic bacteria. The protective effect may spread from one area to another on the same leaf or from one leaf to another. This phenomenon is called induced resistance, or acquired resistance.

A similar phenomenon is observed in many other combinations of hosts and microorganisms. The role of elevated activity of peroxidases as well as biosynthesis of peroxidase isozymes has been postulated as the mechanism of induced resistance, although some contradictory evidence has been reported. It has been documented that the systemic induction of protease inhibitors is involved in plant resistance to insects, and abscisic acid has a key role in the induction as the systemically mobile signal. The signal substance for systemic induction of disease resistance has not been clarified. Ethylene was once suggested as the possible signal, it is currently considered more as a symptom than a signal.

## 8.2.2 PREFORMED RESISTANCE

### *Resistance by Physical Factors*

*Surface Structure*

Bacterial infection through stomata is affected not only by the number of stomata per unit area but also by their size and structural differences. In citrus leaves, the horn-like projections or outer ledges over the front cavity develops as the leaves mature. The open stomata with undeveloped short ledges are always found on young leaves within one month of unfolding, whereas closed stomata with fully developed ledges appear in mature leaves (Fig. 8.10). Rainwater containing citrus canker organism (*X. campestris* pv. *citri*) is prevented from entering the stoma by micro air bubbles in the front cavity. In stoma with open ledges, water can easily enter into the front cavity. Therefore, citrus canker infection through stomata occurs in severe forms on young leaves, but mature leaves about one month after unfolding are essentially free of it.

In twigs the stomata with open ledges distribute in the shoot tip. The infection belt of citrus canker, therefore, moves upward as shoots elongate. When longitudinal growth stops, infection through stomata ceases completely. Infection occurs for a longer period on fruits on which open stomata continuously develop until maturity. Because this type of resistance originates only in the physical blocking of ingress of the pathogen,

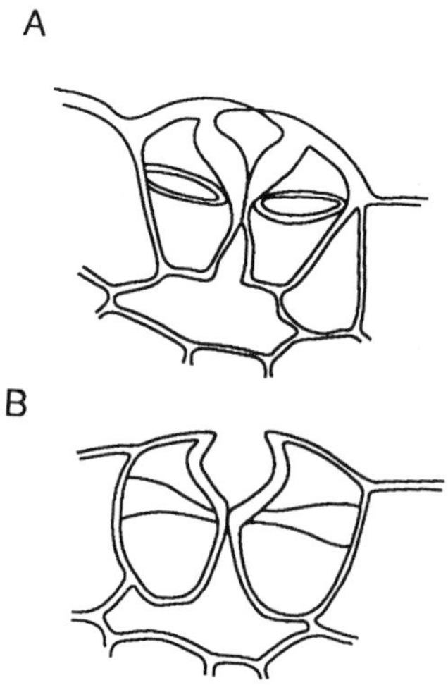

A

B

**Fig. 8.10** Structural alterations of stomata along with maturation of citrus leaves. A, mature leaves; B, young leaves.

mature leaves can be easily infected when pathogenic bacteria are forced to enter into the stomata by pressing between rubber stoppers without injuring the leaf surface. Likewise, differences in the susceptibility of *Leersia* spp. to bacterial leaf blight pathogen (*X. campestris* pv. *oryzae*) are explained by the structural differences of hydathodes.

*Postinfectional Defense Structures*

The moderate resistance of Satsuma orange (*Citrus unshu*) to citrus canker reflects the restricted development of the lesions on twigs of spring shoots which serve as a source of inoculum in the next season. The lesions formed on new shoots are dark green with a water-soaked appearance. As twigs grow and mature, lesions turn reddish brown to dark brown, and eventually stop enlarging after reaching about 0.5 to 2 mm in diameter. A thick defensive cork layer consisting of several- to 10-cells are formed between the lesion and bark parenchyma. Subsequently, these lesions become desiccated and separate from the bark, leaving a healed surface. The causal bacterium dies quickly in the dried lesions. In susceptible navel orange (*C. sinensis*) or Natsudaidai (*C. natsudaidai*), lesions continue to enlarge, forming a water-soaked zone in the margin without formation of a cork layer underneath the lesions; thus, the bacterium can continue to grown in the border adjacent to healthy tissues (Fig. 8.11).

*X. campestris* pv. *pruni* also induces water-soaked lesions on peach leaves. As these spots enlarge, abscission layers are formed around lesions in a host reaction to infection. The lesions subsequently slough off, leaving holes with healed edges. In this reaction, one or two cell layers around the spot enlarge, forming a hypertrophied zone; the pectic substances in the middle lamellae are subsequently dissolved and the connections with healthy cells via the plasmodesmata are severed. Thus, lesions including the pathogen become dry, shrivel, and eventually fall out of the leaf, leaving a small hole.

***Resistance by Chemical Factors***

Resistance may be attributed to chemical components that normal tissues of host plant contain before infection. It may be explained by a nutritional imbalance or by the presence of inhibitory substances. Pathogenic pseudomonads are characterized by a limited capacity to utilize amino acids as a sole source of carbon in comparison with the saprophytic pseudomonads. The inability of bacteria to catabolize the specific amino acids may reflect (1) the absence of permeases, (2) lack of specific enzymes involved in metabolism, or (3) toxicity of the amino acid to the bacteria.

The toxicity of serine to plant pathogenic bacteria has been well documented. Growth of *X. campestris* pv. *citri* is inhibited by the presence

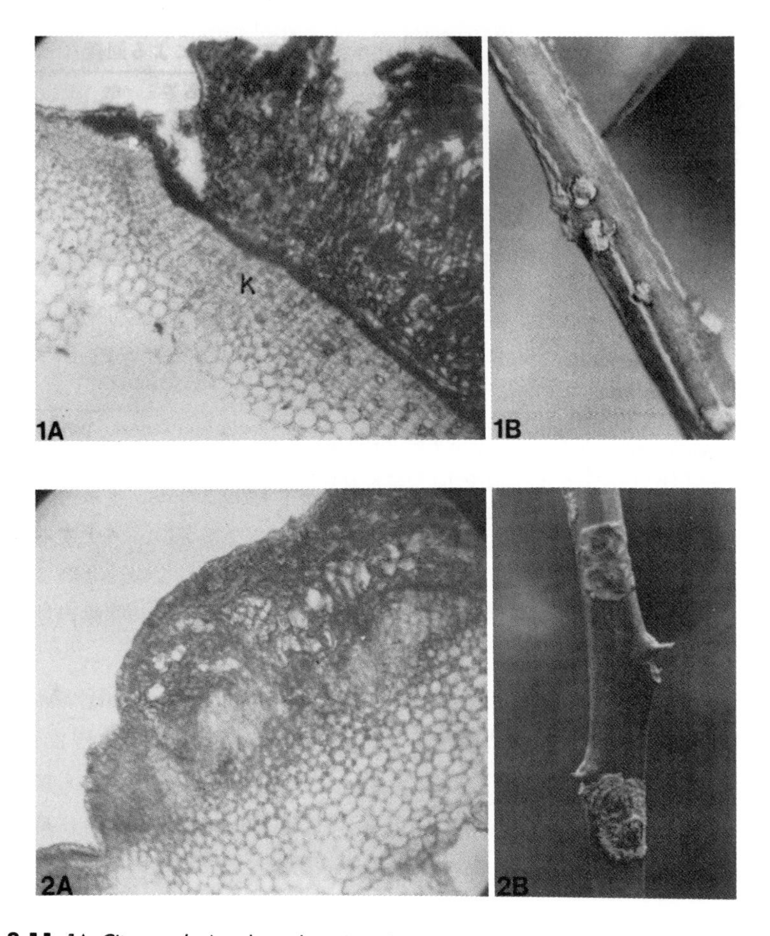


**Fig. 8.11** 1A, *Citrus unshu* (moderately resistant), transverse section of a lesion formed on twig; 1B, Lesions formed on twig of *C. unshu;* 2A, *C. natsudaidai* (susceptible), transverse section of a lesion formed on twig; 2B, Lesions formed on twig of *C. natsudaidai.*

of serine and hydroxylysine. The inhibition due to these amino acids can be reversed by proline in the presence of low concentrations of methionine. The intercellular fluid of citrus leaves contains serine, hydroxylysine, methionine, and proline at concentrations high enough to affect such interactions in amino acid metabolism of the pathogen. It is possible, therefore, that the changes in concentrations of these amino acids in citrus leaves may affect the multiplication of *X. campestris* pv. *citri* during an initial stage of infection. A similar relationship between amino acid metabolism and resistance has also been reported with other plant pathogenic pseudomonads and xanthomonads.

The tissues of *Pyrus* spp. contain arbutin (hydroquinone-β-D-glucoside) and those of *Marus* spp. phloridzin. These phenol compounds have been evaluated as a factor associated with the defense mechanism against the fire blight pathogen (*E. amylovora*). The compounds are hydrolyzed by β-glucosidase into glucose and the aglycone hydroquinone and phloretin. Glucose may be catabolized by the pathogenic bacterium. The aglycones are further oxidized and inhibit growth of the pathogen. They are also polymerized to form brown pigments leading to the formation of defense barriers.

Similar phenol compounds reported in association with resistance include DIMBOA (2,4-dihydroxy-7-methoxy-benzoxazinone) of corn against soft rot *Erwinia* spp., β-D-glucogallin of strawberry against *P. solanacearum*, and phenylacetic acid of rice against *X. campestris* pv. *oryzae*. In addition, diverse kinds of antimicrobial phenol compounds are found in plant tissues. In many cases, however, the role of these compounds in resistance *in situ* has not been precisely elucidated.

### *Resistance by Other Factors*

In general, resistance in plants to bacterial pathogens increases with advance of plant growth. This is referred to as adult plant resistance. Such resistance has been observed with bacterial wilt of tobacco (*P. solanacearum*), bacterial blight of cassava (*X. campestris* pv. *manihotis*), bacterial leaf streak of rice (*X. campestris* pv. *oryzicola*), kresek syndrome of bacterial leaf blight of rice (*X. campestris* pv. *oryzae*), and blister spot of apple (*P. syringae* pv. *papulans*).

There are, however, some instances in which resistance decreases with plant age. In Japan, bacterial soft rot of lettuce incited by soft rot pseudomonads increases as lettuce plants pass through each growth stage of seedling, before heading, heading, and harvest stage. The highest susceptibility of head lettuce just before or at harvest may be associated with the effect of low temperature late in autumn. Low temperature may enhance activities of pathogenic pseudomonads and incite physiological stress on mature lettuce plants.

Distinct differences of disease resistance can be revealed between plant species or cultivars when the density of pathogenic bacteria in splashing raindrops is considerably low. Such differences cannot be detected by inoculation tests with high dose inoculum. For example, the plum and the cherry strains of *P. syringae* pv. *morsprunorum* are clearly distinguished by host specificity when the cross-inoculations are made with the inoculum dose lower than $1 \times 10^5$ cells/ml by stomatal or leaf scar infection. However, such host specificity is not observed by inoculation with an inoculum dose larger than $1 \times 10^6$ cells/ml or by wound inoculation with the low inoculum dose. Similar results have been obtained with bacterial foot rot of rice (*E. chrysnathemi* pv. *zeae*). The susceptible cultivar T65 is attacked at the inoculum dose of $1 \times 10^3$ to $1 \times 10^4$ cells/ml, but most other cultivars require higher doses to be infected.

### 8.2.3 LECTINS AND HOST SPECIFICITY

Lectins are carbohydrate-binding glycoproteins identified in plants, animals, fungi, and bacteria. Their binding specificity is analogous to that of antibody in the antigen–antibody reaction. Therefore, plant lectins may be used for investigating cell surface receptors, typing of bacterial strains, characterizing bacterial cell components, and determining bacteriophage receptors.

It has been postulated that the host specificity of symbiotic strains of *Rhizobium* spp. may be determined by the interaction between carbohydrates at the surface of the bacterial cell wall and lectins of the plant cell wall that bind the carbohydrates, forming a lectin-carbohydrate complex. The determinants of bacterial binding sites are, in general, the terminal residues of polysaccharides on the surface of the cell envelope and are different with the species of lectins. For example, clover lectin specifically reacts with 2-deoxyglucose, which is the terminal residue of the extracellular polysaccharides (EPS) of *R. trifolii.* The possibility has been investigated that the specific interactions between *Rhizobium* spp. and legume lectins may be similar to the mechanisms involved in host recognition of plant pathogenic bacteria.

In bacterial wilt of tobacco, hydroxyproline-rich glycoproteins (HPRGs), originally described as lectins, agglutinate the avirulent strain B-form (butyrous colony form) of *P. solanacearum* but not the virulent F-form (fluidal colony form). The F-form is agglutinated by HPRGs when EPS are removed by washing and the agglutination of B-type is prevented when EPS are added to the reaction mixture. Binding sites for the agglutinin are presumably present in the lipid A portion of lipopolysaccharides (LPS). The initial step of binding is supposed to be the reversible charge–charge interaction followed by irreversible binding by bacterial fimbriae. Although specific recognition with lectins has also been explored in other systems of host–bacteria interactions, it has not been confirmed whether or not such specific agglutinations found *in vitro* take place in plant tissues.

## 8.2.4 RESISTANCE OF CROP CULTIVARS TO PATHOGENIC RACES OF BACTERIA

Some plant pathogenic bacteria differ by strains in pathogenicity to different cultivars of a crop. These strains are termed pathogenic races, or simply races. The genetic interactions between cultivars and races can be interpreted by Flor's gene-for-gene hypothesis. In this hypothesis, virulence is inherited as a recessive characteristic and resistance as a dominant characteristic. Therefore, the host plant phenotype is expressed as resistant only in a combination of dominant genes of resistance (R) and complementary dominant avirulence genes (A).

The resistance governed by R genes that differentially interacts with avirulence genes of races is called vertical resistance. It is controlled by the major genes and stable with the alteration of environmental conditions. However, it may be completely destroyed when new races with complementary virulence genes against the resistant genes arise in the field.

Horizontal resistance, on the other hand, is governed by minor genes or polygenes, and it is evenly spread against all races of the pathogen with no differential interaction. It may be influenced by environmental conditions and not be high enough to completely prevent disease development. However, it has the advantage of stability against different pathogenic races. A few examples of pathogenic races of plant pathogenic bacteria are described below.

### **Pseudomonas syringae** ***pv.*** **glycinea** ***(the Causal Agent of Bacterial Blight of Soybean)***

The pathogenic bacterium attacks leaves, stems, petioles, and pods and induces various symptoms of water-soaked lesions, local chlorotic haloes, and systematic symptoms such as stunting, yellowing, leaf elongation, and chlorotic mottling. The interactions between cultivars and races of this bacterium are shown in Table 8.2. It has been confirmed that race 4 is predominant in the United States and Canada. The genetic background of the race–cultivar interactions was described above.

### **Xanthomonas campestris** ***pv.*** **malvacearum** ***(the Causal Agent of Bacterial Blight of Cotton)***

The disease is sometimes called angular leaf spot, blackarm, or boll rot, depending on the particular phase of the disease. It occurs on cot-

**Table 8.2**
Reactions of Seven Soybean Cultivars to Races of *Pseudomonas syringae* pv. *glycinea*[a]

| Strains (races) | Reactions on cultivar | | | | | | |
|---|---|---|---|---|---|---|---|
| | Acme | Dhippewa | Flambeau | Harosoy | Lindarin | Merit | Norchief |
| K1 (race 1) | S[b] | R | S | R | R | R | R |
| K2 (race 2) | S | S(R) | S | S | S | S | S |
| K4 (race 4) | S | S | S | S | S | S | S |
| K5 (race 5) | R | R | R | S | R | S | R |
| K6 (race 6) | R | R | S | R | R | R | S |
| PgB3 (race 8) | S(I) | S(I) | S(I) | S(I) | R | S(I) | S(I) |

[a] [From Gnanamanickam, S. S. and Ward, E. W. B. (1982). Bacterial blight of soybeans: a new race of *Pseudomonas syringae* pv. *glycinea* and variations in systemic symptoms. *Can. J. Plant Path.* **4,** 73–78].
[b] S, susceptible; R, resistant; I, intermediate; ( ), data described by other authors.

yledons, leaves, stems, or bolls, forming black lesions as well as shoot blights. The differential interactions between races and resistant cultivars have been clearly elucidated, and the variation of virulence has been analyzed genetically.

There are two types of cotton: diploid and tetraploid types. These types contain many species, and each species includes immune, resistant, tolerant, and susceptible cultivars, respectively. The vertical resistance of cotton to bacterial blight is determined by 16 resistant genes (B genes). In contrast, the horizontal resistance of cotton is controlled by the polygene complex in the upland Stonville 2B and Empire, *Bsm,* and those in Deltapine, *Dsm.* The strains of *X. campestris* pv. *malvacearum* are classified into 17 races on a set of differentials consisting of 8 cultivars with different combinations of B-resistant genes and polygene complex (Table 8.3).

### X. campestris *pv.* oryzae *(the Causal Agent of Bacterial Leaf Blight of Rice)*

The disease shows two different symptoms of leaf blight and wilting or "kresek." Leafblight occurs during all growing stages of rice but wilting only in younger stages. Different systems for the classification of races of this bacterium have been independently developed in Japan, the Philippines, and India, and 11 different resistance genes have been identified. The strains isolated in Japan are classified into six races on the

**Table 8.3**

Reactions of 8 Cotton Cultivars and Breeding Lines to 17 Races of *Xanthomonas campestris* pv. *malvacearum*

| Race no. | Differential host and genes for resistance | | | | | | | |
|---|---|---|---|---|---|---|---|---|
| | Acala 44 (none) | Stoneville 2B-S9 (polygenes) | Stoneville 20 ($B_7$+ polygenes) | Mebane B-1 ($B_2$+ polygenes) | 1-10B ($B_{In}$+ polygenes) | 20-3 ($B_N$+ polygenes) | 101-102B ($B_2B_3$+ unknown) | Gregg (unknown) |
| 1 | + | + | − | − | − | − | − | − |
| 2 | + | + | + | − | − | − | − | − |
| 3 | + | + | − | − | + | − | − | |
| 4 | + | + | − | − | − | + | − | |
| 5 | + | + | − | − | + | + | − | |
| 6 | + | + | − | + | + | − | − | |
| 7 | + | + | − | + | + | + | − | |
| 8 | + | + | + | + | + | − | − | |
| 9 | + | − | + | − | − | + | − | |
| 10 | + | + | + | + | + | + | − | |
| 11 | + | + | − | − | − | − | − | + |
| 12 | + | + | + | − | − | − | − | + |
| 13 | + | − | − | − | − | − | − | |
| 14 | + | + | + | − | + | + | − | |
| 15 | + | + | + | − | + | − | − | |
| 16 | + | + | + | + | − | + | − | |
| 17 | + | + | + | − | − | + | − | |

[From Brinkerhoff, L. A. (1970). Variation in *Xanthomonas malvacearum* and its relation to control. *Ann. Rev. Phytoph.* **8,** 85–110.

**Table 8.4**
Reactions of Five Rice Cultivars to Six Races of *Xanthomonas campestris* pv. *oryzae*

| Differential cultivars | Resistance genes | Races distributed in Japan | | | | | |
|---|---|---|---|---|---|---|---|
| | | I | II | III | IV | V | VI |
| Kimmaze | | S | S | S | S | S | S |
| Kougyoku | $X_{a-1}$, $X_{a-kg}$ | R | S | S | S | R | R |
| Te-tep | $X_{a-1}$, $X_{a-2}$ | R | R | S | S | R | S |
| Waseaikoku 3 | $X_{a-w}$ | R | R | R | S | S | R |
| Java 14 | $X_{a-1}$, $X_{a-w}$, $K_{a-kg}$ | R | R | R | S | R | R |

[Adapted from Horino, O. (1987). Specialization of races in *Xanthomonas campestris* pv. *oryzae* and method of screening for horizontal resistance. Plant Protection **41,** 366–370 (Tokyo). (*In* Japanese).

basis of reactions to five differential cultivars with four resistance genes individually or in combinations (Table 8.4). It has been confirmed that race I is predominant in Japan, followed in order by races II and III. The *avr10* corresponding to the resistance gene *Xa10* of the cultivar Cas209 was recently identified on 2.5 Kb fragment of chromosomal DNA. On the other hand, the genes conferring horizontal resistance were identified in the cultivar "Asaminori," and breeding of durable resistant cultivars with the genes is anticipated.

## Further Reading

Ballio, A., Barra, D., Bossa, F., Collina, A., Grgurina, I., Marino, G., Moneti, G., Paci, M., Pucci, P., Segre, A., and Simmaco, M. (1991). Syringopeptins, new phytotoxic lipodepsipeptides of *Pseudomonas syringae* pv. *syringae*. FEBS **291,** 109–112.

Ballio, A., Bossa, F., Collina, A., Gallo, M., Iacobellis, N. S., Paci, M., Pucci, P., Scaloni, A., Segre, A., and Simmaco, M. (1990). Structure of syringotoxin, a bioactive metabolite of *Pseudomonas syringae* pv. *syringae*. FEBS **269,** 377–380.

Binns, A. N., and Thomshow, M. F. (1988). Cell biology of *Agrobacterium* infection and transformation of plants. *Ann. Rev. Microbiol.* **42,** 575–606.

Chatterjee, A. K., and Vidaver, A. K. (1986). "Genetics of Pathogenicity Factors: Application to Phytopathogenic Bacteria." Academic Press, New York.

Collmer, A., and Keen, N. T. (1986). The role of pectic enzymes in plant pathogenesis. *Ann. Rev. Phytopathol.* **24,** 383–409.

Dixon, R. A., and Lamb, C. J. (1990). Molecular communication in interactions between plants and microbial pathogens. *Ann. Rev. Plant Physiol. Plant Mol. Biol.* **41,** 339–367.

Djordjevic, M. A., Gabriel, D. W., and Rolfe, B. G. (1987). Rhizobium—the refined parasite of legumes. *Ann. Rev. Phytopathol.* **25,** 145–168.

Durbin, R. D., and Langston-Unkefer, P. J. (1988). The mechanisms for self-protection against bacterial phytotoxins. *Ann. Rev. Phytopathol.* **26,** 313–329.

Fukuchi, N., Isogai, A., Yamashita, S., Suyama, K., Takemoto, J. Y., and Suzuki, A. (1990). Structure of phytotoxin syringomycin produced by a sugar cane isolate of *Pseudomonas syringae* pv. *syringae. Tetrah. Letters* **31,** 1589–1592.

Gabriel, D. W., and Rolfe, B. G. (1990). Working models of specific recognition in plant–microbe interactions. *Ann. Rev. Phytopathol.* **28,** 365–391.

Goodman, R. N., Kiraly, Z., and Wood, K. R. (1986). "The Biochemistry and Physiology of Plant Diseases." University of Missouri Press, Columbia, Missouri.

Horsfall, J. G., and Cowling, E. B. (eds.) (1979). "Plant Disease, An Advanced Treatise. Volume IV. How Pathogens Induce Disease." Academic Press, New York.

Horsfall, J. G., and Cowling, E. B. (eds.) (1980). "Plant Disease, An Advanced Treatise. Volume V. How Plants Defend Themselves." Academic Press, New York.

Isogai, A., Fukuchi, N., Yamashita, S., Suyama, K., and Suzuki, A. (1990). Structures of syringostatins A and B, novel phytotoxins produced by *Pseudomonas syringae* pv. *syringae* isolated from lilac blights. *Tetrah. Letters* **31,** 695–698.

Mitchell, R. E. (1984). The relevance of non-host-specific toxins in the expression of virulence by pathogens. *Ann. Rev. Phytopathol.* **22,** 215–245.

Mount, M. S., and Lacy, G. H. (eds.) (1982). "Phytopathogenic Prokaryotes. Volume 1. Part IV. Prokaryote Interactions within the Plant." Academic Press, New York.

Nutkins, J. C., Mortishire-Smith, R. J., Packman, L. C., Brodey, C. L., Rainey, P. B., Johnstone, K., and Willams, D. H. (1991). Structure determination of tolaasin, an extracellular lipodepsipeptide produced by the mushroom pathogen *Pseudomonas talaasii* Paine. *J. Am. Chem. Soc.* **113,** 2621–2627.

Sequeira, L. (1978). Lectins and their role in host–pathogen specificity. *Ann. Rev. Phytopathol.* **16,** 453–481.

Van Alfen, N. K. (1989). Reassessment of plant wilt toxins. *Ann. Rev. Phytopathol.* **27,** 533–550.

Van der Plank, J. E. (1978). "Genetic and Molecular Basis of Plant Pathogenesis." Springer-Verlag, New York.

CHAPTER
**NINE**

# *Life Cycles and Dispersal of Plant Pathogenic Prokaryotes*

Plant pathogenic bacteria that have remained alive for long unfavorable survival periods are liberated from their reservoir and infect new plant growth forming foci of infection. This is called the *primary cycle* or *primary infection,* and the source of inoculum is referred to as *primary inoculum.* The pathogens increase their populations in the primary foci and spread to other parts of the field or to other fields. This is called the *secondary cycle* or *secondary infection.*

Under favorable environmental conditions, secondary cycles occur effectively in succession, resulting in an *epidemic* or *epiphytotic.* The bacteria again enter the survival phase when annual plants start senescence or perennial plants enter dormancy. Because the density of inoculum in the primary cycle is much smaller than in the secondary cycle, effective control of primary infection is a principal procedure for prevention of epidemics of bacterial plant diseases.

The chain of events occurring with a pathogenic bacterium in relation to disease development is termed the *disease cycle* and includes survival, dissemination, and infection. The science of the spread of diseases and factors affecting outbreaks of diseases is termed *epidemiology.*

## 9.1 Survival

Plant pathogenic bacteria surviving in nature, even those in diseased plant tissues, are usually exposed to a variety of saprophytic microorganisms. To survive in coexistence with these microorganisms, plant pathogenic bacteria are assumed to free themselves from competition and/or antagonism by entering in hypobiosis, by lying concealed in their own unique habitats, or by establishing a synergistic association with some other organisms. The presence of plants, irrespective of whether they are host or nonhost, exerts a great influence on the interactions between pathogenic bacteria and saprophytic microorganisms in nature, generally providing a selectively beneficial effect on pathogenic bacteria. The response of plant pathogenic bacteria to such a selection pressure may vary among different phagovars, serovars, or biovars of a species, resulting in the predominance of certain groups in a given field or area.

In general, bacterial populations in nature gradually decline during the long unfavorable survival period. Therefore, the specificity and accuracy of detection techniques for particular pathogens are of critical importance in determining their ecological behavior, i.e., natural habitats, densities, or length of survival.

## 9.1.1 PLANT-DEPENDENT SURVIVAL

### *Parasitic Survival*

#### *Survival in Lesions*

This form of survival is usually found in the pathogens of perennial crops, including fruit trees, e.g., *Erwinia amylovora* and *Xanthomonas campestris* pv. *pruni.* The bacteria are carried over in the diseased tissues of twigs or leaves and form the most important source of inoculum in primary cycles. On citrus trees, the population of *Xanthomonas campestris* pv. *citri* in canker lesions of spring shoots markedly declines by the end of growing season on account of active multiplication of *E. herbicola* or *Fusarium* spp. These microorganisms enter into new lesions as the secondary invaders or contaminants and proliferate in a rainy season and a subsequent hot summer. In canker lesions developing late in the growing season, however, the population of the pathogen can be maintained at a high level throughout the winter because of the relative absence of these microorganisms.

Plant pathogenic bacteria may be able to survive in the parasitic form on annual and perennial weeds or volunteers. For example, the long-term survival of *Pseudomonas avenae* in Florida is attributed to the association with a perennial grass, vaseygrass (*Paspalum urvillei*), through repeated infection of its vegetative growth as well as seed transmission. Bacterial leaf blight of corn may have its origin in the infected Vasey grass distributed in corn fields.

### *Survival by Latent Infection*

Latent infection refers to the conditions in which pathogenic bacteria may survive for a long time in plant tissues without development of visible symptoms. *P. syringae* pv. *syringae* or *X. campestris* pv. *citri* can survive in apparently healthy bark tissues of their tree hosts. The systemic existence of the former in the bark of peach trees has been demonstrated and seems to explain the lack of effective control from protective bactericides applied to the surface of trees.

*X. campestris* pv. *citri* can be detected in the discolored bark tissues of trunks, low scaffold limbs, and lateral branches of adult citrus trees. The bacterium may also be isolated with high frequency but in low populations from 6-month to 3-year-old apparently healthy green twigs of Satsuma orange (*Citrus unshu*), although visible symptoms are not subsequently produced. Latent infection in citrus canker is also observed in younger leaves near the top of angular shoots when they are infected late in the growing season. These infections are symptomless during winter and produce minute, water-soaked spots in early spring which serve as a source of rain splash infection for several months.

*Xylella fastidiosa,* the causal agent of Pierce's disease of grape and leaf scorch disease of various fruit and ornamental trees infect diverse kinds of weeds without developing visible symptoms. Because these weeds are usually favorable habitats for the vector insects, latently infected weeds become an important source of the carrier insects.

*Agrobacterium tumefaciens* may infect nursery plants late in the growing season through wounds at the basal portion of the stems produced by grafting or those on roots produced by digging or transplanting. These young plants do not exhibit tumor development until the next season at the onset of active growth of the host plants. The pathogen perpetuates on the surface of root system of such plants as well.

### *Survival in Seeds and Planting Materials*

Seed may be externally infested or internally infected by plant pathogenic bacteria during the course of development and maturation in fruit or pod (Table 9.1). Most seed-borne bacteria survive as long as the seed remains viable; *P. syringae* pv. *tomato* has been shown to survive in dried tomato seed for 20 years, *X. campestris* pv. *phaseoli* and *Curtobacterium flaccumfaciens* pv. *flaccumfaciens* in bean seed for 15 years, and *P. syringae* pv. *phaseolicola* in bean seed for 10 years.

Seed may be infected or contaminated in several ways: (1) invasion of seed through flower infection (e.g., *P. syringae* pv. *glycinea*), (2) invasion of seed through lesions formed on pods (e.g., *P. syringae* pv. *phaseolicola*),

**Table 9.1**
Examples of Plant Pathogenic Bacteria that Survive in Seed

| Bacteria | Disease name |
|---|---|
| *Clavibacter michiganensis* subsp. *michiganensis* | Bacterial canker of tomato |
| *Clavibacter michiganensis* subsp. *nebraskensis* | Goss's bacterial wilt and blight of corn |
| *Curtobacterium flaccumfaciens* pv. *flaccumfaciens* | Bacterial wilt of bean |
| *Pseudomonas avenae* | Bacterial brown stripe of rice |
| *Pseudomonas glumae* | Bacterial grain rot of rice |
| *Pseudomonas syringae* pv. *glycinea* | Bacterial blight of soybean |
| *Pseudomonas syringae* pv. *lachrymans* | Angular leaf spot of cucurbits |
| *Pseudomonas syringae* pv. *phaseolicola* | Halo blight of bean |
| *Pseudomonas syringae* pv. *pisi* | Bacterial blight of pea |
| *Xanthomonas campestris* pv. *campestris* | Black rot of crucifers |
| *Xanthomonas campestris* pv. *malvacearum* | Bacterial blight of cotton |
| *Xanthomonas campestris* pv. *phaseoli* | Common blight of bean |

(3) invasion through the vascular system (e.g., *X. campestris* pv. *campestris, Clavibacter michiganensis* subsp. *michiganensis*), (4) invasion through stomata of seed coat (e.g., *P. glumae,* Fig. 9.1), and (5) contamination during threshing (e.g., many seedborne bacteria). The infected seed may or may not exhibit distinguishable symptoms. In halo blight of bean caused by *P. syringae* pv. *phaseolicola,* about 50% of the total disease transmission originated in symptomless seeds.

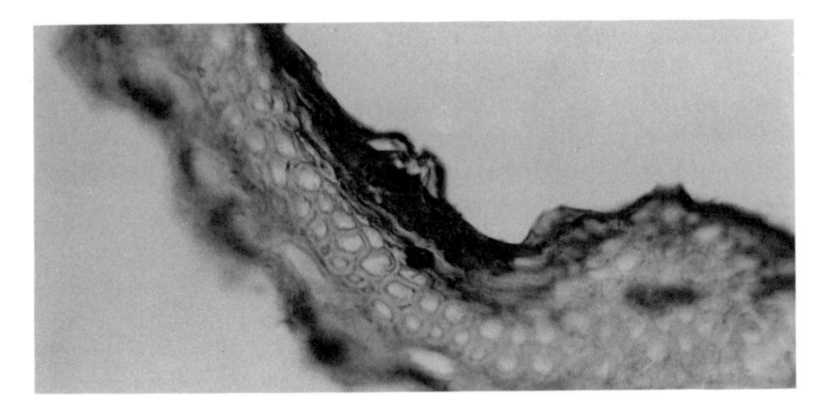

**Fig. 9.1** *Pseudomonas glumae* invaded glume of rice grain through stoma and multiplied in parenchyma. [Courtesy of Dr. H. Tabei.]

Many plant pathogenic bacteria also survive in or on vegetatively propagated planting materials such as bulbs, corms, tubers, rhizomes, or cuttings (Table 9.2). *C. michiganensis* subsp. *sepedonicus* and *P. solanacearum* may be present in the vascular system of potato tubers. Lenticels of potato tubers may carry soft rot *Erwinia* at the maximum level of 100 cells/lenticel, although the infested tubers do not necessarily develop soft rot in the field.

### *Survival as Residents*

Plant pathogenic bacteria have the capacity to grow on the surface of host and nonhost plants utilizing the small amount of nutrients that are secreted on the plant surface. When the environment becomes unfavorable for multiplication, they may survive through hypobiosis. The epiphytically resident bacteria generally show log normal population for any given plant canopy at any given time, irrespective of the total epiphytic bacterial population or selected components of the populations.

Plant surfaces that are subject to microbial colonization can be differentiated into phyllosphere or phylloplane, gemmisphere, and rhizosphere or rhizoplane.

#### *Phyllosphere and Phylloplane*

The outer walls of epidermal cells have fine *ectodesmata* extending from the cuticle through the wall to the lumen of epidermal cells (Fig. 9.2). Ectodesmata are similar to plasmodesmata in function but differ in

**Table 9.2**
Examples of Plant Pathogenic Bacteria that Survive in Planting Materials

| Bacteria | Disease name |
|---|---|
| *Clavibacter michiganensis* subsp. *sepedonicus* | Ring rot of potato |
| *Curtobacterium flaccumfaciens* pv. *oortii* | Silvering of tulip |
| *Pseudomonas caryophylli* | Bacterial wilt of carnation |
| *Pseudomonas gladioli* pv. *gladioli* | Gladiolus scab |
| *Pseudomonas solanacearum* | Bacterial wilt of potato and ginger |
| *Pseudomonas syringae* pv. *photiniae* | Bacterial leaf spot of *Photinia glabra* |
| *Xanthomonas albilineans* | Leaf scald and white streak of sugarcane |
| *Xanthomonas campestris* pv. *begoniae* | Bacterial leaf spot of begonia |
| *Xanthomonas campestris* pv. *hyacinthi* | Yellows of hyacinth |
| *Xanthomonas campestris* pv. *tardicrescens* | Bacterial blight of iris |

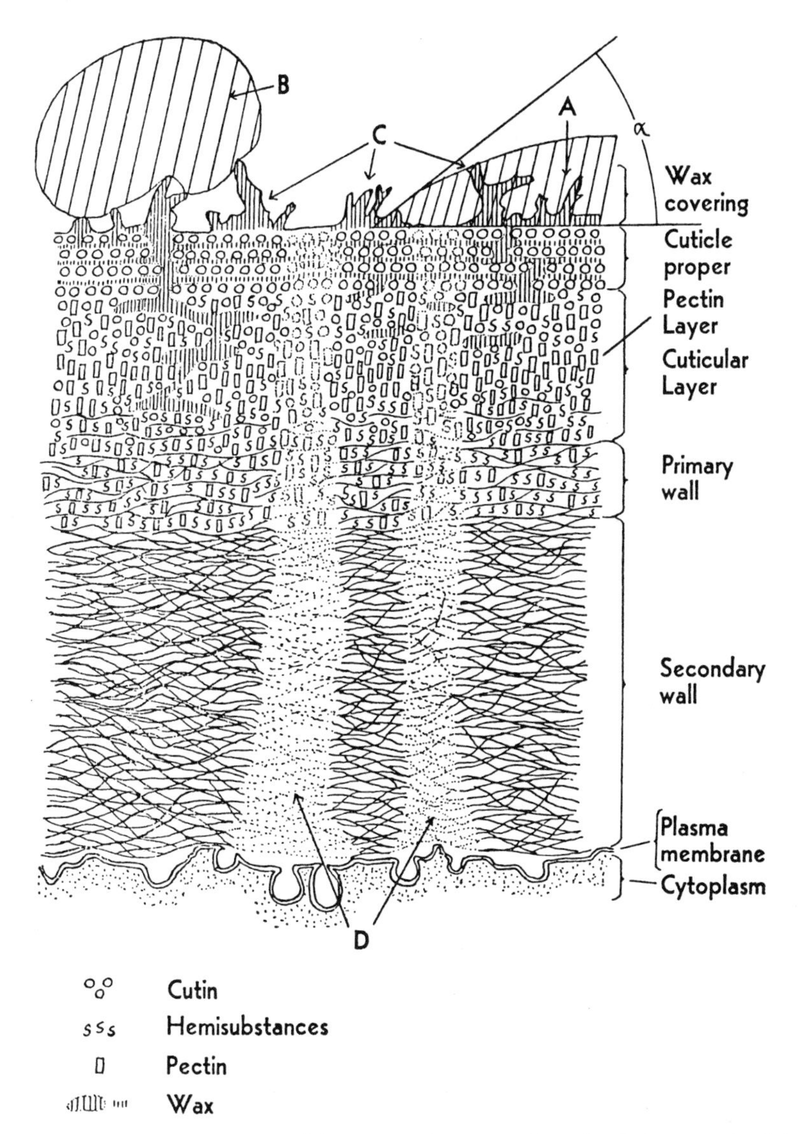


**Fig. 9.2** Schematic illustration of the outer wall of an epidermal cell. A, water droplet with detergent; B, water droplet without detergent; C, wax rodlets; D, ectodesmata; $\alpha$, contact angle. [Franke, W. (1967). Mechanisms of foliar penetration of solutions. *Ann. Rev. Plant Physiol.* **18,** 281–300. Reproduced, with permission, from the Annual Review of Plant Physiology.]

that they lack protoplasm. Ectodesmata are mainly confined to special sites such as trichomes, the epidermal cells surrounding trichomes, anticlinal cells, cells around veins, and particularly guard cells. The major functions of ectodesmata include foliar absorption and foliar excretion of aqueous organic and inorganic substances and cuticular transpiration. Phylloplane bacteria preferentially colonize the leaf surface, where ectodesmata are concentrated, because these are the areas where nutrients are excreted from epidermal cells.

Survival as residents in the phyllosphere has been demonstrated with many plant pathogenic bacteria (Table 9.3). These bacteria maintaining their life cycle on the leaf surface are called *epiphytes*. Populations of epiphytes fluctuate under the influence of various environmental factors such as precipitation, humidity, temperature, and sunlight. Humidity is of particular importance because the resident bacteria readily multiply under high humidity. Rainfall has the greatest effect: the epiphytic population initially decreases by wash-off and then rapidly increases after 12 to 24 hr. The wetness alone is not the factor involved in this rapid multiplication of epiphytes. The momentum of raindrops seems to be the key factor, although its mechanism has not been elucidated.

Multiplication of *X. campestris* pv. *phaseoli* in leaves of navy beans resembles the standard growth curve *in vitro*. The mean generation time in the log phase is 19 hr, and the stationary phase starts with the population at a level of $10^7$ to $10^8$ cells per 20 $cm^2$ of leaf area. The population is maintained at this level until the leaves senesce. In rainfall, at least 10% of the population may be removed by rain run-off becoming the source of inoculum.

**Table 9.3**
Examples of Plant Pathogenic Bacteria that Survive in Epiphytic Form

| Bacteria | Disease name |
|---|---|
| *Erwinia amylovora* | Fire blight of apple and pear |
| *Erwinia carotovora* subsp. *carotovora* | Soft rot of Chinese cabbage |
| *Pseudomonas glumae* | Bacterial grain rot of rice |
| *Pseudomonas syringae* pv. *glycinea* | Bacterial blight of soybean |
| *Pseudomonas syringae* pv. *lachrymans* | Angular leaf spot of cucurbits |
| *Pseudomonas syringae* pv. *morsprunorum* | Bacterial canker of stone fruits |
| *Pseudomonas syringae* pv. *syringae* | Bacterial brown spot of bean |
| *Pseudomonas syringae* pv. *tomato* | Bacterial speck of tomato |
| *Xanthomonas campestris* pv. *malvacearum* | Bacterial blight of cotton |
| *Xanthomonas campestris* pv. *manihotis* | Bacterial blight of cassava |
| *Xanthomonas campestris* pv. *phaseoli* | Common blight of bean |

The epiphytic population of *P. syringae* pv. *syringae* on trifoliate leaves is useful for the prediction of brown leaf spot of beans because symptoms develop when the bacterial density elevates higher than $10^4$ cells per leaflet. The epiphytic phase of *P. syringae* pv. *morsprunorum* is established on peach leaves soon after unfolding and remains at a level of $10^4$ to $10^7$ per leaf until the end of the growing season. Elevation of the epiphytic population always precedes the development of bacterial canker.

*Gemmisphere*

Buds provide a most favorable site for the growth of bacteria because they are protected from the outer environment by scales and hold high moisture and suitable nutrients. The gemmisphere generally holds many saprophytic bacteria, and plant pathogenic bacteria usually survive in coexistence with these saprophytes. The majority of gemmisphere bacteria exist internally, the external population being only 10% or less of the total population. Internal infestation of buds may occur at the very early stage of bud formation where only a few primitive scales are formed. The high population continues until leaves unfold. In general, saprophytic bacteria inhabiting plant buds are limited to certain species, depending on the plant species. For example, more than 50% of gemmisphere bacteria of tea (*Thea sinensis*) consist of irregular, nonspore-forming gram-positive bacteria.

Survival in gemmisphere has been documented with many plant pathogenic bacteria such as *P. syringae* pv. *lachrymans*, *P. syringae* pv. *syringae* and *X. campestris* pv. *glycines*. When new shoots develop, these gemmisphere bacteria may transfer to leaf surface and multiply as the phyllosphere bacteria or may vanish as soon as leaves unfold. The population of *P. syringae* pv. *syringae* perpetuating in buds of sweet cherry rapidly increases when it rains but quickly returns to the normal low or undetectable level when rain ceases. Such dynamic population change associated with rainfall is a common phenomenon in bacteria colonizing the surface of aerial parts of plants.

*Rhizosphere and Rhizoplane*

Plant pathogenic bacteria can saprophytically survive or actively multiply in the rhizosphere or rhizoplane of healthy host and nonhost plants. Growth of these bacteria can be specifically stimulated by nutrients such as amino acids or sugars that are secreted by roots to enhance their ability to compete with other microorganisms. This selective stimulation effect in the rhizosphere is called the *rhizosphere effect.* The substances that are responsible for the rhizosphere effect have rarely been defined chemically.

Populations of rhizosphere bacteria are highly variable and may be a factor of 10 to 50 within a given set of root systems. The capacity of plant pathogenic bacteria to colonize on root systems is not necessarily specific to the host plant. The root colonization is also found in foliage pathogens such as *P. syringae* pv. *tabaci* and *X. campestris* pv. *vesicatoria*, which overwinter in the rhizosphere of nonhost and weed species.

*E. carotovora* subsp. *carotovora* multiplies in the rhizosphere of many cruciferous plant species, where the population can readily increase from $10^2$ cells/g in fallow soil to $10^4$ to $10^6$ cells/g in soil subjected to the rhizosphere effect of Chinese cabbage. In the rhizosphere of Chinese cabbage, 90% or more cells of soft rot bacteria reside on the outer part of soil aggregates, indicating that the rhizosphere effect of this crop does not extend to the inner part of soil aggregates.

The rhizosphere effect differs in magnitude by not only plant species but also growth stages. For example, in Chinese cabbage the effect appears 30 days after planting and becomes pronounced in the head-forming stage. The stimulative effect of rhizosphere affects not only soft rot bacteria but also total bacteria. In particular, gram-positive bacteria subsequently increase and dominate the bacterial flora of rhizosphere, bringing about a gradual decline of soft rot bacteria.

*P. glumae*, the causal agent of bacterial grain rot of rice, remains on rhizosphere and/or rhizoplane of the rice plant from germination through tillering stage. The bacterium moves upward as the booting stage approaches, multiplies rapidly on the surface of young panicles in leaf sheaths, and attacks flowering panicles just after emergence.

Swarming or attraction of bacteria toward the root surface seems to be an important step in the establishment of a resident phase in the rhizoplane. *X. campestris* pv. *oryzae* and *P. solanacearum* are attracted toward root hair of rice and tobacco plants, respectively, when they are immersed in a bacterial suspension. The swarming effect may be variable between host and nonhost plants or susceptible and resistant cultivars. It is conceivable that swarming is principally a chemotactic response toward some unknown substances secreted by roots. The same activation mechanism may operate with other bacteria which have overwintered in a hypobiotic state.

### 9.1.2 SAPROPHYTIC SURVIVAL

Saprophytic survival means here the persistence of plant pathogenic bacteria in soil in the absence of growing plants or in dead plants or plant residues. Plant pathogenic bacteria surviving in the saprophytic phase seem to be in a hypobiotic state with a markedly reduced metabolic activity.

### *Soil*

Many bacteria show prolonged survival in sterilized soil, whereas they rapidly decline in numbers in nonsterilized soil because of their intense interactions with other organisms. *P. solanacearum* and species of *Agrobacterium* are best known with a prolonged soil phase which can be regarded as the true soil-borne pathogens. When these pathogens once become established in a field, the bacteria are maintained at a certain level for decades even in the absence of susceptible host plants.

Okabe (1969, 1971) found that *P. solanacearum* grew more actively in a dry (15–20% water content) than in moist soil (40–50% water content) and in a soil of weakly acidic range (pH 5.4). The environment of such dry and acidic soils does not allow the growth of microorganisms which may compete for nutrients or produce inhibitory substances. *P. solanacearum* is commonly recovered from field soil at 30 cm depth but can be detected even at depth of 80–100 cm. Persistence at such depths, which may be regarded as protected sites, could also be favored by the absence of or a very low population of competing microorganisms. *P. solanacearum* differs from *E. carotovora* subsp. *carotovora* in that it can grow in the inner part of soil aggregates.

*E. carotovora* subsp. *carotovora* has also been considered to survive in soil. However, some recent studies have shown that soft rot *Erwinia* cannot persist for a long time in fallow soil. For example, the pathogen of tobacco hollow stalk rot can be detected from field soil in California after harvest in November but not in May, indicating that *E. carotovora* subsp. *carotovora* cannot overwinter in soil except in decomposing root crown. Another study indicated that soft rot *Erwinia* spp. can be isolated from soil with growing crops, or immediately after crops are harvested, but not from soil fallowed for 6 months. However, the effect of fallowing has rarely been noticed on the population of soft rot bacteria in Japan. Such disagreements in the survival of soft rot *Erwinia* in soil may be derived from the differences in soil types, frequency, and amount of precipitation and vegetations around fields, although there is the possibility of contamination by long distance dispersal through aerosol.

There are many papers reporting that in common scab of potato caused by *Streptomyces scabies,* infected seed tubers are not important as the source of inoculum. This is based on the observation that the progeny tubers are free from the disease even when heavily infected seed tubers are planted. The ineffectiveness of infected seed tubers as inoculum sources has been explained as follows: *S. scabies* usually multiplies in lesions on seed tubers and increases its population, but its translocation in soil is so slow that it cannot reach the progeny tubers. In addition, *S.*

*scabies* is a common soil inhabitant and its distribution is more or less localized.

Recent studies in Japan, however, emphasized the importance of infected seed tubers as the source of inoculum because common scab always occurs in a more severe form with diseased seed tubers than with symptomless seed tubers. In addition, the importance of symptomless seed tubers as the important inoculum source was implied, because common scab occurred in severe form when they are planted in a field disinfected by chloropicrin (Table 9.4).

### *Plant Residues*

Plant residues, dead plants, straw, or hay placed in and on the ground as well as in storage provide favorable survival sites for plant pathogenic bacteria. For example, *X. campestris* pv. *alfalfae* has been isolated for up to 8 years from hay in storage. *X. campestris* pv. *citri* dies out rapidly in diseased leaves buried in soil as the leaves start to decompose. However, the bacterium shows long-term survival on the surface of rice

**Table 9.4**
Effect of Disinfection of Soil and Tubers on the Incidence of Common Scab of Potato[a]

| Disinfection of soil | Scab area of seed tuber (%) | Disinfection of seed tuber[b] | Diseased progeny (%) | Grade of disease severity[c] |
|---|---|---|---|---|
| No | 50 | Yes | 7 | 3 |
| | | No | 52 | 21 |
| | 5 | Yes | 1 | 1 |
| | | No | 17 | 7 |
| | 0 | Yes | 5 | 2 |
| | | No | 11 | 4 |
| Yes (Chloropicrin, 30 liters/10 a[d]) | 50 | Yes | 95 | 69 |
| | | No | 99 | 79 |
| | 5 | Yes | 42 | 19 |
| | | No | 94 | 63 |
| | 0 | Yes | 30 | 13 |
| | | No | 81 | 49 |

[a] Adapted from Tashiro, N. and Matsuo, Y. (1986). Tuber-borne infection in potato common scab disease and effect of soil sterilization and seed treatment on the control of the tuber-borne infection. *Proc. Assoc. Pl. Prot. Kyushu* **32,** 24–27.

[b] Copper compound

[c] $\text{Grade of disease severity} = \dfrac{\Sigma\text{ (disease severity indices} \times \text{number of tubers)}}{\text{Maximum index value} \times \text{total number of tubers examined}} \times 100$

[d] Unit area ($100 m^2$).

straw or grass that is used as mulch in citrus groves, particularly when mulch remains dry. Similar experiences have been reported with many other bacteria such as *C. michiganensis* subsp. *michiganensis, P. syringae* pv. *glycinea, X. campestris* pv. *glycines* and *X. campestris* pv. *oryzae.* In general, the longevity of plant pathogenic bacteria is largely dependent on the capacity to produce extracellular polysaccharide slime. Bacterial cells that are unprotected by slime are killed much faster than those with a thick slime layer.

Many plant pathogenic bacteria persist in diseased host plant residues thrown on the surface of the ground or ploughed into soil but not in the form of free cells that were released from diseased plant tissues into soil. Free cells of *X. campestris* pv. *campestris* introduced into unsterilized soil die within 60 days in winter when the activity of soil microorganisms is markedly reduced, but they become extinct much faster in summer. In contrast, when protected by cabbage stem tissue, the pathogen was recovered in large numbers from plant debris in soil for up to 244 days. The survival time of these bacteria is therefore dependent on the ease of decomposition of the host residues in and on soil. Woody stem tissues resist decomposition in soil, so that vascular pathogens in pieces of stem ploughed into soil may survive for a long time.

Flooding of fields or transforming of upland fields to paddies is often recommended for the control of soil-borne diseases. However, it should be remembered that woody stems of tomato or tobacco remain sometimes undecomposed in deep soil up to the next crop season under relatively anaerobic conditions. The incidence of bacterial wilt caused by *P. solanacearum* or bacterial canker caused by *C. michiganensis* subsp. *michiganensis,* in which the primary inoculum sources originated in this form, has repeatedly been recorded in cultivation of tobacco and tomato in rotation with lowland rice. In contrast, *P. solanacearum* that is free in field soil readily vanishes when raw plant materials such as grass, rape cake, or soybean cake are ploughed immediately before flooding. This is attributed to a rapid increase of certain antagonistic bacteria in soil.

### *Agricultural Materials*

*C. michiganensis* subsp. *michiganensis* has been shown to survive in air-dried conditions for 7–8 months on the surface of wooden stakes and boxes or wires or for 15 months in air-dried tissues of diseased tomato plants. Bacterial exudate produced on the diseased plant surface usually contains a large amount of polysaccharides, which protect bacterial cells from desiccation. Therefore, infestation of agricultural equipment and implements with bacterial exudate facilitates long-term survival of pathogenic bacteria.

### *Surface Water*

*E. carotovora* subsp. *carotovora* was detected from water from drains, ditches, streams, rivers, and lakes in mountainous upland and arable areas of Scotland and Colorado throughout the year. In general, populations were low (less than 1 cell/ml) in winter when temperatures were below 5°C and increased through the spring and summer months, reaching a maximum population of more than 100 cells/ml in late summer and early autumn when water temperatures usually rise to over 20°C. This bacterium was also detected from marine water. All these bacteria in surface free water are assumed to constitute the inoculum source. *X. campestris* pv. *oryzae* and *E. chrysanthemi* pv. *zeae* can survive in irrigation water at a density lower than $10^4$ cells/ml. In a paddy, survival of these bacteria is often enhanced by the presence of some blue-green algae in irrigation water.

# 9.2 Dispersal of Plant Pathogenic Prokaryotes

Plant pathogenic bacteria in diseased tissues are embedded in abundant polysaccharides from which individual bacteria can be separated and released into the air only with difficulty. For effective dispersal, these bacteria need various vectors such as rain splash, wind, surface water, insects, and humans.

On the other hand, in fire blight of apple and pear caused by *E. amylovora,* aerial strands formed on lesions may have the potential for long-range dispersal. Epiphytic population of *P. syringae* pv. *syringae* can also enter the atmosphere from the surface of bean plants in dry weather, implying that plant canopies may constitute a major source of bacteria in air. Nothing is known, however, on the practical importance of these airborne inoculum in disease development.

## 9.2.1 DISPERSAL BY WATER

When lesions occupied by bacterial plant pathogens are wetted, the polysaccharide matrix in which bacterial cells are embedded dissolves readily, and bacterial cells are released to the surface of the lesion.

Bacterial cells thus suspended in water can readily gain access into natural openings or wounds. As long as the bacterial cells are immersed in water, they are protected from the deleterious effects of desiccation and can potentially be available in transport various distances, through air, on plant surfaces, or on the ground to reach new sites of infection. Dissemination by water is the primary means of dispersal for most bacterial plant pathogens.

### *Dispersal by Rain Splash*

During rainfall bacteria emerge from diseased tissues into surface water, providing a source of inoculum for as long as the rain lasts. Large numbers of bacteria are liberated for extended periods of time, particularly from young lesions. For example, young lesions of citrus canker or those of wildfire of tobacco discharge bacterial cells at concentrations of $10^5$ to $10^7$ cells/ml throughout the period of rainfall in water running over the surface of diseased leaves. In these lesions, bacterial multiplication is active enough to supply the discharged bacterial cells.

Water droplets carrying bacterial cells can be dispersed long distances during stormy rains as in a typhoon. However, dispersal of inoculum under storm conditions, which results in subsequent infection, appears not to occur over a long distance. In an experiment with bacterial leaf blight of rice in a rainstorm with maximum wind velocity of 28 m/sec, *X. campestris* pv. *oryzae* was recovered at a distance of 64 m from the source of inoculum. However, actual disease development was detected at a distance of at most only 4 m. A similar result was obtained in an experiment with citrus canker.

The abrupt onset of epidemics of bacterial plant diseases requires, in general, elevation of bacterial populations above the inoculum potential or threshold. Although dispersal directly related to infection is limited to relatively short distances, bacteria scattered by splash dispersal in rainstorms can establish an epiphytic phase on their host plants over a wide area in the field and there become a source of inoculum in subsequent storms. In bacterial brown spot of snap bean caused by *P. syringae* pv. *syringae,* rain primarily contributes to epiphytic multiplication of the pathogen but not to infection through splash dispersal, because this pathogen requires a relatively high inoculum dose for infection.

### *Bacterial Aerosol*

The bacterial aerosol mode of dissemination was discovered in Scotland in the mid-1970s during a study on the recontamination of the highest grade of certified seed potatoes with the black-leg pathogen, *E.*

*carotovora* subsp. *atroseptica*. Since then bacterial aerosol has become widely known as a common mode of distant dispersal of plant pathogenic bacteria. The drops of the aerosol have a diameter of 4–8 $\mu$m and can remain in the air for 60 to 90 min. Although the sources of the bacteria may or may not be confirmed, it has been shown that bacterial aerosols generated by rainfall are important in establishing an epiphytic phase on potato leaves from which seed and progeny tubers can become contaminated by subsequent washing off by rain.

*E. amylovora* can survive longer than soft rot *Erwinia* spp., retaining viability for 2 hr in aerosols that are exposed to the atmosphere at 40–90% relative humidity. Discharge of bacterial aerosol has also been detected in soybean fields where bacterial blight occurs. Aerosols contained an average of 151 cells/m$^3$ of *P. syringae* pv. *glycinea* during rain and 94 cells/m$^3$ during overhead sprinkler irrigation, and most aerosol particles measured 2.1–3.3 $\mu$m in diameter.

### *Irrigation Water*

Irrigation water is an important mode of dissemination in bacterial diseases of lowland rice. *X. campestris* pv. *oryzae* multiplies on a perennial weed, *Leersia oryzoides,* growing along the irrigation canals, producing symptoms on the leaves early in spring. The bacterium is subsequently released from the diseased leaves into irrigation water and introduced into nurseries or paddies, where primary infection of rice seedlings occurs. In the initial stage of disease development, guttation fluid that is turbid with bacterial cells is continuously released into paddy water early in the morning. A drop of guttation fluid contains about 10$^6$ cells of the bacterial pathogen, providing a heavy source of inoculum in irrigation water.

Similarly, the concentrations in paddy water of *E. chrysanthemi* pv. *zeae,* the cause of bacterial foot rot of rice, remains at less than 10 cells/ml during periods of fine and clear weather. However, when rain starts, the population rapidly rises to 10$^3$ to 10$^4$ cells/ml and remains at this level during the period of rainfall, declining to the base level soon after rain ceases. This work shows that rain has an important role in wash-off of the causal bacteria from diseased rice leaves into irrigation water, which then becomes a source of inoculum to other healthy plants.

The recycled ebb and flow irrigation systems are becoming prevalent in contemporary protected horticulture to decrease groundwater pollution. However, in this modern technology, there is a possibility that pathogenic bacteria can be disseminated from a small number of infected plants to whole plants in the system through a plant nutrition solution that is pumped from a storage reservoir and delivered to every potted plant in circulation. This type of dissemination has been documented with *P. solanacearum, X. campestris* pv. *begoniae,* as well as some fungal pathogens.

### 9.2.2 DISPERSAL BY SOIL

Plant pathogenic bacteria surviving in soil can be distantly disseminated through running surface water or agricultural practices, but dispersal by their own movement or by the help of insect or nematode vectors is restricted to a very short distance. The typical soil-borne bacteria include *A. tumefaciens, P. solanacearum,* and *E. carotovora* subsp. *carotovora.* These bacteria can survive for a long time in soil in the absence of host plants.

Most other plant pathogenic bacteria are unable to survive in soil unless there are growing plants or undecomposed plant residues. In these bacteria, undecomposed tissues of diseased plants buried in soil become a potential source of inoculum. For example, *C. michiganensis* subsp. *michiganensis, X. campestris* pv. *campestris, X. campestris* pv. *malvacearum,* and *P. syringae* pv. *tabaci* are disseminated from this type of inoculum source. These bacteria can also survive in the rhizosphere of nonhost plants and infect the lower leaves of host plants by rain splash.

### 9.2.3 DISPERSAL BY SEEDS AND PLANTING MATERIALS

Many plant pathogenic bacteria are disseminated by internally infected or externally infested seeds and vegetatively propagating materials such as tubers, bulbs, tuberous roots, rhizomes, cuttings, budwood, and small plants. Seed-borne diseases are characterized by long-distance dispersal, frequently across continents. This mode of dissemination is of great practical importance, even if the rate of contamination is small in the bulked seed. Under favorable environmental conditions, disease may readily spread from the diseased plants derived from a small number of infested seeds to adjacent healthy plants and finally result in an epidemic.

The lowest detection level of cabbage seeds infected with *X. campestris* pv. *campestris* is about 0.01%, which is an acceptable level for direct planting of cabbage for head production because an epidemic is unlikely to result in the field at this level of infection. However, zero tolerance is required for transplant cultivation because effective dissemination can occur in the seedbed from a very small number of infected seedlings.

*C. michiganensis* subsp. *michiganensis* may be transmitted in nurseries to a bulk of tomato seedlings through seedling harvest practices and/or repeated clipping practices with a rotary mower that is conducted to obtain plant uniformity and vigor. Because of the effective transmission in nurseries through such cultural practices, the contamination of infected seeds at the rates of 0.01 to 0.05% has been sufficient for bringing about outbreaks of bacterial canker in tomato production fields. The shipment of such tomato transplants often create serious problems in fields because these transplants usually remain symptomless for a few weeks or more.

In some vegetables, fruit trees, and flower plants, the majority of pathogenic bacteria can transmit through vegetatively propagated organs. Dispersal of *P. solanacearum* in potato tubers and rhizomes of banana and ginger to areas nearby or to distant countries has frequently occurred. *P. caryophylli* disseminates mainly through cuttings. Scions are also an important source of infection. In a bud propagation trial in peach, about 27% of the scions obtained from apparently healthy trees and grafted onto seedlings died in the first year, and 60% in the second year, because of bacterial canker caused by *P. syringae* pv. *syringae*.

Dissemination through budwoods was also confirmed in the pathogens of deep canker disease of English walnut (*E. rubrifaciens*) and citrus canker (*X. campestris* pv. *citri*). Grafted young plants or seedlings are readily infected by *A. tumefaciens* because nursery soil is likely to be infested with the bacterium and dense plantings are repeated in the same nursery. The spread of foliar diseases in nurseries occurs in rapid and severe form because in general a large number of young plants are raised in very close proximity. Infection in nurseries is particularly important in perennial crops such as fruit trees. Many bacterial pathogens that are found on mature fruit trees in orchards used to be introduced in this way.

Dissemination of bacterial plant diseases across continents may occur by transportation of infected nursery plants, as has been reported in citrus canker and fire blight of apple and pear.

### 9.2.4 DISPERSAL BY INSECTS

All mycoplasma-like organisms (MLO) multiply in their leafhopper and plant hopper vectors. *X. fastidiosa* and clover clubleaf bacteria are also transmitted by their leafhopper vectors. These organisms are the parasites propagating both in insects and in plants and are disseminated by a persistent or circulative transmission but not through transovarial passage. MLO taken up by the insect vector enter into the gut and propagate in its cells. The MLO then systemically spread in the insect body through its blood, and further propagate in the salivary glands. They are injected into phloem of the host plant together with saliva when the vector feeds on the plant by means of its stylet. Consequently, latent periods from several days to several weeks, depending on the MLO–vector combination, are required before the vectors become infective. *X. fastidiosa* attaches and colonizes at the floor of the cibarium and the apodemal groove of the diaphragm of the vectors. In subsequent feeding the bacteria are discharged from the foregut into xylem of host plant by egestion of the sucking pump.

In nonfastidious prokaryotes, with a few exceptions, the vector–bacterium relationship is nonpersistent and in many cases strictly accidental. The first bacterial disease shown to be disseminated by insects was fire blight of apple and pear. Flower-visiting insects have an important role in the development of blossom blight in areas where precipitation is rare in the flowering period. Long-range spread of bacterial wilt (Moko disease) of banana has been shown to occur in parts of Central and South America. In this case, insects convey *P. solanacearum* together with pollen to inflorescences.

The important role of insect transmission in the epidemiology of potato blackleg and soft rot has also been well documented. In this case, dipterous insects naturally infested by soft rot bacteria fly from cull pile or decayed potato seed pieces to broken stems and are attracted to the exudate secreted by broken stems of potato to transmit the bacteria resulting in soft rot of potato foliage.

Another well-known example of insect transmission is *E. stewartii,* the causal bacterium of Stewart's disease of corn. The bacterium overwinters in the body of corn flea beetles and is transmitted when the insects feed on corn leaves. The disease occurs in severe form where and/or when the winter is warm enough to allow survival of the corn flea beetles.

### 9.2.5 DISPERSAL BY FARMING PRACTICES

A wide range of human activities can inadvertently increase the spread of disease in the field or the risk of introduction of pathogens to new countries or continents where the pathogen was previously unknown.

Various cultural practices can increase the spread of diseases. Pathogens can be carried on infested agricultural implements and machinery, and many bacterial diseases are disseminated during pruning or disbudding. The vascular pathogen *P. solanacearum* and *C. michiganensis* subsp. *michiganensis* can be effectively transmitted from infected to healthy plants by pinching of axillary and terminal buds with an infested knife or nail. Fire blight may be transmitted during pruning, and dissemination of ring rot of potato caused by *C. michiganensis* subsp. *sepedonicus* occurs when seed tubers are cut into pieces with an unsterilized knife. *P. avenae* and *X. campestris* pv. *campestris* are carried on the wheels of tractors with alternate weed hosts as the source of inoculum.

## 9.3 Spatial Distribution of Diseased Plants

The spatial and temporal distribution of diseased plants in a unit area and the distribution of lesions on an individual plant are not random or uniform but highly aggregated. In a study of black rot of crucifers, the patterns were best described by the negative binomial distribution with a common $K$ value of 0.15 for lesions per plant and 0.50 for infected plants per unit area, respectively. These values are much smaller than those for most insects (1–8), indicating extreme aggregation. Similar results were also obtained with bacterial blight of soybean and bacterial leaf blight of corn.

In bacterial tan spot of soybean caused by *Curtobacterium flaccumfaciens* pv. *flaccumfaciens*, patterns of disease distribution were not random but were best described as a tendency for the disease to move at a more rapid rate within rows than across rows or diagonal to rows.

These data indicate that the distance of effective transmission of the causal bacteria under normal conditions is very short, i.e., new infections occur very close to previously infected plants. Dissemination by wind or tractor may occur, forming aggregates in a limited area adjacent to the source of infection.

## Further Reading

Baker, K. F., and Snyder, W. C., ed. (1965). "Ecology of Soil-borne Plant Pathogens. Prelude to Biological Control." University of California Press, Berkeley, Los Angeles.

Dommergues, Y. R., and Krupa, S. V., ed. (1978). "Interactions Between Nonpathogenic Soil Microorganisms and Plants." Elsevier Scientific Publishing Company, Amsterdam.

Gray, T. R. G., and Parkinson, D., ed. (1968). "The Ecology of Soil Bacteria. An International Symposium." Liverpool University Press, Liverpool.

Hirano, S. S., and Upper, C. D. (1990). Population biology and epidemiology of *Pseudomonas syringae*. *Ann. Rev. Phytopathol.* **28,** 155–177.

Hopkins, D. L. (1989). *Xylella fastidiosa:* xylem-limited bacterial pathogen of plants. *Ann. Rev. Phytopathol.* **27,** 271–290.

Krupa, S. V., and Dommergues, Y. R., ed. (1979). "Ecology of Root Pathogens." Elsevier Scientific Publishing Company, Amsterdam.

Leben, C. (1981). How plant-pathogenic bacteria survive. *Plant Disease* **65,** 633–637.

Mount, M. S., and Lacy, G. H., ed. (1982). "Phytopathogenic Prokaryotes. Volume 1. Part III. Prokaryote Interactions on Plant Surfaces." Academic Press, New York.

Mount, M. S., and Lacy, G. H., ed. (1982). "Phytopathogenic Prokaryotes. Volume 2. Part I. Epidemiology and Dispersal." Academic Press, New York.

Tarr, S. A. J. (1972). "Principles of Plant Pathology." The Macmillan Press, London.

CHAPTER
**TEN**

# *Infection and Disease Development*

Infected plants develop visible symptoms after certain latent periods. Symptom expression may be influenced by the interactions between pathogenic bacteria and other microorganisms.

## 10.1 Infection

Unlike many fungal pathogens, plant pathogenic bacteria enter plants only via wounds and natural openings and not by direct penetration of noninjured tissues.

### 10.1.1 WOUNDS

All plant parts are likely to be injured by a variety of causes such as insects, wind, farming practices such as pruning, and mechanical damage due to agricultural machinery. These injuries are usually visible and constitute the major portals of entry for most bacterial plant pathogens. The minimum inoculum doses for these wounds are generally at the level of $10^2$ to $10^3$ cells/ml. This level is equivalent to 1/100 to 1/1000 of the population that is required for the infection through natural openings.

Wounds that may not be visible can also become important sites of bacterial infection. For example, when the outer ridges of stomata of citrus leaves are removed when leaves are rubbed against each other in strong winds, stomatal infection may occur through these stomata because the exposed front cavities are no longer different functionally or morphologically from the immature stomata of young leaves. Such injuries are obviously too small to be recognized by the naked eye.

The prevalence and importance of injuries of the underground parts of plants are often overlooked. However, injuries produced by insects or nematodes, the breaking of small roots by rapid movement of aboveground parts, and secondary root openings and damage to roots during cultivation also provide portals of entry for soil-borne bacteria.

The probability that wounds will serve as a pathway for bacterial infection is greatly affected by environmental conditions such as temperature, humidity, and light as well as the water potential of host plants. For example, when citrus leaves are inoculated with *Xanthomonas campestris* pv. *citri* under dry, cool weather conditions in early spring, tissues around the wound become necrotic quickly and localize the bacteria in dead tissues. The reaction causes either significant delay of infection or failure of symptom development. Such phenomena do not occur in warm weather later in the growing season, however.

The susceptibility of injuries to bacterial infection changes as a function of time after wounding. In a warm and dry atmosphere, the infection rate of wounded citrus leaves rapidly decreases within 24 hr, whereas injuries left in a cool and humid atmosphere remain susceptible for about a week (Fig. 10.1). In rice leaves, infection of *X. campestris* pv. *oryzae* occurs for 21–24 hr after wounding, but no infection occurs through injured rice roots that elapsed only 2–4 hr after wounding. Protection takes place much faster than the development of wound periderm. Such rapid

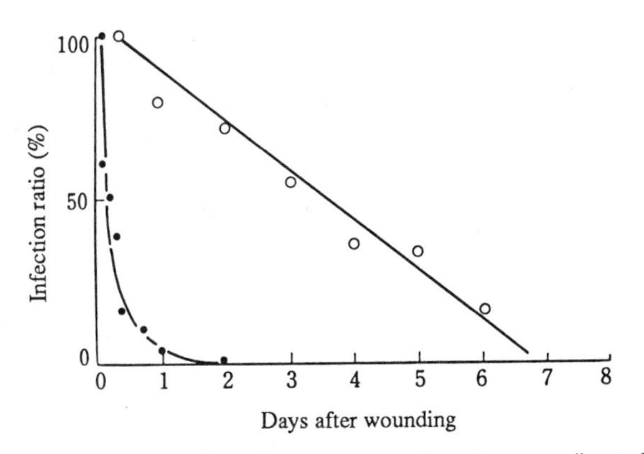


**Fig. 10.1** Effect of environmental conditions on wound healing as well as infection ratio of citrus canker. ●, plants were maintained after wounding in greenhouse at 25°–30°C; ○, plants were maintained after wounding in a humid chamber under illumination at 20°C. [From Goto, M. (1981). "New Bacterial Plant Pathology." Soft Science Co. Ltd., Tokyo.]

protection reactions are usually associated with an activation of the phenylpropanoid pathway, inducing biosynthesis of phytoalexins and lignins.

Wound healing reactions vary with the type of injury. Wounds accompanied by extensive damage to cells quickly induce the production of wound healing hormones, followed by the development of callose and subsequent suberization. In contrast, wound healing reactions are significantly delayed when cells are not substantially damaged, e.g., in feeding injuries by citrus leaf minor (*Phyllocnistis citrella*). Larvae of the insect eat only the middle lamella between the epidermal cells and mesophyll, leaving the separated dead epidermis on the surface. The formation of callus in the mesophyll is markedly delayed when the dry films of epidermis remain unbroken. Such feeding injuries inflicted by leaf minors remain susceptible to citrus canker for 10 to 14 days. Therefore, citrus canker occurs in severe form on shoots flushing in summer through early autumn when feeding damage due to the insect becomes pronounced.

## 10.1.2 NATURAL OPENINGS

The entrance of bacterial cells into healthy plant tissues occurs through natural openings such as stomata, water pores, lenticels, and nectaries.

### *Stomata*

Stomata generally enumerate 50 to 1000 (100–400 in most cases) per 1 mm$^2$ of leaf area. A mature corn plant has about 200 million stomata that are $2 \times 26$ $\mu$m in individual size; the total pore area is estimated at about 1.5% of total leaf area. Stomata are the most common portals of entry for leaf spot bacteria. They multiply in the space immediately below the stoma and then migrate into the surrounding intercellular spaces.

With the exception of soft rot *Erwinia*, cells of most phytopathogenic bacteria are nonmotile when discharged from lesions into water film on the leaf surface. Stomatal infection, therefore, occurs with nonmotile cells under natural conditions, indicating that motility plays no practical role in the infection. Cells of *Pseudomonas syringae* pv. *phaseolicola* and *Erwinia amylovora* which are nonmotile in diseased host tissues become motile in water placed on diseased plant tissues after 30–60 min and 24 hr, respectively.

In general, however, infection through stomata takes place much faster than the appearance of motile cells. For example, when susceptible citrus leaves are sprayed with suspension of the citrus canker pathogen, the stomatal infection does occur even under such conditions that water films on the leaf surface are allowed to dry immediately after spraying. In such case, motility and chemotaxis of the bacterium may have only a secondary role in infection. In contrast, soft rot bacteria are actively motile in diseased pulpy tissues, and loss of motility generally results in reduced spreading of macerated areas, although the infection rate through wounds is not affected.

The minimum inoculum dose for stomatal infection is around $1 \times 10^5$ cells/ml, irrespective of the plant or bacterial species (Fig. 10.2). This inoculum level is approximately equivalent to about one bacterial cell per stoma. Thus, one bacterial cell that successfully enters a stomatal cavity may be sufficient for infection. However, it is evident that only a limited number of these single cells can initiate infection because a very small number of lesions are produced by the inoculum at the level of $1 \times 10^5$ cells/ml.

The number of lesions in relation to inoculum doses may be significantly influenced by air humidity. Moist conditions increase the number of lesions by facilitating the initial growth of bacterial cells that have just entered stomata. Extensive stomatal infection occurs when leaves of host plants become water soaked in a moist atmosphere. For example, severe

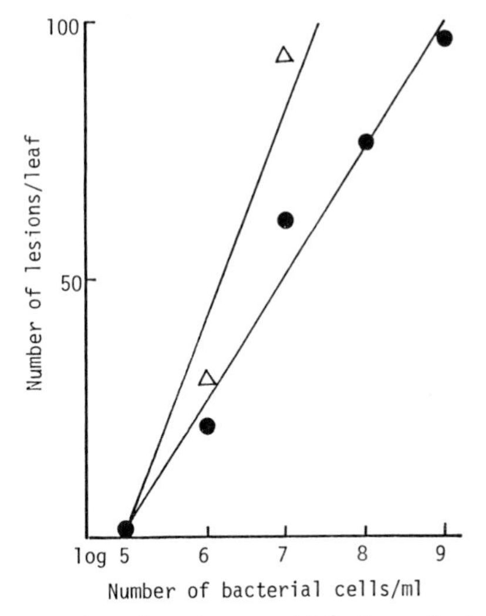


**Fig. 10.2** Relation between inoculum doses and infection rates in citrus canker. ●, spray inoculation; △, inoculated by pressing method.

infection of *P. syringae* pv. *lachrymans* on cucumber leaves occurs only under such conditions. This is attributed to the direct connection of the stomatal cavity and associated intercellular spaces by free water. Thus, it is not surprising that most epidemics of bacterial plant diseases occur in the rainy season or during periods of heavy rains.

### *Hydathodes*

Hydathode, or water pore, locates along the leaf edge and connects with the terminal end of the xylem vessel through a thin walled, chloroplast-deficient parenchyma called *epithem.* In high relative humidity early in the morning or during rain, excess tracheid water moves through the epithem to the water pores, where water is discharged in droplets. This process is called *guttation.*

Guttation water contains a low concentration of various salts, sugars, and a variety of other organic substances. When the concentration of these substances increases, guttation droplets may become injurious to tissues surrounding hydathodes. These physiological injuries may serve as an infection site for some bacteria such as the pseudomonads, causing soft rot or V-shaped lesions along leaf margins. This type of infection may be observed in vegetables and ornamental plants grown in plastic houses or greenhouses.

*X. campestris* pv. *campestris* and *X. campestris* pv. *oryzae* are well-known examples of plant pathogenic bacteria that mainly infect host plants through water pores. These bacteria initially multiply in the epithem and then invade xylem, where they grow rapidly and then block the water transport system. Bacteria subsequently break vessel walls and invade the other adjacent xylem vessels or tracheids. When the vascular tissues and adjacent parenchyma cells are extensively colonized, V-shaped lesions appear along the leaf margins of dicotyledonous plants (Fig. 13.6); wilting symptoms or wave-shaped lesions may also appear on the leaf edges of monocotyledonous plants (Fig. 13.1).

### *Lenticels*

When periderm is formed in stems at the termination of primary growth, the lenticels are simultaneously differentiated underneath stomata. The lenticels are the relatively loose arrangement of complementary (or filling) cells and function in gaseous exchange. Although infection through the lenticel has been suggested in fire blight of apple, it has a only minor importance as the portal of entry in most bacterial plant pathogens except *Streptomyces scabies,* the causal agent of common scab of potato.

*S. scabies* infects potato tubers via lenticels. A potato tuber is formed by expansion of a stolon at the distal end, and axillary buds continue to separate from the apex, forming eyes. The internode is the region between a consecutive pair of eyes. New internodes formed at the apical end of a stolon are resistant to *S. scabies* for 6–8 days at the initial stage of tuber formation where stomata distribute on the epidermis. Young tubers become susceptible to infection when lenticels initiate to develop underneath the stomata with unsuberized complementary cells. The susceptible period continues for about a week. After this period, the complementary cells are progressively suberized, becoming resistant to infection thereafter. The susceptible zone is always located in the third to fourth internode from the apex. Therefore, scab infection usually occurs in the form of a ring when the infection period is limited to a short time by controlling soil moisture.

### *Flower*

*E. amylovora* is an example of the plant pathogenic bacteria that readily invade flowers. The bacterium is transmitted from canker lesions on twigs and trunks by insect vectors and invades various flower parts such as nectary, stigma, anther, and sepal. Infection through the nectary is most common because the nectary surface is highly vulnerable to invasion. Nectar can also serve as the chemoattractant and nutrients for the pathogenic bacterium. A single cell of *E. amylovora* can successfully multiply in the nectary and establish infection via this route. The stigma of the pistil also can become a portal of infection. Bacterial masses travel down the intercellular spaces in the pistils and penetrate the embryos. Bacteria that have invaded and destroyed the flower move into the flower stem and then into the peduncle and spur, causing blossom blight. In severe infections, the disease further extends from spurs to leaves and branches, and then into the trunk, where girdling cankers can eventually kill affected trees. Bacterial pink disease of pineapple is also initiated via infection of *Gluconobacter, Acetobacter,* or *Enterobacter* through the flowers.

### *Foliar Trichome*

*Clavibacter michiganensis* subsp. *michiganensis* commonly enters its host through various morphologically distinct types of trichomes that are distributed on the surface of foliage and stems of tomato plants. These trichomes include glandular and nonglandular short hairs and nonglandular long hairs. The infection rate is much higher on long, septated nonglandular hairs with bulbous bases than short or glandular hairs.

The susceptibility of leaves is generally correlated with the density of trichomes; young leaves with the greatest density of hairs are most susceptible to the disease. The presence of trichomes is not essential for infection by the bacterial canker pathogen because it can also invade plants via stomata and wounds. However, infection through the trichomes significantly increases disease incidence; it is particularly important in the infection of young fruits because they lack stomata.

Entry through trichomes is also observed in *E. amylovora.* The trichomes distributing on the upper surfaces of apple leaves may serve not only as natural portals of infection but also as the potential route for passage of the bacterium into the sieve element of the phloem.

## 10.2 Incubation Period

The time interval from infection of a plant to appearance of disease symptoms on the plant is called the incubation period. In natural infection by bacterial plant pathogens, incubation periods are generally 5–7 days with a few exceptions. The length of the incubation period is governed by various factors of pathogenic bacteria (species, mode of entry, inoculum density, and type of disease), host plant (species, age, and organ), and environment (temperature and humidity).

In leaf spot diseases, water-soaked symptoms generally appear when the bacterial population reaches $10^6$ cells per inoculation site, and the site of bacterial multiplication coincides with the site of water congestion. In vascular diseases such as black rot of cabbage or bacterial leaf blight of rice, however, the latent lesions may be formed before the development of visible lesions (Fig. 10.3).

The incubation period is generally shorter in the actively growing parts of plants than the mature parts. The trend is pronounced in gall-forming diseases. For example, in bacterial gall of Japanese wistaria caused by *E. milletiae,* gall formation may be initiated 5 to 10 days after inoculation on the young growing shoots, whereas it may take several months on thick trunk.

When infection occurs late in autumn, the bacteria as well as plant tissues quickly become dormant because of the low temperatures. Consequently, the incubation period is extended to the following spring, resulting in a latent infection. This type of latent infection can be frequently

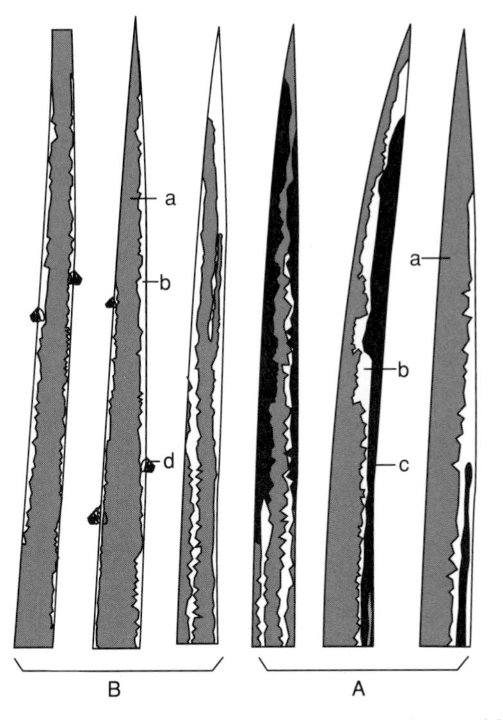


**Fig. 10.3** Latent lesions and appearance of visible lesion in bacterial leaf blight of rice. A, diseased leaves; B, apparently healthy leaves; a, healthy portion; b, latent lesion; c, visible lesion; d, bacterial ooze.

observed in bacterial diseases of fruit trees such as citrus canker, bacterial canker of stone fruit trees, and bacterial shot hole of peach.

## 10.3 Host Range

The host range of plant pathogenic bacteria varies greatly, depending on the bacterial species. It ranges from only one plant species to a large number of species belonging to unrelated taxonomic groups. The bacteria with a wide host range include *Agrobacterium tumefaciens* (643 species of 93 families), *P. solanacearum* (270 species of 3 families), *E. carotovora* subsp. *carotovora* (140 species of 61 families), and *E. amylovora* (180 species of rosaceae).

*P. andropogonis* was first described as the causal agent of bacterial stripe of sorghum and subsequently of maize. However, recent investigations have revealed that the bacterium has a wide host range, invading diverse kinds of plants such as Leguminosae (clover), Nyctaginaceae (bougainvillea), Liliaceae (New Zealand flax, tulip, *Ruscus* spp., *Crinum* spp.), Musaceae (*Strelitzia), Iridaceae (Iris* spp.), and Caryophyllaceae (carnation and gypsophila). The example of *P. andropogonis* indicates that the restricted host ranges of any existing bacteria may be expanded in the future through more extensive surveys.

*P. syringae* and *X. campestris* lumped a number of closely related nomenspecies in the revision of the nomenclature of plant pathogenic bacteria. The individual nomenspecies were designated as pathovars of these two species. Therefore, the host ranges of these bacteria are very wide at the species level but limited to only one to a few plant species at the pathovar level. The exceptions are *P. syringae* pv. *tabaci* and *P. syringae* pv. *syringae* in which about 40 plant species are listed as host plants. In many pathovars of *X. campestris* and *P. syringae,* however, the host ranges have not necessarily been studied intensively. It is highly possible that the number of these pathovars will be reduced when it is seen that some are the synonyms of others by extensive inoculation tests.

The host ranges of plant pathogenic bacteria include plant species shown to be the host plants on the basis of natural infections and/or artificial inoculations. Host ranges due to artificial inoculation involve some problems in relation to the determination of pathogenicity. When toxin-producing bacteria are inoculated, some plants may exhibit symptoms that are simply caused by toxins that were incidentally produced in the inoculation sites without appreciable growth of bacteria in the tissues. Such plants should not be included in the host range because active multiplication in host plant tissues is a requisite for determination of pathogenicity in phytopathogenic bacteria except with *Agrobacterium* spp. In leaf spot diseases, water-soaked lesions usually develop where active bacterial growth takes place.

## 10.4 Disease Development and Symptoms

Plant pathogenic bacteria that enter susceptible host plants grow either in parenchyma tissue or vascular tissue and induce different symptoms through the physiological activities of various virulence factors produced by bacteria, such as toxins, enzymes, phytohormones, and polysaccharides.

Symptoms of bacterial plant diseases include wilting, necrotic spot, blight, soft rot, hypertrophy, and malformation. The symptom appearance depends on many factors, such as disease resistance of the host plant, the virulence of bacteria, and environmental conditions.

### 10.4.1 WILTING

Wilting is induced in most cases by the blocking of water conductance resulting from plugging of xylem vessel with bacterial slime and bacterial cell aggregates (Fig. 10.4). Wilting in bacterial wilt caused by *P. solanacearum* is very rapid; an entire tomato plant may wilt within several hours after the initial symptoms appeared on a few upper leaves. In contrast, wilting in bacterial canker of tomato and potato ring rot occurs slowly and usually accompanies yellowing (Fig. 13.8). *X. campestris* pv. *oryzae* commonly infects rice leaves through hydathodes, inciting leaf blight. In

**Fig. 10.4** Bacterial wilt of *Musa Basjoo* caused by the *Strelitzia* strain of *P. solanacearum.*

Indica-type rice, however, the bacterium can also systemically invade young seedlings, resulting in wilting of the entire plant. This symptom is called *kresek* in Javanese (Fig. 13.1).

In the advanced stages of many vascular diseases, vessel walls may be dissolved by enzymes, and lysigenous cavities filled with the causal bacteria are often formed in the vascular tissues. In bacterial wilt of strawberry caused by *P. solanacearum*, the plugging of vascular systems is limited to a relatively small number of vessels, but bacteria actively grow in xylem parenchyma or even in pith forming bacterial pockets.

### 10.4.2 LEAF SPOT

In general, water-soaked lesions are the initial symptoms of bacterial leaf spot diseases. The symptom appears 4–5 days after infection and is particularly conspicuous under moist conditions. The causal bacteria actively grow in the intercellular spaces of the water-soaked lesions and exude in droplets on leaf surface. The lesions gradually turn yellow to dark brown and eventually become necrotic (Fig. 13.9). In old necrotic lesions, water soaking is often still evident along the edge of lesions, where it merges with healthy tissues.

The shape of leaf spots (angular, round, striped, or streaked) is an important feature for diagnosis of the diseases (Fig. 10.5). In severe infections, many individual lesions coalesce with each other, forming large lesions and resulting in death of the leaves or defoliation.

The initial symptom of citrus canker is a small, white or light greenish localized projection accompanied by negligible water soaking. The projection is callus tissue consisting of the hypertrophied cells. Once these lesions are wetted by rain, however, the tissue collapses rapidly as a result of invasion by secondary organisms, forming typical canker lesions with a water-soaked margin (Fig. 13.19).

### 10.4.3 BLIGHT

Shoot blight or die-back symptoms characterize diseases such as bacterial blight of tea (*P. syringae* pv. *theae*), citrus blast (*P. syringae* pv. *syringae*), and bacterial canker of stone fruit (*P. syringae* pv. *morsprunorum*) (Fig. 10.6, 13.17). Infection usually occurs at the basal portion of leaf petioles and buds, or new lesions extend from holdover cankers. Bacterial ooze may be formed on the surface of infected stems.

*P. syringae* pv. *eriobotryae* causes either bud blight of loquat by infect-

**Fig. 10.5** Bacterial stripe of sorghum caused by *Pseudomonas andropogonis.*

ing the terminal buds or canker by infecting twigs. Both symptoms may eventually result in die-back. Leaf spot inducing bacteria may also cause shoot blight when they severely attack young shoots. For example, shoot blight is often produced on young trees of susceptible *Citrus* spp. that were heavily infected by *X. campestris* pv. *citri* or on susceptible peach cultivars by *X. campestris* pv. *pruni.*

### 10.4.4 SOFT ROT

Bacterial soft rot is characterized by watery collapse of affected parenchyma tissues (Fig. 10.7). Soft rot bacteria frequently spread through xylem tissue in the initial stage of disease development. When *E. carotovora* subsp. *carotovora* and *E. chrysanthemi* pv. *chrysanthemi* infect Chinese cabbage and chrysanthemum, respectively, the pathogens can invade xylem initially and spread rapidly both downward and upward. The invaded vessels are ruptured eventually, and the released bacteria cause soft rot of the adjacent parenchyma, entering again into neighboring xylem vessels. Bacteria can invade pith parenchyma causing soft rot, but spread is naturally much slower than through xylem.

**Fig. 10.6** Citrus blast of *C. unshu* caused by *Pseudomonas syringae* pv. *syringae*.

*Bacillus* spp. and *Clostridium* spp. can cause decay of potato tubers in storage at temperatures higher than those favorable to soft rot *Erwinia*. In addition, *Bacillus* spp. preferentially attacks potato tubers in aerobic conditions, whereas *Clostridium* spp. attacks in anaerobic conditions, with the production of a distinctive thick slimy exudate and a disagreeable odor.

In general, soft rot induced by pseudomonads occurs at a temperature lower than that caused by soft rot *Erwinia*. Bacterial rot of vegetables prevalent in cool seasons from late autumn to early spring is often caused by *P. viridiflava* or *P. fluorescens* (= *P. marginalis* pv. *marginalis*). When

**Fig. 10.7** Soft rot of Chinese cabbage caused by *Erwinia carotovora* subsp. *carotovora.*

storage roots such as carrot or Japanese radish (*Raphanus sativus*) are invaded by soft rot *Erwinia* in the field, the above-ground symptoms, or wilting of foliage, develop only after the root has extensively decayed.

### 10.4.5 HYPERTROPHY (GALL)

Crown gall caused by *A. tumefaciens* is generally formed at the crown (collar of the roots) and on roots but less frequently on aerial parts of the plants except grape plants. In other hypertrophic diseases such as bacterial gall of Japanese wistaria (*E. milletiae*), *Myrica rubra* (*P. syringae* pv. *myricae*), *Melia azedarach* (*P. meliae*), and olive (*P. syringae* pv. *savastanoi*), in contrast, gall formation occurs predominantly on the above-ground parts of twigs and trunks.

These pathogens are all wound parasites and similar in symptoms; galls are knobby or knotty and become more irregular as they age. Crown gall tumors sometimes enlarge to about 30 cm in diameter, and bacterial galls formed on the trunk of *Melia azedarach* reach about 1 m in diameter. The size of galls formed on *Myrica rubra* and olive ranges from a few millimeters to about 5 cm in diameter. Insects usually feed on the galls on

**Fig. 10.8** Bacterial gall of Japanese wistaria caused by *Erwinia milletiae.*

Japanese wistaria, and as a result they collapse within a year; otherwise they reach 10 cm in diameter (Fig. 10.8).

In most bacterial galls, the first reaction of the host plant is to develop hypertrophied cells characterized by thin cell walls, increased cytoplasm, and readily-staining large nuclei. Active cell division is subsequently initiated, producing hyperplastic tissues consisting of a number of daughter cells characterized by different sizes. Hyperplastic tissues erupt from wounds 5 to 10 days after inoculation, developing into visible gall. Some hypertrophied cells differentiate into malformed xylem and eventually connect with the vascular tissues of host plant. In general, the hyperplastic cells of bacterial gall are characterized by irregularly divided cells of various shapes and sizes. With the exception of crown gall, masses of bacterial cells can be observed in the intercellular spaces of hyperplastic tissues (Fig. 10.9).

Although tissues of gall formed on woody plants become lignified and hard as they age, those of newly developing young galls are always tender and contain many bacterial pockets.

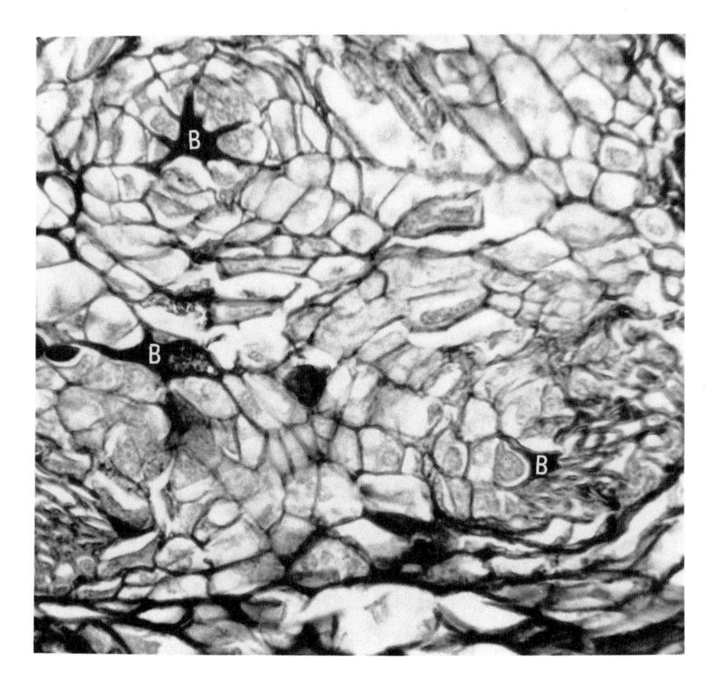


**Fig. 10.9** Hyperplastic tissues of bacterial gall of Japanese wistaria. B, bacterial aggregates.

### 10.4.6 MALFORMATION

Malformed leaves may form on bean leaves that are infected with *P. syringae* pv. *phaseolicola* and on mulberry leaves infected with *P. syringae* pv. *mori,* particularly when lesions are formed along midveins (Fig. 13.24). Malformation is also found when *Rhodococcus facians* and *C. tritici* infect strawberry and wheat, respectively, in association with nematodes.

### 10.4.7 MOSAIC

Mosaic symptom is exhibited in bacterial mosaic of wheat caused by *C. michiganensis* subsp. *tessellarius.* The disease is characterized by yellow lesions with indefinite margins that are formed throughout the leaf and induce the mosaic patterns. The lesions never become water soaked nor exude bacterial stream.

### 10.4.8 YELLOWS AND ROSETTE

Mycoplasmalike organisms and spiroplasmas produce the characteristic yellows-type symptoms which include rosette, leaf yellowing, stunting, apical dwarfing, leaf roll, epinasty, wilting, virescence, and proliferation of axillary shoots.

### 10.4.9 LEAF SCORCH

Leaf scorch induced by infection of *X. fastidiosa* is characterized by discolored or chlorotic areas produced in marginal and interveinal portions of the leaves. Severely infected leaves show brown discoloration and curling and remain attached to the branch.

## 10.5 Interactions between Plant Pathogenic Bacteria and Other Microorganisms

When a plant pathogenic bacterium infects a host plant in the presence of another microorganism, the pathogenicity of the former may significantly be reduced or stimulated, resulting from antagonism or synergism in the process of disease development. Effects of the interactions between two organisms may be expressed as the different types of symptoms, varied rate of infection, length of latent period, and stability of resistance, etc. When a microorganism that interacts with a plant pathogenic bacterium is another plant pathogen, the effect of antagonism or synergism appears on the infection ratio and severity of the two different diseases that they cause.

### 10.5.1 VIRUS–BACTERIA INTERACTION

When potato is infected with the complex of *C. michiganensis* subsp. *sepedonicus* and potato X and S viruses, development of typical symptoms of the virus diseases is inhibited, resulting in a significant decrease in yield compared with the single infection of each pathogen.

In contrast, bacterial pith maceration of cuttings of *Chrysanthemum morifolium* caused by *E. chrysanthemi* pv. *chrysanthemi* is reduced by the infection of chrysanthemum stunt viroid (CSV). The causal bacterium is restricted to xylem vessel in CSV infected cuttings, whereas it spreads from the vessel to the adjacent vascular cells and into pith in virus-free cuttings.

Cowpea seedlings fail to develop tumors when they are first inoculated with *A. tumefaciens* and then 1 day later with the cowpea mosaic virus. The failure of tumor formation by *A. tumefaciens* may be the result of an inhibitory substance that is induced in cowpeas by systemic virus infection and translocated to the crown gall infection site.

### 10.5.2 BACTERIA–BACTERIA INTERACTION

Carnation tissues are rapidly macerated by the simultaneous inoculation of *P. caryophylli* and *Corynebacterium* sp. but not by the inoculation of either organism alone. *Corynebacterium* sp. but not *P. caryophylli* produces endo-pectate lyase, causing the maceration of infected tissues.

*Alcaligenes* sp. (*Achromobacter* sp.) causes systemic latent infection of bean leaves, irrespective of single or mixed inoculations. The number of lesions incited by *P. syringae* pv. *phaseolicola* increase from 2- to nearly 4-fold when inocula are mixed with *Alcaligenes* sp. Enhancement of disease is obtained whether *Alcaligenes* was inoculated simultaneously or before *P. syringae* pv. *phaseolicola.*

Application of *Azotobacter* sp. and *Pseudomonas* sp. to rice as microbial fertilizers results in greater damage by kresek and leaf blight phases of bacterial leaf blight caused by *X. campestris* pv. *oryzae.*

The bacteria–bacteria interactions have been documented mainly from the viewpoint of biological control of bacterial plant diseases. The antagonistic bacteria have been identified in many cases as *Streptomyces, Bacillus,* and *Pseudomonas.* Biological control of crown gall by *A. radiobacter* strain K84 has been used in the commercial base.

### 10.5.3 FUNGI–BACTERIA INTERACTION

Inoculation of *Gibberella cyanogenea, F. oxysporum, F. solani,* and *Verticillium dahliae* on the newly emerged potato plants that were inoculated before planting with *E. carotovora* subsp. *carotovora* and *E. carotovora* subsp. *atroseptica* significantly increases the incidence of black-leg over that caused by the bacteria alone but not seed tuber decay. In particular, *Verticillium dahliae* and *E. carotovora* subsp. *carotovora* are synergistic in the induction of early dying disease of potato: *E. carotovora* subsp.

*carotovora* enhances reduction of plant growth as well as increase of chlorosis and wilting by *V. dahliae,* and *V. dahliae* reciprocally enhances development of stem soft rot by *E. carotovora* subsp. *carotovora.* In wounded citrus fruit, infection of *Penicillium digitatum* is promoted by a coexistence of *P. cepacia* because the bacterium suppresses the increase of phenylalanine ammonia lyase activity and subsequent wound healing.

Fungi–bacteria interactions are of great interest in the production of edible mushrooms. In general, any healthy and clean fruiting bodies irrespective of fungal species carry bacterial microflora, mostly consisting of pseudomonads. Bacteria can attach themselves rapidly and firmly to the surface of fruiting bodies. The attachment rate of the pathogenic bacteria such as *P. tolaasii* and *P. gingeri* is twice as high as that of saprophytic pseudomonads, although the attachment is not a uniform occurrence all over the sporophore surface. It has been documented that some saprophytic pseudomonads stimulate development of fruiting bodies through their synergistic effects.

### 10.5.4 NEMATODE–BACTERIA INTERACTION

Synergism in the combined infection of nematodes and bacteria is relatively common. The tobacco cultivars that are resistant to bacterial wilt caused by *P. solanacearum* often become susceptible by a joint infection with the bacterium and a root knot nematode, *Meloidogyne incognitaacrita.*

Leafy galls are caused by *Rhodococcus fascians* as single infections on strawberry leaves. Similarly, single infections by the ectoparasitic nematodes, *Aphelenchoides fragariae* and *A. ritzemabosi,* on strawberry induces a rosette of small alaminated leaves which usually die, resulting in open-center plants. When strawberry plants are infected by both the bacterium and the nematodes at the same time, proliferation of axillary buds occurs, reducing the crown to the stunted fleshy rosette that is referred to as "cauliflower" disease. Nematodes also can serve as the vectors of *R. fascians* in this disease complex.

In all instances mentioned above, both bacteria and nematodes invade the same part of the host plant at the same time. In bacterial canker of French prune trees incited by *P. syringae pv. syringae,* however, the incidence and severity of the disease is remarkably increased by the attack on the roots by the ring nematode, *Criconemoides xeroplax.*

### 10.5.5 ENDOPHYTE–BACTERIA INTERACTION

It has been known for a long time that apparently healthy potato tubers carry bacteria internally (*Bacillus* spp.). The presence of diverse kind of microorganisms has recently been documented in the actively growing, healthy tissues of various plants. These microorganisms are different from latent infection of pathogens in that they do not elicit diseases in any given environments; hence they are called *endophytic microorganisms*.

A study of the endophytic bacteria inhabiting xylem vessels of citrus trees revealed that they consist of *Pseudomonas* (40%), *Enterobacter* (18%), Gram-positive bacteria (*Bacillus*, coryneforms) (16%), and *Serratia* (6%). Total populations were $1 \times 10^2$ to $1 \times 10^4$ per gram tissue. About 4% of pseudomonads induced hypersensitive reaction on tobacco leaves. Similar results have been documented in storage organs such as cucumber fruit, sugar beet roots, and peanut kernels and the above-ground parts of other plants such as cotton and maples. Some of these endophytes were reported to protect plants from diseases in various degrees: e.g., when a strain of *Bacillus subtilis* that was isolated from xylem vessels of maples was absorbed from roots, 55–80% of maple seedlings were protected from infection of *Verticillium dahliae*.

### 10.5.6 EPIPHYTE–BACTERIA INTERACTION

Phylloplane microorganisms are generally bacteria, yeasts, and fungi. Their populations are mainly affected by sunlight, humidity, and temperature. For example, the naturally occurring populations of microorganisms on citrus leaves are: bacteria, 10–4000; yeasts, 10–500; and fungi, 10–1000 per $cm^2$, depending on leaf age, position of leaves in the tree, and season. Another study on microbial populations on the leaves of onion grown outdoors revealed that the numbers of bacteria ranged from 10 to 5,000 and yeasts and fungi from 10 to 500 per $cm^2$.

The interactions between plant pathogenic bacteria and phylloplane organisms have been studied in several diseases. Infection of cotton leaves with *X. campestris* pv. *malvacearum* always results in an increased number of fungi per unit area, the dominant fungi being *Cladosporium*, *Phoma*, and *Alternaria*. The majority of these fungi are inhibitory to the pathogenic bacterium. Similar interactions have been reported between *Fusarium* spp. and citrus canker organism (*X. campestris* pv. *citri*) (Fig. 10.10).

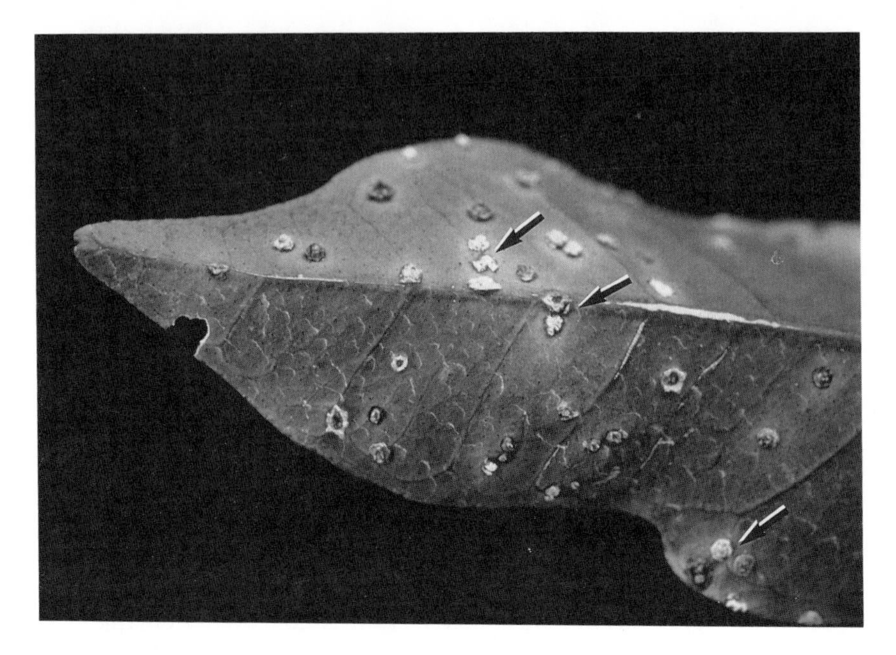

**Fig. 10.10** Citrus canker lesions colonized by *Fusarium* spp (see arrows).

## Further Reading

Carter, W. (1962). "Insects in Relation to Plant Disease." Interscience Publishers, New York and London.

Esau, K. (1953). "Plant Anatomy," 2nd ed. John Wiley & Sons, Inc., New York.

Horsfall, J. G., and Cowling, E. B., ed. (1978). "Plant Disease, An Advanced Treatise. Volume II. How Disease Develops in Populations." Academic Press, New York.

Huang, J.-S. (1986). Ultrastructure of bacterial penetration in plants. *Ann. Rev. Phytopathol.* **24,** 141–157.

Lockwood, J. L. (1988). Evolution of concepts associated with soilborne plant pathogens. *Ann. Rev. Phytopathol.* **26,** 93–121.

Van der Plank, J. E. (1975). "Principles of Plant Infection." Academic Press, New York.

Van der Plank, J. E. (1982). "Host–Pathogen Interactions in Plant Disease." Academic Press, New York.

CHAPTER
ELEVEN

# *Effect of Environment on Disease Development*

In general, epidemics of bacterial plant diseases occur as the result of interactions among three major factors: population of plants, population of pathogens, and the environment. The environmental factors affect the activities of plant and pathogen separately and also the host–pathogen interactions.

It is not uncommon that bacterial diseases of fruit trees or shade trees suddenly break out after complete disappearance for more than a decade. Epidemics of citrus canker (*Xanthomonas campestris* pv. *citri*) and shoot drooping disease of maples (proposed name of the pathogen, *Erwinia acerivora* Takikawa) are such examples in Japan. The former has been occurring in epidemic form every 10 years and the latter occurred in the mid 1980s in severe form after being undetectable for about 50 years since its first reported occurrence. The factors governing epidemics with such long intervals have not been fully elucidated. It is conceivable that the diseases persist in sporadic form at an undetectable level, and that weather conditions, for example, warm temperature and frequent rains in winter, have a significant effect on multiplication of the overwintering pathogens as well as the susceptibility of host plants.

## 11.1 Temperature

### 11.1.1 AIR TEMPERATURE

Temperature can directly affect the multiplication of pathogenic bacteria, influencing the incidence of disease development. The optimum, maximum and minimum growth temperatures are different, depending on the species or strains of bacteria (Table 11.1). The optimum growth temperatures usually fall within the range of 25° to 30°C. The temperature range that bacteria can grow above the optimum is generally narrow but considerably wider below the optimum.

Many pseudomonads are characterized by psychrotrophic traits, with active growth even at 10° to 20°C (Fig. 11.1). Severe outbreaks of citrus blast and bacterial leaf blight of wheat (*Pseudomonas syringae* pv. *syringae*), bacterial canker of stone fruit trees (*P. syringae* pv. *morsprunorum*), and bacterial rot of vegetables (*P. viridifalva, P. fluorescens*) commonly occur late in autumn to early spring, particularly in the period when plants suffer frost damage.

**Table 11.1**
Growth Temperatures of Plant Pathogenic Bacteria

| Bacteria | Minimum | Optimum | Maximum |
|---|---|---|---|
| *Erwinia amylovora* | 0.5°–3.0°C | 28°–30°C | >37°C |
| *E. carotovora* subsp. *atroseptica* | 3 | 27 | 32–35 |
| *E. carotovora* subsp. *carotovora* | 6 | 27 | 35–37 |
| *E. chrysanthemi* pv. *zeae* | 6 | 37 | 40–41 |
| *Pseudomonas andropogonis* | 1.5 | 22–30 | 37–38 |
| *P. avenae* | 0 | 30–35 | 40 |
| *P. caryophylli* | 5 | 30–33 | 46 |
| *P. cepacia* | 6–9 | 30 | 42 |
| *P. solanacearum* | 10 | 35–37 | 41 |
| *P. syringae* pv. *glycinea* | 2 | 24–26 | 35 |
| *P. syringae* pv. *lachrymans* | 1 | 25–27 | 35 |
| *P. syringae* pv. *maculicola* | 0 | 24–25 | 29 |
| *P. syringae* pv. *phaseolicola* | 2.5 | 20–23 | 33 |
| *P. syringae* pv. *pisi* | 7 | 27–28 | 37.5 |
| *Xanthomonas campestris* pv. *begoniae* | 1–3 | 27 | 37 |
| *X. campestris* pv. *campestris* | 7–10 | 28–30 | 36 |
| *X. campestris* pv. *citri* | 10 | 28–30 | 38 |
| *X. campestris* pv. *malvacearum* | 7–10 | 25–30 | 36–38 |
| *X. campestris* pv. *oryzae* | 5–10 | 26–30 | 40 |
| *X. campestris* pv. *pruni* | 7–10 | 24–29 | 37 |
| *Clavibacter michiganensis* subsp. *michiganensis* | 1 | 24–27 | 36–37 |
| *C. michiganensis* subsp. *sepedonicus* | 3–4 | 20–23 | 30–31 |
| *Rhodococcus fascians* | | 25–28 | 37 |
| *Curtobacterium flaccumfaciens* pv. *flaccumfaciens* | 1.5 | 20–30 | 37–40 |
| *C. flaccumfaciens* pv. *oortii* | 5 | 25–30 | 37 |
| *C. flaccumfaciens* pv. *poinsettiae* | 3 | 24–27 | 37–40 |
| *Agrobacterium radiobacter* | 1 | 28 | 45 |
| *A. tumefaciens* | 0 | 25–30 | 37 |

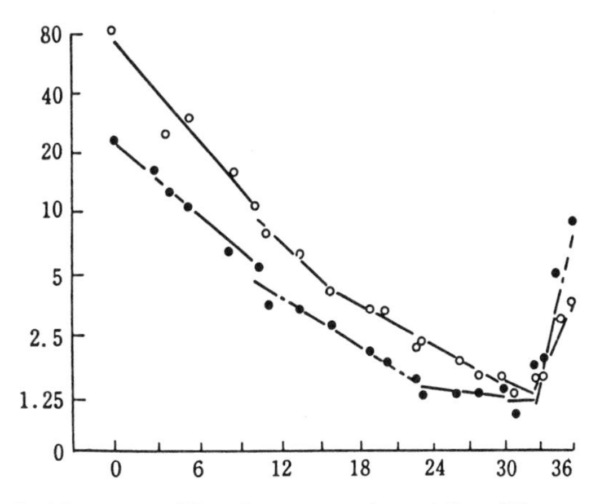


**Fig.11.1** Mean doubling times of *P. syringae* pv. *syringae* (●) and *X. campestris* pv. *pruni* (○) at the temperature range of 0°C–36°C. [Adapted from Young, J. M., Luketina, R. C., and Marshall, A. M. (1977). The effects of temperature on growth *in vitro* of *Pseudomonas syringae* and *Xanthomonas pruni*. *J. Appl. Bacteriol.* **42,** 345–354.]

In psychrotrophic pseudomonads, it has been well documented that the export of certain enzymes such as protease and phosphatase are maximal at temperatures of 15° to 20°C, well below the optimal temperature of 25° to 30°C. Such low temperature specific regulation may also be involved in biosynthesis and secretion of virulence factors and responsible for the prevalence of these diseases in cool weather conditions. For example, *P. syringae* pv. *phaseolicola* preferentially produces phaseolotoxin at the lower temperatures of 15° to 20°C, and this bacterium is protected from its toxin by phaseolotoxin-tolerant ornithine carbamoyltransferase that is also produced at the low temperatures.

In contrast, the plant pathogenic xanthomonads generally favor higher temperature for disease development. For example, bacterial leaf spot of carrot (*X. campestris* pv. *carotae*) occurs in more severe form at 25° to 30°C than at 20°C. This trend is especially pronounced when the vascular system of roots is infected. The natural incidence of bacterial leaf blight of rice (*X. campestris* pv. *oryzae*) is more severe at 25° to 30°C but is rare below 20°C. On the other hand, high summer temperatures ranging from 32° to 33°C are rather depressive to disease occurrence. This is chiefly a result of the less effective dissemination of the causal bacterium under intensive sunshine and dryness rather than of the effect of temperature.

Air temperature also affects the susceptibility of host plants. The resistance of cotton cultivars to bacterial blight incited by *X. campestris* pv. *malvacearum* is lowered by low night (19°C) and high day temperature (36.5°C). The enhanced disease development under this temperature regime is particularly severe in the susceptible cultivars, but the disease reaction of the immune cultivar is much less affected (Table 11.2).

Development of bacterial blight of corn caused by *P. avenae* is accelerated by a lower temperature regime (day/night temperature lower than 22°C/14°C), rather than a temperature regime higher than 22°C/18°C. This relationship is just the reverse in wheat and oat, which become more susceptible at the higher temperatures. Therefore, the damage of bacterial blight of corn becomes less severe where night temperatures are higher than 22°C, and the disease incidence may be decreased on wheat and oat by seeding late in autumun when the temperature is cooler.

In crown gall of radish, growth reduction of host plant is temperature dependent, being more severe in higher temperatures. No growth retardation is obtained with either the above-ground or the underground tissues in the day/night temperature regime of 14°C/10°C, but highly significant retardation occurs at 30°C/26°C to 34°C/30°C.

Pierce's disease of grape (*Xylella fastidiosa*) occurs only in moderate or subtropical climates, and severe winter conditions limit distribution of the disease. The low temperature may indirectly affect recovery from the disease through the physiological change of host plants rather than direct therapeutic effect on the causal bacteria.

**Table 11.2**
Effect of Different Day and Night Temperatures on the Development of Bacterial Blight of Cotton[a]

| Temperature | | Cotton strains | | | |
|---|---|---|---|---|---|
| | | Susceptible | Resistant | | Immune |
| Night | Day | Acala 44 | Acala1517BR−1 | 20−3×4Ac44 | Im2−4−4 |
| 19 | 36.5 | S+ | S | S | Im− |
| 19 | 25.5 | S+ | R | R | Im− |
| 26.5 | 26.5 | S− | R | R | Im |
| 26.5–29 | 36.5 | S− | R | R | Im |

[a] Adapted from Brinkerhoff, L. A. and Presley, J. T. (1967). Effect of four day and night temperature regimes on bacterial blight reactions of immune, resistant, and susceptible strains of upland cotton. *Phytopathology* **57,** 47–51.

## 11.1.2 SOIL TEMPERATURE

The ordinal strains of *P. solanacearum* (improperly designated race 1) usually do not produce wilt symptoms in infected plants when soil temperature is below 21°C. In inoculation tests conducted at variable soil temperatures and fixed air temperature of 22°C, the higher the soil temperature, the more severe and rapid the wilt development. In contrast, bacterial wilt of potato due to the potato strain (improperly desig-

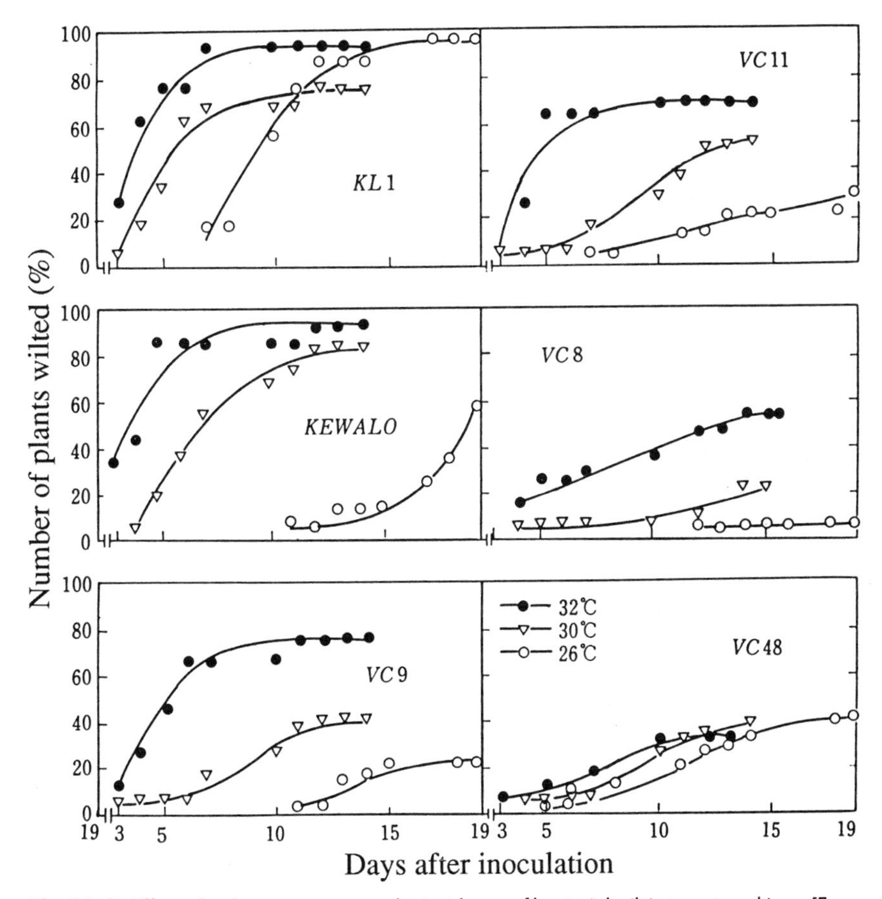


**Fig.11.2** Effect of soil temperatures on the incidence of bacterial wilt in tomato cultivars [From Mew, T. W. and Ho, W. C. (1977). Effect of soil temperature on resistance of tomato cultivars to bacterial wilt. *Phytopathology* **67,** 909–911.]

nated race 3) is characterized by development of more severe symptoms at 21°C than at 23°C.

Resistance of some tomato cultivars to bacterial wilt is soil temperature dependent. The cultivars VC8 and VC11 are resistant at 26°C but become moderately susceptible and susceptible, respectively, at 32°C. The rate of wilting or the time in days for 50% wilting also differed among cultivars at different soil temperatures. It appears that 32°C is the critical soil temperature for separating the two types of bacterial wilt resistance in tomato (Fig. 11.2).

*E. carotovora* subsp. *atroseptica* causes considerably more black-leg disease than *E. carotovora* subsp. *carotovora* in cool soil temperature, i.e., 7° to 26°C for the first 30 days after planting, whereas the reverse is true under high soil temperature, i.e., 21° to 35°C.

## 11.2 Humidity

### 11.2.1 AIR HUMIDITY

Because rain is a major means of dissemination for most plant pathogenic bacteria, an environment with frequent rains and high humidity is most favorable to infection. Relative humidity (RH) affects directly or indirectly the time that free water remains on plant surfaces, whether derived from fog, dew, or rain. Consequently, RH affects the duration of water congestion, amount of bacterial exudation and its desiccation, epiphytic population of pathogens, and interaction between pathogens and saprophytic microorganisms on plant surfaces.

Angular leaf spot of cucumber caused by *P. syringae* pv. *lachrymans* is one of the diseases that require high humidity for disease development. Water-soaked, large angular spots are formed when RH is maintained in saturation for 24 hr after infection or RH of 90–94% saturation is retained until disease development. In contrast, lesions remain in tiny spots with minor damage when the saturated RH lasts within 6 hr or 83–85% saturation is maintained until disease development.

The persistence of free water on plant surfaces is sometimes essential for disease development. Disease severity of *Philodendron selloum* infected by *E. chrysanthemi* pv. *phylodendron* is not affected by the presence of free water on the preinoculation leaf surface but significantly affected by misting for 24 hr after inoculation. Misting for longer than 24 hr after inoculation does not affect disease severity. Development of bacterial blotch of mushroom (*Agaricus bisporus*) caused by *P. tolaasii* is promoted by the persistence of free water on primordium. Hence, frequent overhead watering increases disease incidence as well as disease severity.

In bacterial leaf streak of rice (*X. campestris* pv. *oyzicola*) or fire blight of apples (*E. amylovora*), a large amount of bacterial ooze exudes on leaf surfaces under high humidity. The ooze dries rapidly under a low RH. In black rot of cabbage (*X. campestris* pv. *campestris*) and bacterial leaf blight of rice (*X. campestris* pv. *oryzae*), bacterial droplets appear along leaf edges at the sites of hydathodes early in the morning under high RH but soon disappear as the RH decreases after sunrise.

Plant pathogenic bacteria survive as epiphytes together with saprophytic bacteria on plant surfaces. The RH and the presence of a wet film on plant surfaces substantially affect the epiphytic capacity of these bacteria. In general, bacteria of the *P. syringae* group have a high capacity to reside on leaf surfaces and can retain high populations at a RH of 40%. On cucumber leaves, both *P. syringae* pv. *lachrymans* and saprophytic bacteria increase rapidly for the first 12 hr on a surface wet with dew. After this rise, the population of the pathogen declines precipitously because of a very large increase of saprophytes. When infected leaves are subjected to continuous rain, however, the populations of the pathogen and the saprophytes increase in 8–16 hr and remain at a high level thereafter.

After establishment of infection, rain continues to influence the symptoms. For example, leaf spots enlarge rapidly in rain with formation of a conspicuous water-soaked appearance. In citrus canker, small callus projections that are formed in the dry atmosphere collapse in rain, forming typical large lesions of the disease.

### 11.2.2 SOIL HUMIDITY

Soil humidity significantly influences the distribution and survival of bacteria in soil, the activity of other soil organisms, the process of root infection, as well as the water potential of host plants. These factors affect the resistance of host plants, altering the incidence and severity of bacterial plant diseases. High correlations are generally observed between disease severity and water potential of plant tissues. *E. carotovora* subsp. *carotovora* cannot cause soft rot of parenchyma tissue of Chinese cabbage placed in 30% RH for 2 hr. This condition remains unchanged even if the tissue is transferred to a high RH afterwards. It is assumed that reduced water potential of the tissue causes inhibition of water movement between cells, resulting in deficiency of intercellular fluid and subsequent inhibition of bacterial spread in the tissue.

In the field, *E. carotovora* subsp. *carotovora* multiplies on the root surface of Chinese cabbage that appears on the soil surface as the plant approaches the heading stage. The population of the bacterium in the rhizoplane is naturally higher in humid soil in fields with poor drainage than in dry soils. Soft rot incidence increases when the lower leaves of Chinese cabbage contact these roots or soil that holds these roots.

Soil moisture is a critical factor in the development of potato common scab caused by *Streptomyces scabies*. Infection of the pathogen through lenticels does not occur when the soil humidity is less than 10 cm Hg moisture tension by watering but increases where the soil is allowed to dry to 30–50 cm or more Hg moisture tension. Allowing soil to dry for only 3 days in the tuber initiation period is enough to increase scab incidence. If field soils can be held to 20.32 soil moisture deficit by irrigation for 3 weeks after tuber initiation, economic control can be obtained. *S. scabies* infects tubers through newly formed unsuberized lenticels. However, potato lenticels proliferate rapidly in wet soil and displace infected tissues in the early stage of scab formation or when only small lesions have developed. Furthermore, when the soil become dry again, the proliferated tissues collapse readily and a suberized protective layer is formed underneath.

## 11.3 Wind

Strong wind inflicts injuries to leaves, stems, or roots by swinging, twisting, or waving above-ground parts. These wounds become potential portals of infection when exposed to rain splash or irrigation water carrying pathogenic bacteria. Therefore, the most hazardous wind is that accompanying rain. Many bacterial plant diseases occur in a quite severe form after a typhoon or heavy rainstorms. This undoubtedly reflects multiple wounding of plants as well as effective dissemination of bacterial pathogens.

## 11.4 Sunlight

Sunlight has important effects on the physiological functions of host plants such as photosynthesis and respiration, increasing or decreasing susceptibility to bacterial plant diseases. In a breeding line 1169 of tomato, reduction of light intensity from 19,350 lux to 8,075 lux significantly decreased resistance to *P. solanacearum* at 29.4°C but not at 26.6°C. On the contrary, disease symptoms of bacterial leaf spot of carrots caused by *X. campestris* pv. *carotae* are more severe under high light intensity of 30,140 lux than low intensity of 21,520 lux.

Light also influences symptom development. For example, the toxin-induced chlorosis caused by *P. syringae* pv. *coronafaciens* is light dependent. Distinct halo blight symptoms are produced on rye leaves under full light intensity (1,425 $\mu W/cm^2$) for 3–4 days after inoculation, whereas no symptom or only faint symptoms appear under reduced light intensity (202 $\mu W/cm^2$) for 7 days. A similar relationship is also observed between toxoflavin of *P. glumae* and rice seedlings. In both cases, however, light intensity does not affect the rate of multiplication of pathogenic bacteria in plant tissues.

## 11.5 Chronobiology

Plants have biorhythms in their growth patterns as do human and animals. The genetically fixed rhythmic processes considerably affect host–parasite interactions through the effects on pre- or postinfectional conditions of host plants and pathogens. The effects of illumination and temperature in day/night cycles, of time of day of inoculation, or light intensity on disease development or hypersensitive response and infection rate are postulated to be under such genetically fixed rhythmic potentials of host plants and pathogens. For example, primary leaves of bean plants inoculated with *X. campestris* pv. *phaseoli* show special rhythmic movements that are different from healthy primary leaves during the light span but not in darkness for various period of time before death of the plants. Thus, normal biorhythms of plants seem to be significantly affected by bacterial infection. This is a relatively new field of plant pathology that was termed *chronophytopathology* by B. W. Kennedy (1987).

## 11.6 Nutrition

Plant nutrition affects disease development not only through physiological tolerance to bacterial invasion but also through the capacity of host plants to allow the pathogen to readily colonize on the plant surface. In general, plants show maximum growth and greatest disease resistance when fertilizers such as nitrogen, potassium, magnesium, phosphate, and calcium, as well as the minor elements, are provided in proper proportion. However, the effect of fertilization on disease development is difficult to generalize because it varies depending on plants and pathogenic bacteria or their combinations.

The susceptibility of tomato fruits to *Erwinia* soft rot increases markedly when the nitrogen level is increased twofold at the same level of potassium. The incidence of the disease becomes greater after the application of fertilizers by sidedressing or topdressing rather than broadcasting. Susceptibility after topdress-application is attributable to the temporary calcium deficiency of the fruits. This has been explained on the basis of leaching of calcium from the soil or a competitive inhibition of calcium uptake caused by the other salts leached to the root zone by rain percolation.

The incidence of bacterial diseases is generally is less severe when plants are fertilized with low nitrogen and high potassium, whereas greater incidence is induced in plants that receive high nitrogen and low potassium. This trend can be seen either with annual crops or perennial crops and in greenhouses or in open fields. However, in bacterial leaf blight of philodendron and syngonium (*E. chrysanthemi* pv. *philodendron* and *X. campestris* pv. *syngonii*), and bacterial leaf spot of dwarf schefflera (*P. cichorii*), fertilization with high nitrogen significantly reduces disease severity. In these diseases, high levels of nitrogen accelerate development of young leaves, and these actively expanding young leaves are more resistant than mature leaves.

Physiological disorders preceding bacterial infection are often observed in rapidly growing vegetables. In heart rot of Chinese cabbage incited by calcium deficiency, soft rot pseudomonads often initiate their growth in the injured tissues developed on the marginal parts of the heart leaves, resulting in bacterial rot. For such diseases, correct diagnosis of the primary cause is possible only at the very early stage of disease development.

## 11.7 Soil

The factors of soil such as soil type and pH affect directly or indirectly the growth of host plants and the survival and multiplication of soil-borne pathogens as well.

### 11.7.1 pH

There is a report that the severity of bacterial canker of tomato caused by *C. michiganesis* subsp. *michiganensis* is greatest in soil at pH 8.0, followed by soil at pH 4.0. The damage was least in tomato plants grown at pH 6.5. However, soil pH rarely becomes the critical factor in disease development. For example, bacterial wilt caused by *P. solanacearum* occurs in soils with a wide range of pH value, e.g., from acid soil of pH 4.3–4.5 to alkaline soil of pH 7.5–8.5.

Since Gillespie's report in 1918, potato common scab (*S. scabies*) had been considered to occur with highest severity in soils above pH 5.2. Therefore, lowering the soil pH has been an accepted control practice. In 1953, however, potato scab appeared in highly acid soils with pH as low as 4.5 in Maine, and this strain is referred to as uncommon scab pathogen. This strain is pathogenic on tubers grown in soil at pH 4.5–4.9 where *S. scabies* are not pathogenic. The uncommon scab pathogen is also distinct from *S. scabies* in other morphological and physiological characteristics; hence, it was named *S. acidiscabies* (see chapter 13).

### 11.7.2 SOIL TYPE

The relationship between the incidence of bacterial plant diseases and soil type has not been well documented. It has been reported that bacterial leaf blight of rice is frequently observed in severe form in sandy loam, and clay loam in alluvial plains, and also in areas where rocks are shale, granite, or andesite. On the other hand, bacterial wilt of solanaceous plants occurs in many different type of soil. Severe fire blight occurs on pears and apples growing in any type of soil where it is shallow with underlying rock or where the water table is high.

### 11.7.3 MICROBIAL FLORA

In some soil-borne diseases caused by fungal pathogens, *Fusarium* in particular, the presence of soils where diseases rarely occur in nature has been known. This type of soil is termed *suppressive soil,* and the disease suppression has been mainly attributed to microbial flora of the soil in addition to physical and chemical properties. In bacterial plant diseases, however, little information is available on suppressive soil, with the possible exception of potato common scab.

The occurrence of potato common scab may be significantly influenced by soil microorganisms. The disease often occurs in severe form where microflora are naturally simple or destroyed by soil-disinfection (see Chapter 9, Section 9.1.2). Occurrence of the disease is reportedly prevented when suppressive soil is mixed with conducive soil at the ratio of 1 : 1 and 9 : 1 by the antagonism between the pathogen and microorganisms inhabiting the suppressive soil. The decrease of potato common scab in wet soil may also be attributed to the antagonism and/or competition between the pathogen and saprophytic bacteria that grow abundantly in humid soil, in addition to the lenticel proliferation described above. Decrease of disease incidence where green manure was applied is also attributable to the drastic change of soil microflora.

When seeds or vegetatively propagating materials are treated with cultures of certain bacteria, they colonize on rhizoplane of the treated plants, enhancing subsequent plant growth and increasing crop yields. This treatment is called *bacterization,* and this type of bacteria is termed *plant growth promoting rhizobacteria* (PGPR). Most PGPR so far described belong to *Pseudomonas, Bacillus, Serratia, Arthrobacter,* and *Streptomyces.* There is a report that when potato seed pieces were treated with suspension of *P. fluorescens* and *P. putida* at a density of about $10^9$ cells/ml, tuber yield increased 100% in greenhouse trials and 14–33% in field trials.

The mechanism of such growth promotion by the treatment is explained as follows: the population of these bacteria actively increases on the rhizoplane of treated plants, reaching $4 \times 10^4$ to $4.7 \times 10^7$/mg root. They inhibit colonization of *deleterious rhizobacteria,* DB (quasipathogenic bacteria), such as cyanate bacteria, through competition for nutrients or for colonizing sites on root surface and/or antagonistic effect by siderophores or antibiotic substances. Yield increases of 10–30% have been recorded in various crops such as sugar beet, radish, and rape, in addition to potato. Bacterization is also applied for biological control of plant diseases (see Chapter 12, Section 12.3.4).

## 11.8 Pollution

Industrialization, urbanization, and the motorization of modern society consume huge amounts of fossil energy and have created new hazardous environments for plants through air and water pollution. Air pollution has been attributed to various chemicals such as ozone, sulfur dioxide, hydrogen fluoride, nitrogen oxide, peroxyacetyl nitrate, and acid rain.

These phytotoxic substances either directly injure plants though acute or chronic effects or indirectly affect plant production through interaction with other pests and diseases. Investigations have revealed that most air pollutants except for acid rain appear to prevent rather than enhance the incidence of bacterial diseases.

### 11.8.1 OZONE

Exposure of soybean plants to ozone causes considerable reduction of shoot, root, and plant dry weight. The number, size, and weight of nodules are also reduced by ozone treatment in the most legume–rhizobia combinations. However, a series of studies with several combinations of plants and plant pathogenic bacteria such as kidney bean and *P. syringae* pv. *phaseolicola* indicate that ozone is repressive to subsequent bacterial infection, and in contrast, bacterial infection decreases the degree of injuries caused by subsequent exposure to ozone. The reduction of infection is explained by the production of bacteriostatic or bacteriocidal substances (phytoalexins), such as isoflavonoid compounds and peroxidases, in the leaves treated with $O_3$. On the other hand, ozone tolerance resulting from bacterial infection is explained by a reduction in uptake of ozone due to impairment of stomatal function and/or reduced availability of functional stomata or potential ozone-injury sites.

### 11.8.2 SULFUR DIOXIDE

Investigations on the effect of sulfur dioxide on lesion development by *Clavibacter michiganensis* subsp. *nebraskensis* on corn leaves and *X. campestris* pv. *phaseoli* on soybean leaves indicate that $SO_2$ exposure inhibits lesion development and delays the onset of symptom development in both diseases. The effect was observed regardless of the time of exposure, i.e., preinoculation, postinoculation, or both. The extended latent period and the reduced rate of lesion development result in a decrease of the population of pathogenic bacteria in the field and consequently in a concomitant reduction of disease severity at the end of an epidemic.

### 11.8.3 HYDROGEN FLUORIDE

In bacterial blight of red kidney bean plants caused by *X. campestris* pv. *phaseoli,* continuous postinoculation exposure to hydrogen fluoride causes an extension of the latent period and smaller initial lesion sizes.

Disease development is not affected by the presence of hydrogen fluoride in the atmosphere after the pathogen becomes established. The effect of hydrogen fluoride on disease development mainly depends on the accumulation of fluoride in plants rather than the concentration, frequency, or total dose of the pollutant. The accumulation of fluoride in plants occurs more effectively by intermittent exposure to the pollutant rather than continual exposure.

### 11.8.4 ACID RAIN

Acid rain has been implicated in direct injurious effects on plants, such as induction of necrotic leaf spots, increased nutrient leaching from leaf surfaces, and decreased total plant weight. Acid rain also indirectly induces chemical injuries by $Cu^{2+}$ that is dissolved at phytotoxic concentrations from copper protectant deposited on leaf surfaces. In an experiment with simulated rain acidified with sulfuric acid to pH 3.2 and 6.0 and halo blight of kidney bean caused by *P. syringae* pv. *phaseolicola,* preinoculation exposure to acid rain of pH 3.2 increased disease severity by 4.2%. This may have resulted from increased portals of infection because of visible injuries to the leaf surface. In postinoculation exposure, however, disease development was inhibited by 22%, possibly because reduced susceptibility of host plants resulted from the stress. The possibility of a direct effect of acid rain on the pathogen, however, cannot be eliminated because bacterial suspensions in the acid rain induced no infection.

## 11.9 Controlled Atmosphere

Controlled atmosphere (CA) or gas mixture of $O_2$, $CO_2$, and $N_2$ combined with low temperature is generally used to extend the storage longevity of fresh fruits and vegetables to minimize the incidence of postharvest diseases and ripening. The ratio of $O_2$ and $CO_2$ is critical for reducing deterioration. The effectiveness may differ depending on the plant and the pathogenic organisms. In general, the ratios of 4% $O_2$, 0–2% $CO_2$ and 96–94% $N_2$ at 10° to 13°C seem to be favorable for reducing physiological deterioration and development of postharvest diseases of bacterial and fungal origins in storage.

## 11.10 Ice Nucleation-Active Bacteria and Frost Damage

A small amount of distilled water can be supercooled to as low as −40°C without being frozen where no ice nuclei are present. In the presence of silver iodide, supercooling is broken down at −8°C, and water freezes instantly. In the early 1970s, the capacity to induce ice formation at −2° to −3°C was discovered with some bacteria that reside on plant surfaces. These bacteria are called ice nucleation-active (INA) bacteria. Investigations on INA bacteria have been conducted in relation to frost damage of plants, determination of INA components of bacterial cell, genetic analysis of ice nucleating activity, industrial use of INA bacteria, and their INA proteins for artificial snow or food technology.

Some plants, such as potato and pea, are cooled to temperatures as low as −4° to −5°C for at least several hours with no apparent damage if they are free of ice nuclei. When these plants are gently cooled down to −3° to −5°C, ice crystals are intercellularly formed. When the temperature goes up again, these ice crystals are melted, and water vapor is transpirated through the stomata without any injury to plant cells. When the temperature lowers quickly, in contrast, ice crystals are intracellularly formed, causing substantial damage to plant cells. The INA bacteria that reside on plant surfaces induce quick freezing of plant cells in the temperature range that plants can recover from if INA bacteria are not involved.

Ice nucleating activity of bacteria can be detected by various procedures, e.g., ice formation of a bacterial suspension held in a test tube at −4°C for 5 min, the freezing rate of droplets of bacterial suspensions (10 μl) placed on aluminum foil, or bacterial suspensions held in capillary tubes when cooled to −5°C for 10 to 30 sec. Ice nucleating activity is not always found in every cell of a bacterial population but only in a fraction of cells; this fraction increases with decreasing temperature, however.

The INA bacteria include *E. ananas, E. herbicola, E. stewartii, P. fluorescens, P. viridiflava,* many, but not all, of the pathovars of *P. syringae,* epiphytic strains of *P. syringae, X. campestris* pv. *translucens,* and an epiphytic strain of *X. campestris.*

It has been confirmed in various combinations of INA bacteria and plant species that destructive frost damage occurs when the suspensions of INA bacteria at concentrations of $10^8$ cells/ml or more were sprayed on plant surfaces and subjected to −3° to −5°C for 0.5 to 1 hr (Table 11.3). It has also been consistently observed that if the density of bacterial cells is reduced to $10^6$ cells/ml, no detectable frost injuries occur.

**Table 11.3**
Reduction of Frost Injuries by Chemical Spray on Broccoli that Was Previously Sprayed with Ice Nucleation-Active *Xanthomonas campestris*[a]

| Chemicals | Concentration | Percent frozen leaves | |
|---|---|---|---|
| | | Treatment A[b] | Treatment B[c] |
| Wettable copper (Kocide)[d] | 1,000 | 76.5 | 83.3 |
| Copper hydroxy nonylbenzenesulfonate (Yonepon) | 1,000 | 55.9 | 97.3 |
| Streptomycin-oxytetracycline (agrimycin-100) | 1,500 | 47.2 | 55.8 |
| Kasugamycin (kasumin) | 1,000 | 8.3 | 17.9 |
| Control | | 100 | 94.9 |

[a] From Goto, T., Inaba, T., and Goto, M. (1988). Effect of chemical treatments on the frost damage of broccoli sprayed with ice nucleation-active bacterium, *Xanthomonas campestris* subsp. *Proc. Assoc. Pl. Prot. Shikoku,* **23,** 57–59.

[b] Supercooled to −5°C for 60 min on the day of chemical spray.

[c] Supercooled 2 days after chemical spray.

[d] Commercial names in parenthesis.

Frost damage induced by INA bacteria sprayed on plant surfaces can be remarkably reduced by spraying bactericides (Table 11.3). On the basis of these observations, a concept that INA bacteria have an important role in the initiation of frost damage has been proposed. However, a considerable number of articles that have subsequently been published indicate that this is not always true.

In tea plant (*Thea sinensis*), young shoots with unfolded terminal buds hold INA bacteria at populations of $5 \times 10^6$ or lower per bud. In most cases, the INA bacteria consist of an epiphytic and nonpathogenic strain of *X. campestris* or sometimes *E. ananas.* The species of INA bacteria and their population as well as ratio of tea buds carrying INA bacteria vary, depending on localities and fields. In field trials of the artificial generation of first damage, any positive correlations have not been obtained between the shoots suffering frost injuries and those carrying INA bacteria and/or their population; frost injury of sprouting buds occurred independently from the presence of INA bacteria and their populations. This may be explained by the specific cold sensitivity of young tea shoots derived from cell membrane components such as phosphatidylglycerol. The results obtained with tea plant cannot be generalized to other plants, however, because the frost–INA bacteria relationship may be different, depending on the plant species and/or plant organs.

Frost injuries incited by INA bacteria at restricted sites of the plant surface may become excellent avenues of infection for the epiphytic plant pathogen. It is assumed that a freeze-thaw cycle may create water soaking in plant tissues and promote the ingress and spread of bacteria. Such enhancement of disease severity has been implicated with many diseases that occur in winter to early spring, i.e., soft rot of vegetables caused by *P. viridiflava,* pear blossom blight, bacterial canker of poplar, apricot, and peach caused by *P. syringae* pv. *syringae,* and bacterial blight of pea caused by *P. syringae* pv. *pisi.*

## Further Reading


Coakley, S. M. (1988). Variation in climate and prediction of disease in plants. *Ann. Rev. Phytopathol.* **26,** 163–181.

Horsfall, J. C., and Cowling, E. B., ed. (1978). "Plant Disease, An Advanced Treatise. Volume III. How Plants Suffer from Disease." Academic Press, New York.

Kennedy, B. W., and Koukkari, W. L. (1987). Chronophytopathology. *Adv. Chronobiol.,* Part A, 95–103.

Lindow, S. S. (1983). The role of bacterial ice nucleation in frost injury to plants. *Ann. Rev. Phytopathol.* **21,** 363–384.

Perombelon, M. C. M., and Kelman, A. (1980). Ecology of the soft rot erwinias. *Ann. Rev. Phytopathol.* **18,** 361–387.

Reinert, R. A. (1984). Plant response to air pollutant mixtures. *Ann. Rev. Phytopathol.* **22,** 421–442.

CHAPTER
TWELVE

# *Diagnosis and Control of Bacterial Plant Diseases*

Because of the absence of sufficiently effective chemicals, control of bacterial plant diseases relies largely on an integrated control which consists of improved methods of cultivation, field hygiene, plant quarantine, and use of resistant cultivars.

## 12.1 Diagnosis

The correct and rapid determination of plant diseases is the first step in their effective control. The process of identifying plant diseases is called *diagnosis,* and the science of theoretical and practical diagnosis is termed *diagnostics.*

### 12.1.1 NAMES OF PLANT DISEASES

Each plant disease has its own name in different languages. When several different names have been given to a particular disease in the past, one of them is generally selected and approved as the common name by each phytopathological society on the basis either of priority or of popularity in use. These names are available in the lists of common names of plant diseases, which are published by phytopathological societies or equivalent organizations. For example, all names of plant diseases occurring in Japan are listed in five volumes of *Common Names of Plant Diseases in Japan*, published by The Phytopathological Society of Japan.

Plant diseases are named to express the characteristics of symptoms in the most preferable way and to aid practical diagnosis of diseases from symptoms. Diseases caused by the same bacteria are usually given the same names, irrespective of the species of host plant. An exception is the Japanese name of bacterial wilt of tobacco caused by *Pseudomonas solanacearum*. It has been given a different name from a disease of other plants due to the same bacterium mainly on the basis of popular use in the past. In reverse, the same name may be given to diseases caused by different pathogenic bacteria, e.g., bacterial leaf spot is generally adopted regardless of bacterial species or host plant when the symptoms are characterized by the formation of restricted necrotic lesions.

Disease names do not always express the characteristics of the diseases correctly because symptoms may progressively change depending either on the process of disease development or plant organs affected. For example, bacterial grain rot of rice caused by *P. glumae* is primarily referred to as the disease of rice panicles, but the bacterium also causes another syndrome of seedling rot in nurseries. Similar cases are also found in bacterial leaf blight of rice (*Xanthomonas campestris* pv. *oryzae*) that induces leaf blight and wilting, and bacterial blight of tea (*P. syringae* pv. *theae*) that shows shoot blight and leaf spot, respectively. In the United States, these different symptoms may be incorporated into one name, e.g., "Goss's bacterial wilt and blight of corn." As found in this name, the name of a person or locality is sometimes used in disease names indicating first discovery or description of the diseases, e.g., "*Stewart*'s disease of corn" and "*Granville* wilt of tobacco."

### 12.1.2 FIELD DIAGNOSIS

The first step in diagnosis of a plant bacterial disease is to examine the perspective of the disease and farming practices. Careful examination in this step may bring significant information that may not be obtained from the examination of the diseased plant in the laboratory. Particularly, information on distribution of the disease in the field, record of the problem in the past, cultural managements such as application of fertilizers and pesticides, crop rotation and cultivars are sometime valuable for identification of the disease under consideration. For example, the distribution of diseased plants along a furrow indicates that pathogens may be disseminated by farming practices. When a disease spreads around a focus in the field, it may have originated in the infected seed or seedling.

## 12.1.3 PLANT DIAGNOSIS

Diagnosis of disease from individual plants or plant parts includes macroscopic examination of symptoms and signs, microscopic examination of bacteria streaming out of the diseased tissues, isolation of the bacteria, inoculation test, phage-typing, and serotyping.

### *Symptoms and Signs*

Symptoms of bacterial diseases can be classified into wilting, necrotic spot, blight, soft rot, hypertrophy, malformation, and yellows. Although similar symptoms may be found in fungal and viral diseases, those in bacterial diseases can be determined in many cases, but not all, by the appearance of water soaking in the affected tissues and formation of bacterial ooze or exudate on the surface of lesions. The amount of bacterial ooze varies greatly depending on the nature of the disease and environmental conditions; it is most conspicuous on fresh lesions under humid conditions. Bacterial ooze turns to flakes or scales in a dry atmosphere, often becoming detectable only with the help of a magnifying glass, especially when they are sparse.

In bacterial wilt, the surface of the transverse section is moister in appearance than that in fungal wilt. White droplets of bacterial exudate usually appear on the cut surface with or without squeezing.

### *Diagnosis by Microscope*

Plant parts may be examined under the microscope for the presence of a mass of bacteria streaming out of the cut margin of diseased tissues or lesions. Phase contrast microscopy is preferable in order to distinguish bacteria from particles of starch and/or latex, which may also stream out of tissues. Fresh lesions at an early stage of disease development are best for observing clouds of bacteria.

In aged lesions, pathogenic bacteria may already be dead or replaced by saprophytic bacteria. Pathogenic bacteria that diffuse out of a lesion are rarely motile except in the case of soft rot bacteria. Diffusion of a small number of bacteria is commonly found in moribund plant tissues affected by other diseases as well as chemical and mechanical injuries. However, pathogenic bacteria are clearly distinguished from saprophytes by their massive streaming.

A portion of the lesion is cut from the border adjacent to healthy tissue and is mounted in a drop of water on a slide and gently overlaid with a coverslip for microscopic examination. Clouds of bacteria streaming out of the periphery of the mounted tissues can be observed. In bacterial wilt, the cortical tissue is peeled off to expose the woody tissue or xylem. A small piece of discolored xylem is excised and mounted in a water drop on a slide. In hyperplastic diseases other than crown gall, a gall is cut into halves, and a small piece of tender, hyperplastic tissue with a water-soaked appearance is removed with a needle or knife and mounted for examination of bacterial streaming.

In soft rot diseases, a small piece of tissue at the junction between firm and softened tissues is taken with a needle or knife and examined under the microscope. A number of actively motile bacteria can be observed. In some diseases caused by fungi such as *Sclerotium, Pythium,* or *Phytophthora,* the primary symptom may also be soft rot. When the possibility of fungal disease is suspected, diseased samples are held in a plastic bag for a few days at room temperature. Fungal diseases are easily recognized by the development of white mycelium at the surface of the specimens.

In diseases caused by fastidious prokaryotes such as the mycoplasma-like organisms *Spiroplasma* and *Xylella fastidiosa,* the vascular bundles of the sectioned tissues are examined under an electron microscope.

### *Diagnosis by Isolation of Plant Pathogenic Bacteria*

A few square millimeters of a lesion between diseased and healthy tissues is cut out with a knife or scissors, put in a small test tube containing 1–2 ml of 1% peptone water, and macerated with a glass rod. A loopful of this suspension is streaked out on agar plates. Plant pathogenic bacteria may be determined by their characteristic growth on the plates. It is not likely to be the bacterial disease if colonies developing on the plates are significantly small in number and/or vary in color and size.

In general, isolation of plant pathogenic bacteria from lesions on roots is not an easy task because of heavy contamination with saprophytes. This is particularly true when the specimens are deteriorated or growth of pathogenic bacteria is slower than that of saprophytes. In such a case, diagnosis by inoculation tests can be applied as follows.

### *Diagnosis by Inoculation*

Host plant tissue is practically the best selective medium for separation of plant pathogenic bacteria from saprophytic microorganisms. Various inoculation methods can be applied for diagnosis, depending on the bacterial species and/or the type of disease: spray inoculation, tissue slice inoculation, leaf infiltration, multineedle inoculation, and seedling inoculation. Bacterial soft rot is rapid in development of symptoms and contamination of rotted tissue by secondary invaders as well. In a heavily contaminated specimen, a small amount of rotted pulp is placed on sliced potato or carrot and examined for the development of soft rot within 24 hr.

Plant galls can be caused by various agents such as bacteria, fungi, insects, and physiological disorder. When the causal agent is not certain, gall tissues may be macerated in a small amount of water and its sap inoculated on a host plant. When typical symptoms develop, the causal bacterium is isolated on agar plates for confirmation.

### *Diagnosis by Bacteriophage*

The phage technique may be useful in the examination of infected seed because it has an advantage that the presence of host bacteria can be demonstrated in several hours. In principle, seeds are macerated in a liquid medium to release bacterial cells, a certain number of phage particles are added, and incubated for several hours to overnight. A significant increase in the number of phage particles indicates the presence of host bacterium. The factors involved in successful application of the technique have been described in Chapter 5, Section 5.1.

### *Diagnosis by Serological Methods*

Serological methods are also advantageous for the detection of plant pathogenic bacteria because they are rapid and specific. However, the procedures must be carefully evaluated with regard to sensitivity, i.e., the lowest level at which bacteria can be detected and specificity, i.e., inter- and intraspecies cross reactions. Among various procedures, immunofluorescence and ELISA have been most widely used for the diagnosis of phytopathogenic bacteria.

### *Gene Diagnosis*

Plant pathogenic bacteria may be identified rapidly and accurately by nucleic acid hybridization by the dot-blot or colony hybridization techniques. The probe may be the synthesized oligonucleotides or cloned chromosomal or plasmid DNA fragments that specifically hybridize to DNA of the target bacteria. The probes may be labeled with radioactive nucleotide or with enzyme-linked antibodies.

Single-stranded target DNA molecules are immobilized on a nitrocellulose or nylon filter surface. The filters are prehybridized with denatured salmon sperm DNA to block nonspecific DNA-binding sites. The denatured probe is added to the filters allowing hybridization with the complementary sequences of target DNA. After excess unbound probe is washed off, the hybridized sequences are assayed either by radioautography or by a colorimetric method. The colony hybridization technique was described in Chapter 6, Section 6.4. Successful identification has been documented with many plant pathogenic bacteria.

## 12.2 Disease Loss Assessment

One of the most difficult tasks in plant pathology is to assess crop loss due to disease in terms of yield or money. Although many bacterial plant diseases cause severe yield losses every year, methods of measuring crop loss have been established with only a few diseases.

### 12.2.1 BACTERIAL LEAF BLIGHT OF RICE

Yield reduction becomes serious when the disease occurs at the younger growing stages of rice plants. The earlier the appearance of the disease, the larger the yield loss. Influence on yield is practically negligible when the disease develops on foliage one week after panicle emergence. A positive correlation is observed between disease severity indices of leaf blight and yield reduction in percent, the ratio in percent of empty husks, low quality grains, broken grains, and weight of kernels. The weight of a panicle or a number of grains per panicle does not correlate with the severity of leaf blight symptoms. Although several methods had been proposed for crop loss assessment of the disease, the formula devised by Tokai-Kinki Agrigultural Experimental Station in 1951 has been widely accepted. This is:

$$\text{Yield loss in \% } (Y) = \frac{1\text{I} + 3\text{II} + 4\text{III} + 5\text{IV} + 7\text{V}}{20}$$

where I is the ratio in percent of the leaves with less than 20% leaf area affected; II, 33%; III, 50%; IV, 60%; and V, above 61%, respectively.

The maximum crop loss calculated by this formula is 35%. When disease occurs at a growing stage later than the milky stage, the maximum crop loss is equivalent to 50% of the above. Yield reductions by natural infections in fields were calculated as 35% for the most severe form of the disease and 19.7% for the moderately severe form, indicating that the formula is applicable to crop loss assessment of the disease.

## 12.2.2 BACTERIAL GRAIN ROT OF RICE

The causal bacterium (*P. glumae*) causes seedling rot in nurseries and grain rot in the field. In the latter case, the bacterium infects inflorescence only during several days of the flowering stage, resulting in under-developed, low quality, and broken grains. When it occurs in severe form, panicles turn white and keep standing erect even in maturing stage.

The correlation between the disease severity indices ($X$) and the weight of kernels ($Y$) is shown by the correlation coefficient $r = -0.8965$ and a regression curve of $Y = -2.632X + 424.9$. According to this formula, a 50 in disease severity index corresponds to 44% of yield reduction (Fig. 12.1).

## 12.2.3 BACTERIAL STREAK OF WHEAT AND BARLEY

Bacterial streak of wheat and barley, caused by *X. campestris* pv. *undulosa* and pv. *translucens*, is characterized by dark green to dark brown water-soaked streaking of leaves, culms, glumes, and awns. Small droplets of bacterial exudate are observed along streaks. The correlation between the disease severity indices on the flag leaves ($X$) and the grain weight ($Y$) was shown by the regression curve of $Y = B_0 + B_1X$ in which the parameters $B_0$ and $B_1$ in the different field sites are shown in Table 12.1. According to this formula, a 50% disease severity on the flag leaves results in an 8–13% loss in kernel weight and a 100% disease severity in

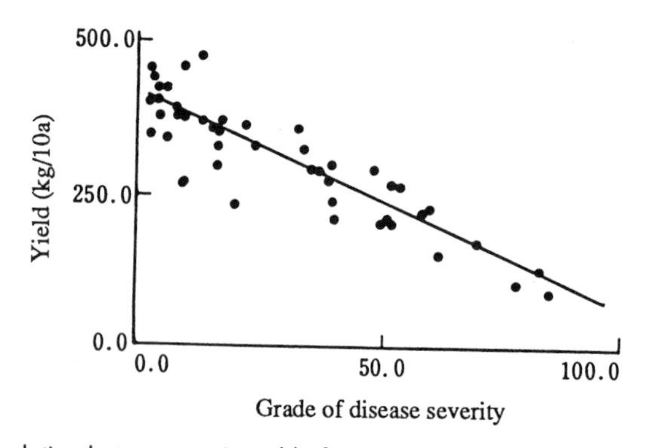


**Fig. 12.1** Correlation between grain yield of rice ($Y$) and severity of bacterial grain rot ($X$). [From Isogawa, Y., Kato, S., Uebayashi, Y., Kato, N., and Amano, T. (1986). Studies on environmental factors and control methods of bacterial grain rot of rice, caused by *Pseudomonas glumae*. (1) Ecology and crop loss assessment. *Res. Bull. Aichi Agric. Res. Ctr.* **18,** 55–66.]

**Table 12.1**
Correlation between Grain Yield (kg/ha) and Disease Severity (%) of Flag Leaves of Angus Wheat[a]

| | Regression parameter[b] | |
|---|---|---|
| Field site | $B_0$ | $B_1$ |
| Morris east | 3,550 | −10.3 |
| Morris west | 4,511 | −18.4 |
| Rosemount | 2,410 | −56.4 |

[a] Adapted from: Shane, W. W., Baumer, J. S., and Teng, P. S. (1987). Crop losses caused by Xanthomonas streak on spring wheat and barley. *Plant Disease* **71,** 927–930.

[b] Regression model: $Y = B_0 + B_1X$, where X is disease severity (%) and Y is plot yield in kg/ha.

13–34% loss. It was confirmed that the protein contents of grains are also correlated with disease severity of flag leaf.

## 12.3 Control

### 12.3.1 PLANT QUARANTINE

The first step in disease control is to prevent movement of the pathogenic bacterium from infested to noninfested areas by restricting the transfer of seeds, vegetative propagules, plants, plant products, soil, or packing materials. Both international and domestic transfer may be potentially controlled by plant quarantine regulation and legislation.

#### *Global Spread of Bacterial Plant Diseases*

Many plant pathogens introduced from abroad have caused serious epidemics in new areas where they were previously unknown. A few examples of bacterial diseases that illustrate this type of epidemics are cited below.

*Fire Blight of Apple and Pear*

Fire blight of apple and pear, which was prevalent in the United States as early as the 1780s, was introduced into England in the 1950s, either on infected plant materials or wooden boxes of apple fruits, causing the first outbreak in 1958. The disease was subsequently introduced into continental Europe sometime in the 1960s and rapidly spread on the coastlines on the English Channel via migratory birds. Many European countries undertook eradication campaigns to eliminate infected fruit trees and other hosts such as hawthorn or pyracantha that are common as hedges around orchards.

*Citrus Canker*

In the 1910s, the pathogen of citrus canker was introduced from Asia to the Gulf States of North America through shipment of nursery stock and caused a severe epidemic in oranges and grapefruit, especially in Florida. A cooperative federal–state eradication program from 1914 to 1934, one of the most intensive programs in the history of plant pathology, was undertaken in Florida, Alabama, Georgia, Louisiana, Mississippi, South Carolina, and Texas.

The complete elimination of citrus canker was announced in 1952. It has been cited as an outstanding example of successful eradication of a bacterial plant disease after its establishment in a favorable environment. During the campaign, approximately 260,000 orchard trees and 3,000,000 nursery plants were destroyed at a cost of about 6 million dollars. Because of such circumstances, the states of Florida and California have had stringent restrictions on the importation of citrus fruits from areas where citrus canker is endemic.

*Bacterial Leaf Blight of Rice*

In the early 1960s the dwarf rice cultivar IR8 was developed by the International Rice Research Institute and distributed in many countries of Southeast and South Asia. This cultivar had many promising properties such as high yields, nonlodging, and response to high nitrogen, and was cited as a green revolution. Unfortunately, IR8 was highly susceptible to bacterial leaf blight, and widespread introduction of this cultivar to these regions resulted in an epidemic of the disease by the pathogen that had previously persisted on the native cultivars in an endemic form.

### *International Plant Quarantine*

The hazardous means of introduction of phytopathogenic bacteria are on the infested or infected seeds, budwoods, and young plants, as well as in soil or hay. Any living plants should be considered as potential carriers of bacterial plant pathogens, even in the absence of visible symptoms. Bacterial pathogens are generally highly adaptable to any new environments in which their host plants thrive.

The legal basis for quarantines controlling plant introduction is in the legislation issued by national governments in the form of regulations and rules. The regulations for international quarantine are promulgated among the member countries of the plant protection convention of the Food and Agricultural Organization of the United Nations.

Some plant pathogenic bacteria may be disseminated over long distance by insects or birds from outside a country. For this type of disease, control measures could only be adopted by exchange of information on disease incidence at the earliest stage among the member countries of the biogeographical regions.

Different plant pathogenic bacteria are not necessarily equally hazardous or threatening. In the adoption of appropriate and effective quarantine measures, it is rational to categorize plant pathogens by the potential danger that they present to specific crops.

The degree of hazard is usually classified by the capacity of a pathogen to cause an outbreak of disease in the newly introduced area. The practical bases for categorization include (1) the pathogenic potential of the pathogen, (2) the presence or absence of the pathogen under consideration, and (3) the potential of the pathogen to cause epidemics. Examples of categorization of plant pathogenic bacteria on the basis of level of hazard are found in the classification of European and Mediterranean Plant Protection Organization (EPPO) and the classification of the International Seed Testing Association (ISTA) (Table 12.2).

### *Domestic Plant Quarantine*

Major legislation for domestic plant protection in Japan includes (1) control of the most important diseases and pests designated by the government that are targets of the national forecasting projects, (2) emergency control of abrupt outbreaks of diseases and pests, (3) inspection of specified seeds and seed plants, and (4) legislative inspection of agricultural chemicals.

### *Disease Forecasting*

The major criteria that have been used in forecasting disease epidemics include ecological behavior of the pathogen, meteorological data, and health of host plants. These factors are usually used in combination to some extent. Forecasting systems based on only one of these criteria stand on the premise that other factors proceed in average conditions preferable to the disease epidemics. If changes occur in other factors, the system must be reworked.

**Table 12.2**
Categorization of Seed-Borne Bacteria on the Basis of Potential Hazard in the International Quarantine[a]

| | |
|---|---|
| Category A | |
| Definition: | Dangerous plant pathogens not present in the region of introduction that have a high or considerable epidemic potential. |
| Precautions: | Complete prohibition against introduction from areas with infections. |
| Examples: | *Curtobacterium flaccumfaciens* pv. *flaccumfaciens* (French bean, soybean)<br>*Clavibacter rathayi* (cocksfoot)<br>*Xanthomonas campestris* pv. *papavericola* (poppy)<br>*Xanthomonas campestris* pv. *sesami* (sesame) |
| Category B | |
| Definition: | Dangerous plant pathogens not present in the region of introduction that have a moderate epidemic potential. |
| Precautions: | The seed may be tested based on adequate sampling. The tested samples must be found completely free of infection or contamination. |
| Examples: | *Pseudomonas syringae* pv. *glycinea* (soybean)<br>*Xanthomonas campestris* pv. *campestris* (crucifers)<br>*Xanthomonas campestris* pv. *vesicatoria* (tomato) |
| Category C | |
| Definition: | Other plant pathogens of importance to the planting value of seed. |
| Precautions: | Testing by adequate procedures of representative samples is advisable. |
| Examples: | *Clavibacter michiganensis* subsp. *michiganensis* (tomato)<br>*Pseudomonas syringae* pv. *lachrymans* (cucumber)<br>*Pseudomonas syringae* pv. *phaseolicola* (French bean)<br>*Pseudomonas syringae* pv. *pisi* (pea)<br>*Xanthomonas campestris* pv. *phaseoli* (French bean) |

[a] Excerpted from Neergaard, P. (1977). Quarantine policy for seed in transfer of genetic resources. *In* "Plant Health and Quarantine in International Transfer of Genetic Resources" p. 309–314. (W. B. Hewitt and L. Chiarappa eds). CRC Press, Inc. Cleveland, Ohio.

To improve these limitations in the present forecasting systems, computer-based simulation modeling has recently been activated for predicting plant epidemics. In this system, the factors that influence epidemics are subdivided into various components of the disease cycle, such as behavior of the pathogen, susceptibility of host plant, and environmental conditions. All data of the variables constituting these components are computerized to resynthesize the whole epidemic system by linking them in a model. The model system so constructed reserves the possibility that the data of any given environmental factors input into the computer can provide a prediction of epidemics in the near future.

Simulation approaches for predicting epidemics of bacterial plant diseases are currently in progress.

The existing forecasting systems of bacterial plant diseases are described with three examples.

*Bacterial Leaf Blight*

In bacterial leaf blight of rice, the time of initial disease incidence and the extent of disease development in the early stage of rice growth significantly influence disease severity later in the season and directly affect yield. A forecasting system has been developed on the basis of meteorological factors such as temperature, precipitation, the number of rainy days, and the occurrence of typhoons or flooding in the early growth season.

Another forecasting system is based on seasonal assays of the irrigation water or paddy water for populations of the bacteriophages that attack the pathogen. Although the number of bacteriophages in paddy water does not indicate the actual number of causal bacteria, the population of bacteriophages fluctuates in proportion to that of the pathogen because the phages propagate in paddy water on bacterial cells released from diseased rice plants. The increase in phage population usually occurs well in advance of the onset of epidemic.

*Fire Blight of Apple and Pear*

*E. amylovora* grows epiphytically on flowers and increases its population in advance of the onset of blossom blight. A prediction system developed in California is based on monitoring such changes in population of the pathogen on flowrs. In Michigan, however, monitoring populations of *E. amylovora* on the surface of cankered branches is more reliable than those on flowers as the means for predicting outbreak of the disease. In this region, an increase in the bacterial population on flowers is too rapid to monitor during the humid blossom period.

The fire blight warning system developed in England is primarily based on the potential daily doubling of the pathogen estimated by temperature, the rain score calculated from daily precipitation, and additional data on flower infection rates as well as insect activity. This system has shown a high correlation with the fire blight epidemics in England in the past that occurred when the maximum temperatures exceeded 18°C or the mean temperatures exceeded 14.5–16.5°C and it rained or the relative humidity exceeded 60% in the spring blossom period.

Forecasting systems of fire blight vary between the United States and Europe and even in the United States between California and Michigan. Such variation mainly originates in the differences in such local factors as the major means of dissemination, blossom period, insect activity, and weather conditions that influence the incidence of spring infection.

*Stewart's Disease (or Bacterial Wilt) of Corn*

Both overwintering and dissemination of *E. stewartii* are dependent on the corn flea beetle, *Chaetocnema pulicaria,* although the pathogen may also be seed-borne. The incidence of the disease is related to numbers of the insects that will carry the bacterium internally. The numbers of insects that overwinter is related to the temperatures during December, January, and February. Thus, in Pennsylvania a computerized forecasting program has been developed on the basis of a winter temperature index derived from the sum of the average monthly temperatures for December, January, and February.

## 12.3.2 DISEASE CONTROL BY FARMING PRACTICES

### *Eradication*

Eradication methods are applied directly against the pathogen, to the host plants or alternate hosts. Practical eradication procedures include fumigation of storage houses and machinery, heat treatment, solarization or flooding of soil, and burning or removal of plant residues.

A good example of effective host eradication was the citrus canker eradication campaign mentioned previously. Similar eradication programs have been implemented in Australia, Mozambique, New Zealand, South Africa, and South America. Eradication of this type usually requires the total elimination of all infected host plants from the geographical area affected. Such a drastic measure has no value if the elimination is incomplete or infected nursery plants are replanted after elimination of infected trees. Moreover, extensive eradication of infected host plants may not be successful unless there is full support by the growers, government agencies, and the public at large.

### *Field Hygiene*

Removal of diseased plant or plant parts from field is important to reduce the density of inoculum. Defoliated leaves of pruned twigs of fruit trees and straw of cereals are preferably burned. Infected woody stems such as tomato, eggplant, and tobacco should not be ploughed into soil because they may remain partly undecomposed in northern or temperate climates. Disinfection of tools and equipment is important to control the spread of bacterial pathogens through field practices such as pruning or harvesting. Solutions of 10% sodium hypochlorite are widely and safely used for the purpose of disinfection.

### *Disinfection of Seeds and Planting Materials*

Use of pathogen-free propagative materials is the primary and key step in control of bacterial diseases. In perennial crops such as fruit trees, new diseases are often introduced by infected nursery stocks. Crown gall, citrus canker, and fire blight pathogens have been spread in this manner. Because bacterial wilt of carnation is mostly spread by cuttings, hygienic treatment is required throughout the entire process of propagation.

In seed-borne diseases, the frequency of primary infection through infected seeds is generally low. However, pathogens can readily disseminate from the primary infection centers by wind-driven rain droplets, aerosols, or farming practices. The disinfection of infested or infected seeds by heating or chemicals is, therefore, an important and effective measure to eliminate the primary inoculum and to prevent disease establishment in fields.

### *Disease Escape*

Plants can escape from bacterial infection by a change in their cultivation period. For example, severe damage of bacterial leaf blight of rice has been avoided in Japan by the combination of planting the early maturing cultivars and setting the cultivation periods early enough to harvest before the typhoon season starts. Similarly, an earlier cropping system of tobacco plants has been successfully adopted in Japan in order to escape severe damage due to bacterial wilt that usually occurs in mid-summer after the rainy season.

### *Rotation*

In phytopathogenic bacteria with narrow host ranges, disease development may be controlled by rotation if seed-borne infection can be successfully prevented. However, rotation may not be effective when the pathogens have wide host ranges or are well adapted for long-term survival in the field.

When pathogens survive epiphytically on weeds or volunteers, field hygiene should be accompanied by rotation for successful disease control. In general, however, rotation is ineffective with soil-inhabiting bacteria such as *P. solanacearum, A. tumefaciens,* and soft rot *Erwinia* spp.

### *Irrigation*

In potato common scab caused by *Streptomyces scabies*, irrigation for several weeks after tuber initiation is effective in reducing scab. Practical control of the disease at the Rothamstad Experiment Station is accomplished by irrigation that maintains a soil moisture deficit of 0.6 for 3–4 weeks in the tuber initiation period. In Japan common scab in the autumn crop season can be effectively prevented by irrigation. However, the effect of irrigation is not obvious in spring cultivation because tuber initiation continues for a long time and soil temperature ranges around 20°C, which is favorable for infection. Therefore, irrigation must be continued for a long time.

### *Nutrition and Plant Density*

Adequate plant density and fertilization are obviously important for growing healthy plants. High nitrogen levels and dense spacing may increase disease susceptibility of host plants and resident populations of the pathogen because proliferated dense foliage reduces air movement and drying of leaf and stem surfaces. In dense spacing, a between-plant infection may preferentially occur, whereas a within-plant infection occurs in low plant densities.

Balanced nutrition is, of course, necessary for the healthy growth of plants. Deficiency or excess of a nutritional element may itself become the cause of physiological disorder, but it may also reduce the resistance of plants to bacterial diseases.

### *Grafting*

Muskmelon and watermelon are usually grafted on pumpkin root stocks to prevent infection of *Fusarium oxysporum* f. sp. *melonis* and *F. oxysporum* f. sp. *niveum*. Severe infection due to *P. syringae* pv. *lachrymans* and *X. campestris* pv. *cucurbitae* may occur on these plants through pumpkin stocks that are raised from infected seeds. *Agrobacterium tumefaciens* and *A. rhizogenes* usually infect nursery plants through wounds of grafting. In these diseases, latent infections become potentially important sources of dissemination.

Use of healthy young plants is the first step in citrus canker control. Both root-stock seedlings and scions grafted on them are readily infected in densely planted nurseries, even if the primary inoculum is very low. Such infected nursery plants become, in most cases, the major source of epidemics in citrus orchards.

To control bacterial wilt, tomato and eggplant may be grafted on root stocks of KNVE, LS-89, *Solanum torvum*, *S. intergrifolium*, or *S. mammosum* that are highly resistant to *P. solanacearum*. Because the susceptibility of the scions is not enhanced by the resistant root stocks, the scion plant may be exposed to infection when farming practices are not hygienic.

### 12.3.3 BREEDING OF DISEASE-RESISTANT CULTIVARS

Use of disease-resistant cultivars is economically and technically the most practical method of disease control. A number of breeding programs have been highly successful, and commercially acceptable cultivars are available now for the following diseases: bacterial wilt of peanut and tobacco, Stewart's disease of corn, bacterial blight of cotton, black-rot of cabbage, cauliflower, and broccoli (Fig. 12.2).

Breeding resistant cultivars starts with a survey for resistant genes that can be incorporated into breeding lines. To aid this process the International Board of Plant Genetic Resources is fostering the collection, conservation, documentation, evaluation, and utilization of plant

**Fig. 12.2** Incidence of black rot of broccoli on resistant (left) and susceptible (right) cultivars.

germ plasm under the network of various national and international organizations in the world.

In potato, resistance of *Solanum brevidens,* the diploid nontuber-bearing wild species, to *Erwinia* soft rot bacteria was successfully transferred to a tetraploid potato cultivar (*S. tuberosum*) by protoplast fusion. In this case, the resistance thus incorporated into the somatic hybrid could be sexually transferred to the progeny. The success in the interspecific transfer of resistance has revealed the practical usefulness of somatic hybridization to incorporate disease resistance from distantly related plants.

In general, monocropping of a resistant cultivar for a long period results in the development of new races of the pathogen that can affect the cultivar. When a cultivar carries a high level of vertical resistance governed by a major gene and a low level of horizontal resistance governed by polygenes, disease may occur in the destructive form. This is called the *vertifolia effect.* Therefore, current breeding programs aim at the durable resistance that is derived from a balanced combination of major genes and polygenes. In cotton, for instance, the cultivars or breeding lines that have high levels in yield, quality, and resistance to bacterial blight (*X. campestris* pv. *malvacearum*) have been developed by incorporating the major genes ($B_2$, $B_3$, $B_6$, and $B_7$) and the polygenes ($B_{sm}$ and $D_{sm}$) in combination.

## 12.3.4 BIOLOGICAL CONTROL

In general, biological control of plant diseases involves the deliberate use of one organism to control or eliminate a pathogenic organism. Biological control has attracted great interest in plant pathology because the unnecessarily frequent use of pesticides is increasingly causing concern in modern society in terms of human toxicity and hazardous effects on natural environments.

Biological control of fungal diseases with the use of bacteria has been documented in many diseases such as Take-all of wheat (*Gaeumannomyces graminis* var. *tritici*) with fluorescent pseudomonads and Fusarium wilt of bottle gourd (*F. oxysporum* f. sp. *lagenariae*) with *P. gladioli,* etc. Most of these techniques are, however, still in the stage of empirical application, either in the greenhouse or the field, with a few exceptions. Bacteria that have an antagonistic effect against pathogens and the potential as biological control agents include *Argrobacterium, Alcaligenes, Aerobacter, Bacillus, Erwinia, Flavobacterium, Lactobacillus, Pseudomonas,* and *Streptomyces.*

A practical method of commercial biological control is found in crown gall and has been developed in Australia with the strain K84 of *A. radiobacter*. This is undoubtedly one of the most innovative and important advances in biological control of bacterial plant diseases. When peach seedlings were dipped in a suspension of *A. radiobacter* strain K84 at a concentration of $10^8$ cells/ml and potted in soil infested with *A. tumefaciens*, galling was reduced to 31% in contrast to 79% in the untreated control. The effectiveness of the method has been confirmed in many laboratories in the world with various host plants such as *Prunus, Rubus, Malus, Salix, Vitis, Libocedrus, Chrysanthemum, Crataegus, Rosa, Pyrus,* and *Humulus.*

Biological control from *A. radiobacter* strain K84 seems to be a complex effect involving the bactericidal effect of a bacteriocin, Agrocin 84, the competitive inhibition of attachment of *A. tumefaciens* to the receptor sites on host cell wall, and competitive and exclusive uptake of nutrients that are common to these bacteria. Therefore, effective control can be obtained only when the plants are treated with *A. radiobacter* strain K84 before infection of *A. tumefaciens* takes place, i.e., the employment of the bacterium as the preventive agent in a preplanting dip is necessary.

The proportion of *A. radiobacter* strain K84 cells to *A. tumefaciens* cells is important in the application of the bacterium; effective control is generally obtained at the ratio of $>1:1$. Failures in biological control with strain K84 mostly originate when the population of the antagonist is significantly lower than the pathogenic strains.

Because damage due to crown gall is particularly severe on young trees, control with the strain K84 is effective in preplanting dip of seeds, cuttings, or grafted nursery plants. Dipping of plants carrying latent infections or sprays of the bacterial suspension over an infected area are not effective.

Biological control with *A. radiobacter* strain K84 is ineffective in control of *A. vitis*. This bacterium specifically attacks grapevines and differs from *A. tumefaciens* in physiological properties and resistance to agrocin 84.

The plasmid governing production of agrocin 84 and resistance against it may be transferred *in planta* by conjugation from strain K84 to *A. tumefaciens* in the presence of opines that are produced by the transformed tumor cells. The conjugative transfer of the plasmid makes biological control with this strain ineffective. To get away from the risk, a mutant strain in which the transfer gene (*tra*) of the plasmid is inactivated has been developed by gene manipulation and is currently in commercial use.

Biological control may be exerted by antagonists screened by chemicals on infection sites under natural conditions. Some fungicides that are not inhibitory to bacteria *in vitro* sometimes show effectiveness to bacterial diseases *in situ* by unknown mechanisms. Benomyl, pyroquilon, and probenazol are this type of chemical, and they are effective in controlling bacterial grain rot of rice caused by *P. glumae.*

Seed dressing with benomyl is particularly effective in control of seedling rot phase of the disease. Benomyl exerts a selective pressure on saprophytic microflora on rice germlings. As a result, some bacteria that are antagonistic to the pathogen become dominant and suppress growth of the pathogen under the level of infection threshold for some time when rice seedlings become resistant to infection. The mechanism is essentially biological control with antagonistic bacteria screened by benomyl.

## 12.3.5 CHEMICAL CONTROL

### *Chemotherapy*

Application of practical and effective chemotherapy is rather limited in bacterial plant diseases.

#### *Trunk Injection Method*

The canes of kiwi (*Actinidia chinensis*) are infected in winter with *P. syringae* pv. *actinidiae* through pruning of unnecessary canes where some of them are infected. The canes are most susceptible to infection from late autumn to early spring and in most cases are infected in this manner. Subsequently, leaves are infected in spring with the pathogen that exudes from diseased canes and trunks. Because of this disease cycle, either streptomycin or kasugamycin compounds in solution can be applied in late autumn, just before defoliation, by a systemic injection from the basal part of trunks. A problem with this method is the appearance of antibiotic-resistant strains, particularly in streptomycin compounds, after repeated applications.

#### *Coating Method*

Chemotherapy can be applied to crown gall with "Bacticin," which is a mixture of aromatic hydrocarbons in an oil–water emulsion (Table 12.3). Bacticin can be applied on the exposed crown gall tumors with some overlap onto the surrounding healthy tissues. The chemicals pene-

**Table 12.3**
Composition of Bacticin[a]

| Ingredient | Concentration | Ingredient | Concentration |
|---|---|---|---|
| 1,2,3,4-tetrahydronaphthalene | 3% | m-cresol | 0.4% |
| diphenylmethane | 3% | paraffin oil | 60% |
| dimethylnaphthalene | 3% | triethanolamine | 6% |
| 2,4-xylenol | 0.4% | stearic acid | 3% |

[a] From Schroth, M. N., Weinhold, A. R., McCain, A. H., Hildebrand, D. C., and Ross, N. (1971). Biology and control of *Agrobacterium tumefaciens*. *Hilgardia* **40,** 537–552.

trate selectively into tumor tissues, killing the tumor cells and the causal bacterium but not cells in healthy tissues. Large tumors may be removed with a hatchet before the treatment.

### *Chemical Spray*

Chemical compounds that are currently registered in Japan for controlling bacterial plant diseases include streptomycin, kasugamycin, oxytetracycline, novobiocin, copper compounds, dithianon, polycarbamate, oxolinic acid, tecloftalam, and pyroquilon as foliar sprays, sodium hypochlorite and oxolinic acid as seed disinfectants, and chloropicrin and metam-ammonium (NCS) and thiram as soil disinfectants (Fig. 12.3).

#### *Copper Compounds*

Wettable copper compounds are widely used for the leaf spot type of bacterial plant diseases. Effective ingredients include either copper chloride basic [$CuCl_2 \cdot 3Cu(OH)_2$], copper sulfate basic [$CuSO_4 \cdot 3Cu(OH)_2$], copper hydroxide [$Cu(OH)_2$], or 8-quinolinol that is the metal chelating compound 8-hydroxy quinoline-Cu.

Bordeaux mixture is still employed by farmers in spite of its complex method of preparation. The major effective ingredients of Bordeaux mixture are the complex salts of copper sulfate, copper hydroxide, and calcium hydroxide, with a formula of $CuSO_4 \cdot 3Cu(OH)_2 \cdot nCa(OH)_2$.

#### *Sulfur Compounds*

Sulfur compounds are available in the form of lime sulfur and wettable sulfur powder. Lime sulfur is sometimes applied against citrus canker in the dormant season. Wettable sulfur powder is not used against bacterial plant diseases.

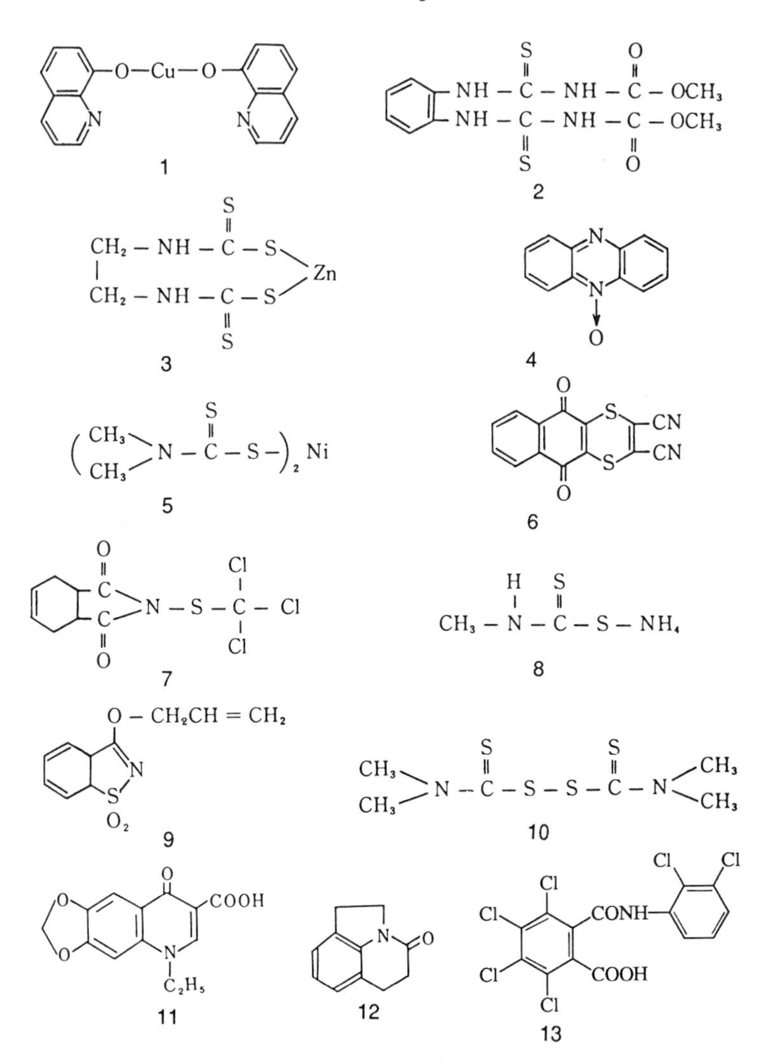

**Fig. 12.3** Structures of chemical compounds registered in Japan for controlling bacterial plant diseases (1990). 1, Oxine copper; 2. Thiophanatemethyl; 3, Zineb; 4, Phenazine oxide; 5, Nickel dimethyldithiocarbamate; 6, Dithianon; 7, Captan; 8, Metam-ammonium; 9, Probenazole; 10, Thiram; 11, Oxolinic acid; 12, Pyroquilon; 13, Tecloftalam.

*Other Compounds*

Nickel dimethyldithiocarbamate, probenazole, tecloftalam, and phenazine oxide have been applied to control bacterial leaf blight of rice. Probenazole and pyroquilon, which are primarily used for controlling rice blast disease (*Pyricularia oryzae*) have also bee registered against bacterial grain rot of rice based on its proven effectiveness in field trials. Polycarbamate is applied to bacterial leaf spot of cucumber. Oxolinic acid is a new compound recently developed in Japan against bacterial grain rot of rice. It has also been shown to be effective against soft rot and fire blight. This compound prevents bacterial cell division by inhibiting DNA gyrase. Other chemicals, such as thiophanate methyl, dithianon, or captan, are mixed with copper components or antibiotics in compounds that are registered for both bacterial and fungal diseases.

*Antibiotics*

*Streptomycin sulfate* This is an aminoglycoside antibiotic produced by *Streptomyces griseus* and is available in Japan under the common names phytomycin and agrimycin, and it is also used as a mixture with copper compounds. Streptomycin is an inhibitor of protein synthesis. It prevents binding of formyl-methionyl tRNA (fMet- tRNA) to 30 S subunit and reassociation of 30 S and 50 S subunits to form the 70 S ribosome–mRNA initiation complex.

*Kasugamycin* This is an aminoglycoside antibiotic produced by *Streptomyces kasugaensis*. It is available in Japan under the commercial name "kasumin" and also as a mixture with copper or other compounds. The antibiotic was initially developed for controlling rice blast (*Pyricularia oryzae*), but its effectiveness against various bacterial plant diseases has subsequently been confirmed. Kasugamycin inhibits bacterial protein synthesis by interacting with 30 S subunit of ribosome and prevents binding of fMet-tRNA and hence initiation of protein synthesis.

*Oxytetracycline* Oxytetracycline (terramycin) is an antibiotic produced by *Streptomyces rimosus* and usually used as a mixture with streptomycin sulfate. This antibiotic inhibits protein synthesis by preventing binding of aminoacyl-tRNA to 30S subunit of ribosome.

*Novobiocin* This is a glycoside antibiotic produced by *Streptomyces niveus*, *S. spheroides*, and *S. griseoflavus* and that is more effective against gram-positive than gram-negative bacteria. This antibiotic inhibits bacterial DNA gyrase. Novobiocin is registered in Japan only for tomato plants that are grown for seed production to protect them from *Clavibacter michiganensis* subsp. *michiganensis*.

*Soil Disinfectants*

*Chloropicrin* Chloropicrin is a broad spectrum fumigant with the effective ingredient of trichloronitromethane, $CCl_3NO_2$. It is commonly used for the control of soil-borne bacterial diseases such as bacterial wilt of solanaceous plants. The dispersion into and evaporation from soil of this material vary depending on soil temperature. Therefore, complete evaporation from soil before planting is important to avoid chemical damage. Because this material is a highly penetrating tear gas, great care must be taken in its use near suburban area.

*Other Soil Disinfectants* Thiram and metam-ammonium are registered in Japan for use against only bacterial wilt of tobacco.

*Other Compounds*

*Sodium hypochlorite* This compound is used for sterilizing the surface of such fruits as oranges and apples. It is also widely used for disinfesting seeds, tubers, and agricultural equipment. Calcium hypochlorite may also be used. Successful disinfestation with these compounds requires treatment for 5 min at least, and it should be remembered that the effectiveness of these compounds may be reduced in the presence of organic matter.

*Miscellaneous* In general, practical control of bacterial plant diseases relies mainly on copper compounds and antibiotics (streptomycin and kasugamycin), even if their effectiveness is not sufficient to prevent disease development. Therefore, various trails have been made for control of bacterial diseases with various chemical compounds available.

It has been repeatedly documented that some growth regulatory substances or herbicides alter the interactions between bacterial pathogens and their hosts providing significant suppression of disease development. For example, chloromoquat chloride [(2-chloroethyl)trimethylammonium chloride] reduced 50–70% of bacterial leaf spot of *Hibiscus rosa-sinensis* caused by *P. cichorii*, *P. syringae* pv. *hibisci* and *X. campestris* pv. *malvacearum*. Exposure of tomato plants to dinitroaniline herbicides suppressed the symptom development of bacterial wilt caused by *P. solanacearum*. A single spray of daminozide reduced approximately 50% of the incidence of common scab of potato.

# Further Reading

American Phytopathological Society's Committee on Standardization of Common Names for Plant Diseases: Common names for plant diseases (1985). *Plant Disease* **69,** 649–676.

British Society for Plant Pathology (1962). "Names of British Plant Diseases and Their Causes." Cambridge University Press, London.

Burdon, J. J., and Chilvers, G. A. (1982). Host density as a factor in plant disease ecology. *Ann. Rev. Phytopathol.* **20,** 143–166.

Cook, R. J., and Baker, K. F. (1983). "The Nature and Practice of Biological Control of Plant Pathogens." American Phytopathological Society, St. Paul, Minnesota.

Horsfall, J. G., and Cowling, E. B., ed. (1977). "Plant Disease, An Advanced Treatise, Volume I. How Disease Is Managed." Academic Press, New York.

Jackson, R. D. (1986). Remote sensing of biotic and abiotic plant stress. *Ann. Rev. Phytopathol.* **24,** 265–287.

Kranz, J., and Hau, B. (1980). Systems analysis in epidemiology. *Ann. Rev. Phytopathol.* **18,** 67–83.

Miller, S. A., and Martin, R. R. (1988). Molecular diagnosis of plant disease. *Ann. Rev. Phytopathol.* **26,** 409–432.

Mount, M. S., and Lacy, G. H., ed. (1982). "Phytopathogenic Prokaryotes, Volume 2. Part IV. Strategies for Control." Academic Press, New York.

Mulder, D., ed. (1979). "Soil Disinfestation." Elsevier Scientific Publishing Company, Amsterdam.

Russell, G. E. (1978). "Plant Breeding for Pest and Disease Resistance." Butterworths, London and Boston.

Saylor, G. S., and Layton, A. C. (1990). Environmental application of nucleic acid hybridization. *Ann. Rev. Microbiol.* **44,** 625–648.

Teng, P. S. (1985). A comparison of simulation approaches to epidemic modeling. *Ann. Rev. Phytopathol.* **23,** 351–379.

Weller, D. M. (1988). Biological control of soilborne plant pathogens in the rhizosphere with bacteria. *Ann. Rev. Phytopathol.* **26,** 379–407.

Young, H. C., Jr., Prescott, J. M., and Saari, E. E. (1978). Role of disease monitoring in preventing epidemics. *Ann. Rev. Phytopathol.* **16,** 263–285.

CHAPTER
THIRTEEN

# *Specific Diseases*

In this chapter, the characteristics of individual diseases are described on the basis of brief history of occurrence, symptoms, causal organism, disease cycle, control methods, and diagnosis from similar diseases. Eighteen prokaryote diseases are selected from field crops, vegetables and fruit trees on the basis of economic importance, historical significance, and/or unusual nature of the pathogen.

## 13.1 Bacterial Leaf Blight of Rice

Bacterial leaf blight of rice was first discovered in Fukuoka Prefecture, Japan in 1884. After Bokura's proposal of the invalid name *Bacillus oryzae* in 1911, Ishiyama reinvestigated the pathogenic bacterium and renamed it *Pseudomonas oryzae* in 1922. The disease has been one of the most important diseases of rice and has been intensively studied since the 1950s. From the mid 1970s, however, the incidence of the disease in Japan has declined. The occurrence of bacterial leaf blight was confirmed in South and Southeast Asia in the early 1960s, and the disease is still occurring there in severe form. The disease was also discovered in Texas and Louisiana in 1987.

### 13.1.1 SYMPTOMS

The symptoms of bacterial leaf blight can be divided into leaf blight, wilting (kresek), and pale yellow leaf. The leaf blight symptoms are characterized by longitudinal lesions developed along leaf edges. The lesion starts as small water-soaked streaks a few centimeters from the tip, where water pores distribute. It rapidly enlarges in length and width, forming a yellow lesion with a wavy margin along leaf edges. The lesion turns white to gray later. Lesions may be formed either on one side or both sides of a leaf blade, or sometimes also along the midrib. In young lesions, guttation of white milky ooze is continuously observed early in the morning at the portion of leaf edges where water pores distribute. When the disease occurs in severe form, the entire field is blighted in white (Fig. 13.1). The disease occurs on panicle, producing gray to light brown lesions on glumes and resulting in poor fertility and low quality of grains.

In the tropics, particularly on cultivars of Indica type, two additional symptoms may be observed. When infection occurs on younger plants of susceptible cultivars, the causal bacterium spreads through xylem of the infected leaf to the culm and infects the base of other leaves. The infected plant shows rolling and withering of several leaves of an infected tiller in the early stage. All tillers connected to it are subsequently infected, resulting in quick wilting of the entire plant. This symptom was first called kresek in Indonesia (Fig. 13.1). When the systemic infection occurs late in the tillering stage, pale yellow leaves may develop, depending on cultivars.

### 13.1.2 CAUSAL ORGANISM: *XANTHOMONAS CAMPESTRIS* pv. *ORYZAE* (Ishiyama 1922) Dye 1978

*X. campestris* has the following characteristics in addition to the common properties of Genus *Xanthomonas* given in Chapter 3, Section 3.6.1. Growth on agar plates is generally good, forming colonies of 1 to 2 mm in diameter in 48 hr. Viscid and slimy colonies are formed on nutrient agar containing 5% sucrose or potato dextrose agar. Acid is produced from arabinose, glucose, fructose, mannose, galactose, sucrose, trehalose, and cellobiose. The reactions are positive in aesculin hydrolysis and production of hydrogen sulfide, whereas they are negative in urease production. The capacity to liquefy gelatin and to hydrolyze starch varies depending on the pathovar. Some pathovars require methionine, glutamic acid, and/or nicotinic acid for growth. Growth is inhibited by the presence of NaCl at concentrations of 2–5%.

*X. campestris* pv. *oryzae* is distinct from other pathovars in slow growth on agar plates, requiring 3–4 days to form colonies 1–2 mm in diameter.

**Fig. 13.1** Bacterial leaf blight of rice caused by *Xanthomonas campestris* pv. *oryzae*. Left, leaf blight phase; right, wilting (kresek) phase.

Colony formation from a single cell is poor but can be improved by the addition of beef extract, methionine, or glutamic acid. The ability to hydrolyze starch and gelatin varies depending on the strain.

Recently, the name *X. oryzae* pv. *oryzae* has been proposed for this bacterium based on the phenotypic properties (negative activity of lecithinase, no growth on propionate as a carbon source), fatty acid profiles, whole cell protein profiles, DNA–DNA homology, and reactions to specific monoclonal antibodies. In this proposal, *X. oryzae* is defined to include two pathovars of pv. *oryzae* and pv. *oryzicola,* the pathogen of bacterial leaf streak of rice.

Bacteriophage strains found in nature are greatly different in host ranges, although they are differentiated into three major groups in serological properties, i.e., Op1, Op2, (Wakimoto), and Bp3 (Goto). Phage-typing of the bacterium is complex when bacterial strains are collected from wide geographical areas. The number of strains that can be phage-typed may be 80% at most, even when a number of phage strains are employed.

Natural infection is found on weeds such as *Leersia* spp. and *Zizania latifolia.* Several other weed hosts are also infected weakly through artificial inoculation but not by natural infection. The pathogen infects either through wounds or hydathodes. Bacterial cells that entered through water pores multiply in epithem, followed by active multiplication in xylem vessels. Strains distributing in tropical areas are generally more virulent than those in temperate zones. The bacterium is classified into several races on the basis of pathogenic reactions to the differential cultivars (Table 8.4). An avirulence gene of a race has been demonstrated by cloning.

The strains of *X. campestris* pv. *oryzae* isolated in Australia were different from those distributing in Asia in physiological properties. The strains isolated in the United States in 1987 were also distinguished by monoclonal antibody technique. This information implies that the disease discovered in these areas is not seed-borne or it was not introduced from Asian countries through infected seeds.

## 13.1.3 DISEASE CYCLE

The disease cycle of bacterial leaf blight in Japan is shown in Fig. 13.2. Seed dissemination of the disease has not been confirmed. The major source of inoculum is found on *Leersia* spp., which grow in colonies along canals and ditches. The pathogen overwintered on rhizosphere of these alternative hosts multiplies in spring when new shoots of the weeds

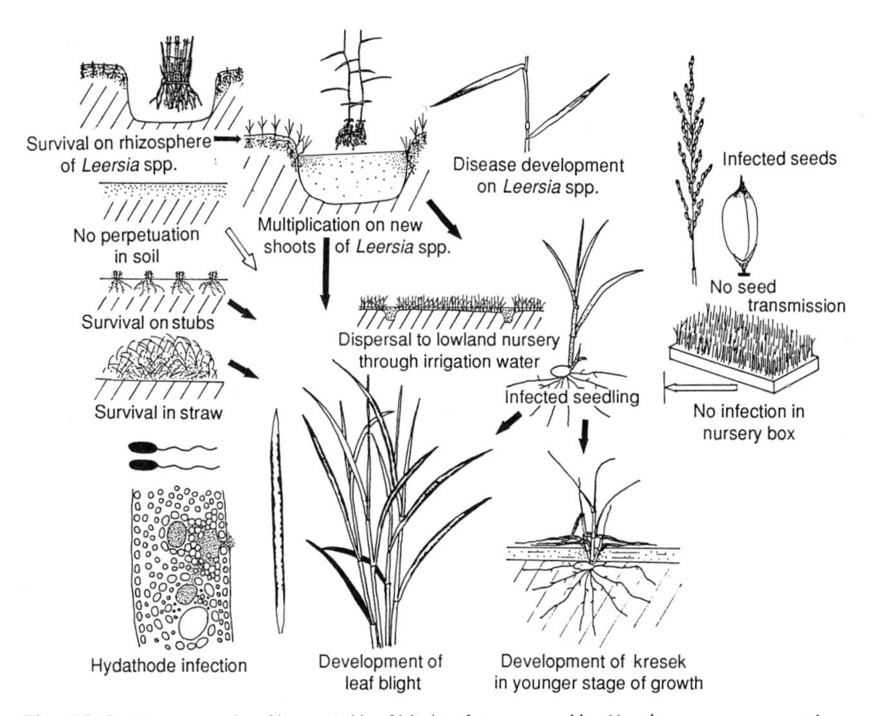


**Fig. 13.2** Disease cycle of bacterial leaf blight of rice caused by *Xanthomonas campestris* pv. *oryzae*.

begin to emerge. The bacterium is dispersed through irrigation water and infects rice seedlings in lowland nurseries or those transplanted in paddy. This is the primary infection.

Secondary spread of the disease in severe form occurs usually in summer to early autumn under humid weather conditions. Epidemics are readily induced by rainstorms or flooding of paddy fields in this season. Excess application of nitrogen fertilizer also increases disease severity.

The noticeable decline of disease incidence in Japan in the past few decades is considered to be attributed to eradication of colonies of *Leersia* spp. by reformation or reconstruction of irrigation systems. Innovation in the way of raising seedlings in nursery boxes for adapting to rice planters eliminated the chance of introducing infected seedlings into paddy fields. Even in these days, therefore, bacterial leaf blight occurs in endemic form in the area surrounding spring of water where colonies of *Leersia* spp. are left growing. In the tropics, however, the disease incidence is still severe because environments and the ways of rice cultivation are entirely different from those in Japan.

### 13.1.4 CONTROL

Alternative hosts growing along canals and ridges are removed to eliminate the natural habitats of the pathogen and its dispersal through irrigation water. Care must be taken for field sanitation by burning or taking out diseased rice straw from fields. Fertilization in either excess nitrogen or in deficiency of potassium and phosphorous should be avoided.

Tecloftalam, phenazine oxide, nickel dimethyldithiocarbamate, and probenazole can be applied for chemical control of bacterial leaf blight. Probenazole (granule form) is applied to the paddy water before or just after transplanting, well in advance of disease occurrence. Other chemicals are sprayed two to three times, depending on the intensity of the disease.

### 13.1.5 SIMILAR DISEASES

Wilting syndrome (kresek) of bacterial leaf blight somewhat resembles that of bacterial foot rot (*Erwinia chrysanthemi* pv. *zeae*). These diseases are distinguishable in that diseased tillers in the latter readily pull out and have a strong odor. The advanced stage of bacterial leaf streak (*X. cam-*

*pestris* pv. *oryzicola* somewhat resembles the leaf blight phase of the disease. Bacterial leaf streak is easily distinguished by translucent and yellow streaks limited by leaf veins and a number of droplets of bacterial ooze formed on their surface. Several other bacterioses of rice can be distinguished by their characteristic symptoms.

## 13.2 Bacterial Grain Rot of Rice

Bacterial grain rot disease was discovered in the Kyushu district of Japan in 1955. It had been noticed as a minor disease until the early 1970s, when disease incidence became increasingly severe in the southern and central regions of Japan. This period coincided with wider use of seedlings raised in nursery boxes rather than in traditional lowland nurseries for adapting to rice transplanters. In the mid-1970s, the first incidence of seedling rot by the same organism was reported in central Japan, revealing that the disease has two different phases of grain rot in panicles and seedling rot in nursery boxes. Since then seedling rot has become a serious problem in nursery stages also. However, seedling rot has scarcely been observed in lowland nurseries, even if infected grains were mixed in bulked seed.

### 13.2.1 SYMPTOMS

***Grain Rot Phase***

Symptoms appear on glumes and kernels but never on rachilla or rachis. The hulls of infected grains on young panicles rapidly fade in color from the base, turning greenish white at first, and then pale pink to yellowish brown. They soon wither and dry up. The affected grains are either sterile or carry brown rice with dark-brown discoloration at the base. In severe cases, almost all grains on a panicle are attacked, so that these white panicles remain standing erect even in the mature stage (Fig. 13.3). Symptoms may vary depending on the temperature and the amount of nitrogen applied before panicle emergence.

***Seedling Rot Phase***

Seedling rot in nursery boxes may occur on seedlings derived from the infected seeds, on those infected during presowing water soaking, or

**Fig. 13.3** Bacterial grain rot of rice caused by *Pseudomonas glumae.* Left, grain rot; right, seedling rot.

on those infected secondarily in nursery boxes. Seeds that are infected in the preemergence stage result in rotting and do not germinate. Young seedlings around these rotted seeds may be infected, forming infection foci in nursery boxes. Seedlings in the foci develop brown lesions on coleoptile and soon rot.

When infection takes place several days after sowing, the disease appears as a brown discoloration of leaf sheaths and chlorosis of newly expanding leaves; a folded youngest leaf may also turn pale yellow to light brown and is sometimes twisted. These seedlings soon turn yellow as a whole and finally decay. When infection occurs in mild form, seedlings exhibit yellowing on a few younger leaves but continue to grow without fatal damages.

### 13.2.2 CAUSAL ORGANISM: *PSEUDOMONAS GLUMAE* Kurita and Tabei 1967

This bacterium has the following characteristics in addition to the general features of Genus *Pseudomonas* described in Chapter 3, Section 3.6.1. Positive reactions are found in the tests of hydrolysis of gelatin, reduction of nitrate, and activities of catalase, lipase, and lecithinase. In contrast, reactions are negative in the tests of oxidase, tylosinase, arginine dihydrolase, hydrolysis of arbutin, salicin, and aesculin, and utilization of sucrose, maltose, and rhamnose as a sole source of carbon. Good growth occurs at 40°C, and crystals of calcium oxalate are produced on agar plates that contain calcium chloride. This bacterium produces several pigments that are the nonpyroverdin type, fluorescent, water soluble, and yellowish green to green in color. Some of these pigments are known to be phytotoxic. Major components of the pigments are toxoflavin with the highest toxicity and feruvenulin with the next highest toxicity. The former is readily transformed to reumycin (Fig. 8.1).

Natural infection of the pathogen has been observed only on rice plant. Several plants of Gramineae such as *Alopecurus aequalis* var. *amurensis* and *Poa annua,* as well as those of Compositae, Umbelliferae, Leguminosae, Cruciferae, and Polygonaceae have been reported to be weakly infected with the bacterium by artifical inoculation.

### 13.2.3 DISEASE CYCLE

The disease is seed-borne. The pathogen invades glumes through stomata located outside lemmas and colonizes in the stomatal cavity as well as parenchymatous tissues surrounding them and survives until the next crop season (Fig. 9.1). These bacteria multiply when seeds are soaked in water for presowing emergence of embryo and infect young germlings. When densely seeded nursery boxes are placed under high temperature and an excess of humidity, the seedlings surrounding these rotted germlings are secondarily affected, forming disease foci in nursery boxes.

On the seedlings that recovered from mild infection, the bacterium epiphytically perpetuates at the base of hills in a relatively low density and gradually moves upward along with plant growth. In the booting stage, the bacterium quickly multiplies on the surface of a young panicle enveloped by leaf sheath and infects inflorescence at the flowering stage. Severe infection usually takes place when heading and flowering of panicles occurs under high temperature and frequent rains (Fig. 13.4).

### 13.2.4 CONTROL

Because this disease is seed-borne, use of healthy seeds is the primary method of control. When contamination of the infected seeds are suspected, bulked seed may be subjected to salt-water selection with a spe-

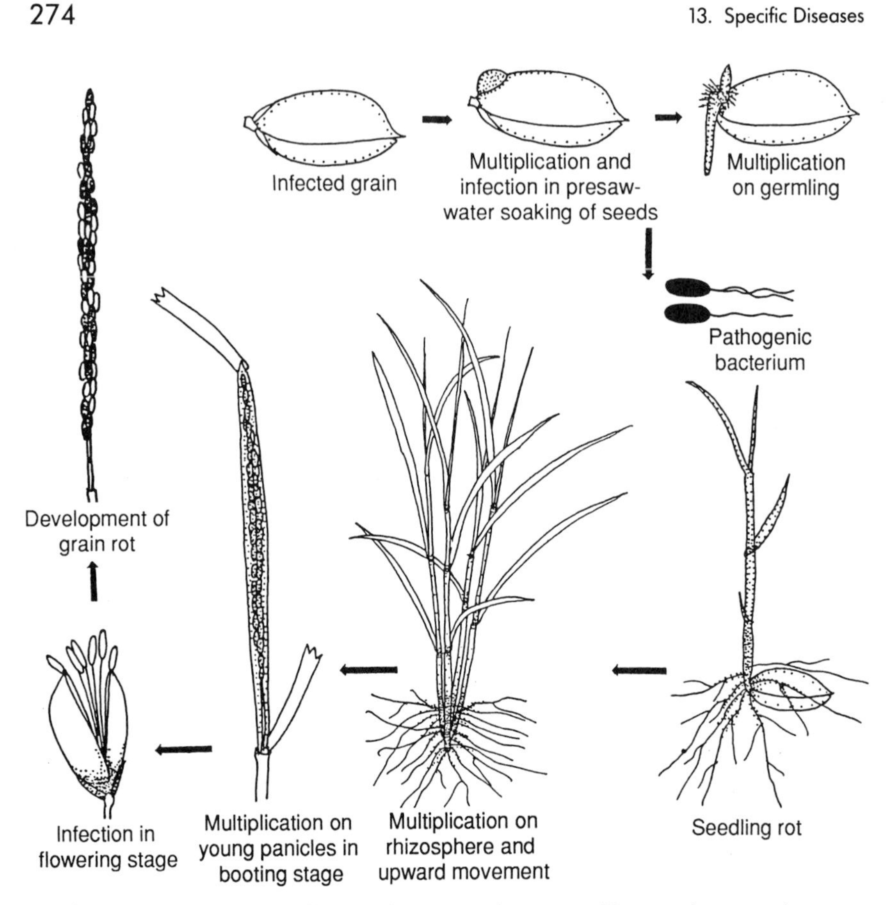


**Fig. 13.4** Disease cycle of bacterial grain rot of rice caused by *Pseudomonas glumae.*

cific gravity of 1.18 to eliminate contaminated seeds. Overwatering nursery boxes must be avoided. The surface of boxes must be kept even and level to avoid localized standing water after watering. The temperature of nursery sites is kept between 15° and 30°C.

For chemical control of grain rot on panicles, oxolinic acid or kasugamycin compounds are sprayed on plants at pre- or postheading stages. Granule-type compounds of pyroquilon and probenazole may be applied to paddy soil a week to several weeks before heading.

The effective control of seedling rot in nursery boxes may be achieved by seed soaking in solutions of oxolinic acid, kasugamycin, or sodium hypochlorite, or by presow-mixing of kasugamycin dust with bed soi.

### 13.2.5 SIMILAR DISEASES

Other bacterial diseases of rice found in nurseries include bacterial brown stripe (*P. avenae*) and seedling blight (*P. plantarii*). The former is distinguishable by the primary symptoms of brown stripes extending along leaf blades and/or leaf sheaths. Extensive rotting or chlorosis of the entire leaf is rarely observed. Seedling blight is distinctive in that it never rots seedlings. Bacterial palea browning (*E. ananas*) is another disease occurring on panicles. It is readily distinguishable by characteristic discoloration of palea, but not lemmas, into dark brown.

## 13.3 Soft Rot

Under natural conditions, soft rot of plants may be caused by various bacteria included in *Erwinia, Pseudomonas, Bacillus,* and *Clostridium.* Although these bacteria have the potential to cause soft rot at any time of the year, the pathogens of *Pseudomonas* generally cause soft rot at lower temperature than those of *Erwinia* and *Bacillus.* In this section, soft rot diseases caused by *E. carotovora* and *E. chrysanthemi* are described.

Soft rot of vegetables caused by *E. carotovora* was first reported by L. R. Jones (1901) in carrots. In Japan, J. Omori reported in 1896 on soft rot of Japanese horseradish (*Eutrema wasabi*) and named the causal organism *Bacillus alliariae,* although this report had totally been overlooked by the scientific community. Soft rot caused by *E. chrysanthemi* was first reported with bacterial blight of chrysanthemum from the United States in 1953. The disease was detected in Shizuoka Prefecture in Japan on chrysanthemum and shasta daisy in 1957. Bacterial soft rot is an important disease of various kinds of vegetables. In addition, the disease also occurs on some crops of Gramineae and woody trees.

### 13.3.1 SYMPTOMS

The first symptom of soft rot of fleshy vegetables or storage organs caused by *Erwinia* soft rot bacteria is the appearance of small water-soaked lesions, no matter what plants or plant parts are infected. The lesions rapidly enlarge and cause extensive maceration of the affected tissues so that the diseased plants finally collapse.

The disease often starts from the plant parts that come in close contact with the soil surface (Fig. 10.7). In root crops such as carrots, the disease may often be recognized by wilting of the aboveground parts in the field. Potato tubers in storage are internally transformed into a soft pulpy mass, with the corky epidermis left intact, and bacterial exudate may be abundantly found on the surface of the skin. Soft rot of mulberry plant is characterized by blackening of bark tissues, soft rot, wilting, or hollow stalks of young shoots.

*E. chrysanthemi* attacks a wide range of host plants such as vegetables, cereals, and woody plants, producing different symptoms depending on the affected plants. For instance, *Erwinia* rusty canker of pear is characterized by staining of affected twigs into rusty red by bacterial ooze and plant sap exuding from affected bark. Bacterial foot rot of rice is characterized by discoloration of the basal portion of culms, with a disagreeable odor, and bacterial stalk rot of corn by fall-down of decayed culms.

### 13.3.2 CAUSAL ORGANISM: *ERWINIA CAROTOVORA* (Jones 1901) Bergey, Harrison, Breed, Hammer and Huntoon, 1923; *ERWINIA CHRYSANTHEMI* Burkholder, McFadden, and Dimock 1953

These bacteria are common in the general characteristics of Genus *Erwinia* described in Chapter 3, Section 3.6.3 but classified into two distinct species by the characteristics listed in Table 13.1. *E. carotovora* is further distinguished into four subspecies according to the features shown in Table 13.1. Some of these subspecies are distinctive in the affected host plants by natural infection, but it is difficult to demonstrate the differences by inoculation tests. *E. chrysanthemi* is classified into six pathovars by host range patterns, but these pathovars are also distinguishable by differences in the ability to use carbon sources.

### 13.3.3 DISEASE CYCLE

The primary sources of inoculum are bacteria perpetuating as a saprophyte in soil, as an epiphyte on phyllosphere of host plants and/or weeds, in residues of diseased plants, or in soil attached on planting materials. In addition, the role of bacterial aerosol and irrigation water has come to researchers' attention these days. Soft rot bacteria are one of the typical wound parasites; hence, disease incidence remarkably increases when host plants are injured by farming practices, wind, plant

**Table 13.1**
Comparison of Bacteriological Characteristics of Subspecies of *Erwinia carotovora* and *E. chrysanthemi* pv. *chrysanthemi*

| Characteristics | Bacteria | | | | |
|---|---|---|---|---|---|
| | Eca[a] | Ecc | Ecb | Ecw | Echch |
| Liquefaction of gelatin | + | + | − | + | + |
| Hydrolysis of casein | − | + | − | +(s)[b] | + |
| Methyl red test | + | + | − | + | − |
| Reducing substance from sucrose | + | d | + | − | + |
| Growth in 5% NaCl broth | + | + | + | − | − |
| Growth at 36°C | − | + | + | − | + |
| Growth in KCN broth | + | + | − | − | + |
| *O*-Nitrophenyl-β-D-galactopyranoside | + | + | + | − | + |
| Production of acetoin | − | d[c] | + | − | + |
| Activities of: phosphatase | − | − | − | − | + |
| lecithinase | − | − | − | − | + |
| Hydrolysis of Tween 80 | − | − | − | − | + |
| Production of indole | − | − | − | − | + |
| Erythromycin sensitivity (15 μg) | R[d] | R | R | R | S[e] |
| Production of acid from: galactose | + | + | + | +(lag) | + |
| lactose | + | + | + | +(lag) | + |

[a] Eca, *E. carotovora* subsp. *atroseptica,* Ecc, *E. carotovora* subsp. *carotovora;* Ecw, *E. carotovora* subsp. *wasabiae,* Ecb, *E. carotovora* subsp. *betavaculorum;* Echch, *E. chrysanthemi* pv. *chrysanthemi*

[b] s, strong reaction.

[c] d, different by strain.

[d] R, resistant.

[e] S, susceptible.

growth itself, or such insects as stripe cabbage flea-beetle, cabbage worm, and cabbage army worm.

When the growing stage of the host plant advances and dense foliage come to cover the soil surface, the humidity under the plant canopy elevates and root systems appear on the soil surface, making the microenvironments favorable to bacterial multiplication. In Chinese cabbage, the disease usually occurs from leaf petioles in contact with the soil surface under such circumstance (Fig. 13.5).

## 13.3.4 CONTROL

1. Resistant cultivars are cultivated if they are available.
2. The density of causal bacteria is reduced by crop rotation and field sanitation.

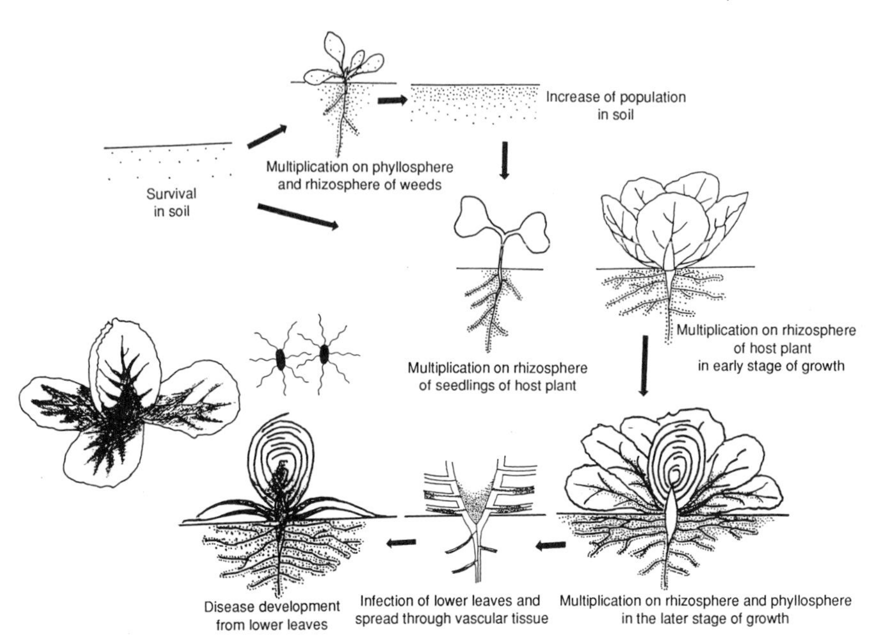


**Fig. 13.5** Disease cycle of soft rot of Chinese cabbage caused by *Erwinia carotovora* subsp. *carotovora.*

3. Farming practices are carefully performed not to injure crops. Control measures are taken against insects.
4. Fields are well drained and ridges are prepared in humid soil.
5. Oxolinic acid and copper compounds can be sprayed every 7 to 10 days when disease appears.

## 13.4 Black Rot of Crucifers

Black rot of crucifers was discovered by Pammel in the early 1890s in Iowa, and the causal organism was named *Bacillus campestris* in 1895. It was subsequently found in Wisconsin and soon recognized to be a serious disease of crucifers throughout the world. In Japan, the disease was first recorded in Hokkaido in 1909. Black rot can be seen all year round where crucifers grow, but severe incidence may occur during later stage of growth, particularly in autumn.

### 13.4.1 SYMPTOMS

The pathogen invades the vascular tissue of leaves, stalks, stumps, and roots. When infected seeds germinate, black discoloration occurs at the top margin of cotyledon around the end of the midvein. The affected cotyledons wither soon and are finally killed. In aged seedlings, infection usually takes place from lower leaves through hydathodes. Typical symptoms on these leaves are large, yellow to dark brown lesions formed along leaf margins. These lesions develop in the shape of a wedge or V around a hydathode. Young yellow lesions are usually surrounded by an indistinct, pale green and withered area. When disease advances, dark brown to black nets of leaf veins develop in the center of lesions on the yellow to light brown background. The margins become defined and surrounded by a narrow yellow halo. When leaves are injured, wound infection takes place in the center of leaf blades, resulting in the same symptoms as those in hydathode infection (Fig. 13.6).

In susceptible hosts as certain cultivars of cauliflower, infection sometimes occurs on the stumps, causing blackening of the vascular tissue, which may result in dwarfing, wilting, or uneven or twisted plant growth. Infected stumps usually appear externally healthy, but the disease is revealed when it is transversely excised.

### 13.4.2 CAUSAL ORGANISM: *XANTHOMONAS CAMPESTRIS* pv. *CAMPESTRIS* (Pammel, 1895) Dowson, 1939

This bacterium is characterized by the following properties in addition to the general features of genus *Xanthomonas* in Chapter 3, Section 3.6.1 and those of *X. campestris* in Chapter 13, Section 13.1.2. Starch is strongly hydrolyzed and gelatin is moderately liquefied. The majority of strains are lysogenic and release temperate phages that are different in host range spectrum.

Natural infection occurs in 34 species of crucifers, of which 20 species belong to *Brassica*. However, severe damage is usually found on susceptible cultivars of cabbage, cauliflower, and broccoli. The bacterium can affect zinnia by artificial inoculation. Investigations in Japan revealed that the strains isolated from cabbage and cauliflower and those from Chinese cabbage and turnip were distinguished from each other in host ranges and in susceptibility to bacteriophages. The xanthan-deficient spontaneous mutants usually exhibit markedly reduced virulence.

**Fig. 13.6** Black rot of cabbage caused by *Xanthomonas campestris* pv. *campestris.*

### 13.4.3 DISEASE CYCLE

The primary infection occurs through internally infected or externally infested seeds. Because the disease may readily be spread from a small number of infected seeds in a seedbed, a very small proportion of infected seeds has the potential to induce epidemics when seedlings raised in a nursery are planted in the field. Young plants grown in nurseries may also be latently infected and become carriers.

The bacterium can survive for long periods in diseased plant residues, particularly in woody tissues remaining in or on the ground. The bacterium surviving in these woody tissues in soil plays a significant role in infection through root systems. The secondary cycle of the disease starts from the foci of primary infection by rain splash or irrigation water. The disease epidemic is facilitated by favorable meteorological conditions as well as injuries caused by insects, wind, or farming practices.

### 13.4.4 CONTROL

1. Resistant cultivars are planted. In cabbage, cauliflower, and broccoli, cultivars with stable resistance to black rot are available. The resistance genes of these cultivars originated in the Japanese cabbage cultivar "early Fuji" (Fig. 12.2).
2. Healthy seeds are planted. When contamination of infested seeds is suspected, bulked seed is treated at 50°C for 25 min for cabbage or at 50°C for 15 min for other crucifers.
3. Field sanitation is well maintained by removing diseased plant residues as well as cruciferous weeds.
4. Control measures are taken against insects such as striped cabbage flea-beetle, cabbage worm, and cabbage army worm.
5. Contaminated field soil may be disinfected by chloropicrin. Foliage may be protected from infection by spraying copper or kasugamycin compounds.

### 13.4.5 SIMILAR DISEASES

The pathogenic bacteria which cause leaf spots on crucifers include *P. syringae* pv. *maculicola, X. campestris* pv. *rhaphani, X. campestris* pv. *aberrans* (cauliflower), and *X. campestris* pv. *armoraciae* (horseradish). These bacteria affect primarily parenchyma but never vascular tissue and produce small spots 2–5 mm in diameter on leaf lamina. Black rot disease is, therefore, rarely misdiagnosed with these bacterioses if its characteristic symptoms of large, brown lesions with black leaf vein nets are examined.

# 13.5 Bacterial Wilt

The first record of bacterial wilt in the world may be tuber rot of potato that was reported by Burrill in 1890. The disease was first reported in Japan in Uyeda's paper on bacterial wilt of tobacco in 1904. However, it is said that the disease had been noticed in Kagoshima Prefecture since the 1880s. Bacterial wilt is one of the most important diseases of plants in terms of its distribution and damage. It is not rare that 100% of solanaceous crops are killed by the disease in the field.

## 13.5.1 SYMPTOMS

Bacterial wilt is characterized by sudden wilting of foliage. The acute wilting is particularly notable in susceptible young plants. Once wilting appears on some leaves, the plants quickly wither and die without recovering (Fig. 10.4). Different symptoms may develop depending on the plant species, cultivar, growth stage, and environmental conditions such as temperature. They include aboveground symptoms such as yellowing, dwarfing, stunting, and development of adventitious roots and underground symptoms such as decay of tubers of potato and ginger. Regardless of the symptoms mentioned above, discoloration of the vascular system from pale yellow to dark brown is the common feature of diseased plants. The surface of a transverse section is moist, and droplets of milky bacterial ooze exude on the invaded tissue. Infected tubers of potato and ginger externally appear healthy when disease has not advanced, but section reveals the discoloration of the vascular region and bacterial exudation from invaded tissue.

## 13.5.2 CAUSAL ORGANISM: *PSEUDOMONAS SOLANACEARUM* (Smith, 1896) Smith, 1914

The bacterium has the following characteristics in addition to the general properties of genus *Pseudomonas* (Chapter 3, Section 3.6.1): fluorescent pigment is not produced; poly-$\beta$-hydroxybutyrate is accumulated; nitrate reduction and denitrification occur; a member of rRNA II group; mol% G + C of the DNA is 66.5–68%; five biovars are differentiated by the use of carbohydrates (Table 13.2). High correlations have been recognized among biovars, pathotypes, phage types, and bacteriocinogenic types.

**Table 13.2**
Biovars of *Pseudomonas solanacearum*

| | Biovar | | | | |
|---|---|---|---|---|---|
| Carbohydrates | I | II | III | IV | V[a] |
| Cellobiose | − | + | + | − | + |
| Lactose | − | + | + | − | + |
| Maltose | − | + | + | − | + |
| Dulcitol | − | − | + | + | − |
| Mannitol | − | − | + | + | + |
| Sorbitol | − | − | + | + | − |

[a] Biovar V: mulberry strain.

Less virulent or avirulent mutants readily appear in subcultures on artificial media; pathogenic wild strains form fluidal opaque colonies on agar plate and cells are nonflagellated, whereas attenuated mutant strains form butyrous transparent colonies and cells are flagellated.

*P. solanacearum* attacks a number of plants, covering more than 270 species of 3 families. Among these plants, disease incidence is particularly common in tobacco, tomato, potato, pepper, eggplant, peanut, and banana. The presence of races has been elucidated in the bacterium attacking resistant root stocks of eggplant such as *Solanum torvum, Solanum integrifolium, S. mammosum,* and KNVF (a root stock cultivar). Buddenhagen and Kelman (1964) defined the strains attacking potato at lower temperature as race 3, those attacking banana and Heliconia as race 2, and others as race 1. However, use of the term *race* in this way should be avoided because races are defined in plant pathology on the basis of the ability to cause disease on certain selected differential cultivars of a host plant. The pathogenic differentiation of *P. solanacearum* is quite complex and needs further study in the future, not only by artificial inoculation tests but also by analysis of naturally infected host plants.

## 13.5.3 DISEASE CYCLE

Major infection routes of the bacterium in tomato, tobacco, banana, and potato are shown in Fig. 13.7. *P. solanacearum* is a typical soil-inhabiting bacteria and can survive for a long time without the presence of host plants. Therefore, wound infection through root systems is the most common mode of entry. However, infection through planting materials or insect transmission may be the major mode of infection, de-

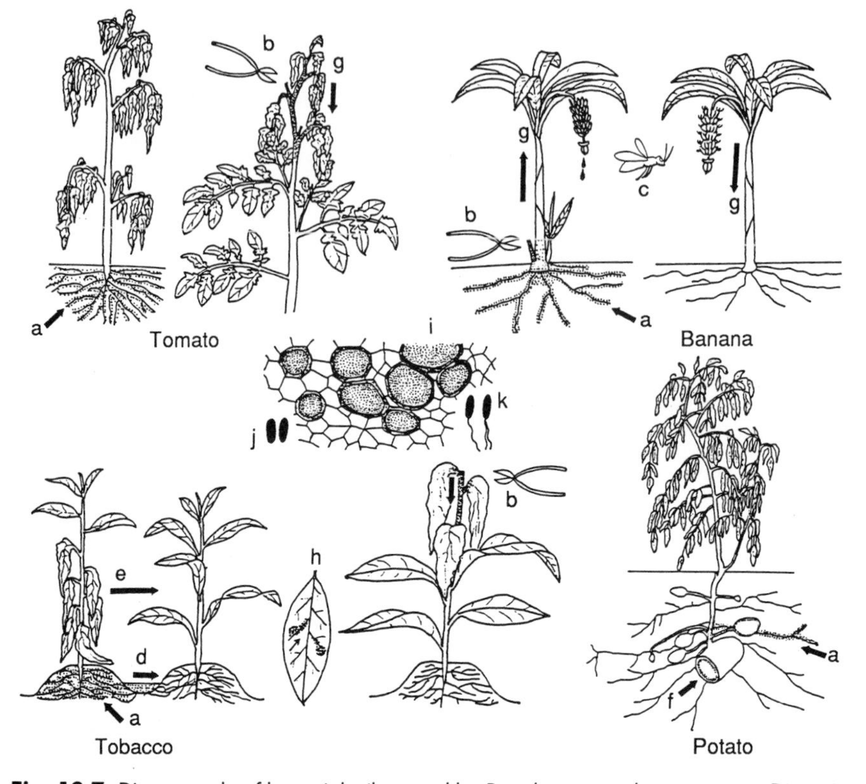


**Fig. 13.7** Disease cycle of bacterial wilt caused by *Pseudomonas solanacearum*. a, Dissemination by soil; b, dissemination by knife; c, dissemination by insects; d, dissemination by surface water; e, dissemination by rain splash; f, dissemination by infected tubers; g, spread of the pathogen; h, lesions developed by rain-splash dispersal; i, multiplication in xylem; j, virulent strain; k, avirulent strain.

pending on host plants. Under intensive cultivation, infection may occur through wounds produced by various farming practices such as harvesting, pinching, and topping. Adult tobacco plants may be infected from wounds of root systems developed on the soil surface through surface water in rain or from wounds of leaves by rain splash.

Bacterial wilt does not occur in areas where mean temperature is below 10°C. The disease usually develops when mean temperature rises above 20°C. Bacterial wilt occurs in practically any type of soil or soil pH where host plants can normally grow.

### 13.5.4 CONTROL

1. Soil of nurseries and fields is disinfected by chloropicrin, steam treatment, or solarization.
2. Resistant cultivars are available in tobacco. Tomato and eggplant may be grafted on resistant root stocks (*Solanum torvum, S. integrifolium,* or *S. mammosum* for eggplant, BF-Okitsu 101, KNVF, or LS-89 for tomato).
3. Disease-free tubers or rhizomes are planted in potato, ginger, and bird of paradise.
4. To eradicate root knot nematodes which increase the incidence of bacterial wilt, field soils are treated with D-D or chloropicrin.
5. The disease may be avoided by planting crops early to achieve maturity while mean temperature is still below the range for disease development.
6. In pruning, knives or fingernails are disinfected with proper disinfectants.

### 13.5.5 SIMILAR DISEASES

Irreversible wilting of plants may be induced by diverse causes such as diseases (fungi, nematodes), insect damage, and physiological disorders. Bacterial wilt is readily diagnosed by formation of macroscopic exudation of bacterial ooze from a cut surface or by conspicuous bacterial streaming from vascular tissue under a microscope.

## 13.6 Bacterial Canker of Tomato

Bacterial canker of tomato was first discovered in Michigan in 1909 and subsequently reported by the 1920s from Europe, Australia, Central and South America, Africa, and Asia. In Japan, occurrence of the disease was first recorded in 1958 in Hokkaido and in 1962 it was detected across the country.

### 13.6.1 SYMPTOMS

Two different symptoms of wilting and spotting are produced, depending on environmental conditions, cultivars or plant organs affected.

### *Wilting*

Wilting is the most common symptom of the disease and is caused by invasion of vascular bundles, cortex, and pith. Wilting starts on the lower leaves in which leaflets often show curling, followed by interveinal pale brown scorch and progressive withering and drying. When the disease advances, brown stripes and/or cracks appear on the surface of stems and leaf stalks. Symptoms sometimes develop on one side of a plant when the foliage of another side remains intact. In general, wilting advances slowly and partially on affected plants; hence, the disease may be distinguishable from bacterial wilt caused by *P. solanacearum.* A longitudinal section of affected stems exhibits browning of the vascular bundle, cortex, and pith, as well as distinct cavities in pith (Fig. 13.8).

### *Spotting*

Round spots, 2–3 mm in size, may form on fruits, stems, and leaves. The spots are white at first, but later show a light brown, raised, and corky appearance. The lesions formed on fruits are characterized by bird's-eye spots carrying a white halo around the spots.

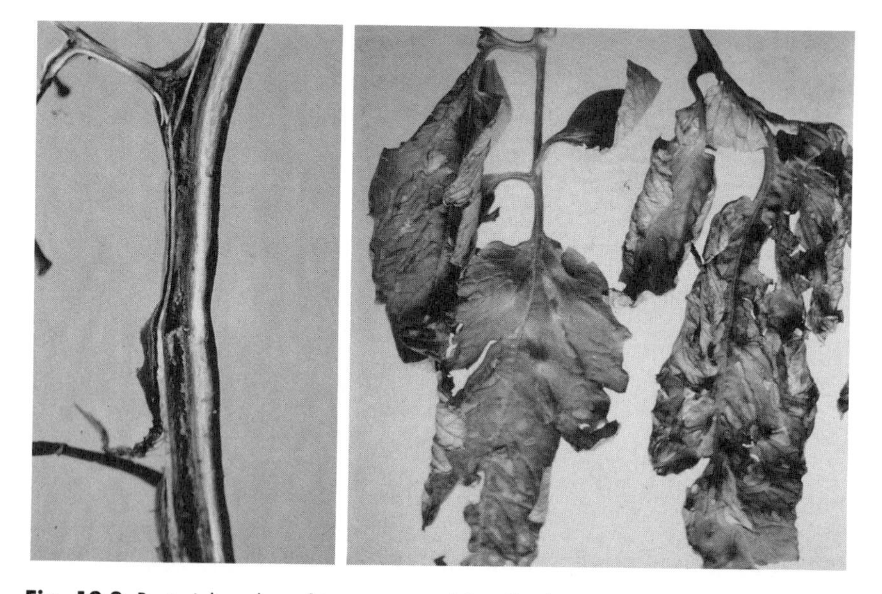

**Fig. 13.8** Bacterial canker of tomato caused by *Clavibacter michiganensis* subsp. *michiganensis.*

### 13.6.2 CAUSAL ORGANISM: *CLAVIBACTER MICHIGANENSIS* subsp. *MICHIGANENSIS* (Smith 1910) Davis, Gillaspie, Vidaver, and Harris 1984

In addition to the general features of genus *Clavibacter* (Chapter 3, Section 3.6.4), *C. michiganensis* has the following common characteristics: usually good growth on nutrient agar containing 5% glucose; acid is produced from fructose and sucrose but not from dulcitol, sorbitol, inulin, and $\alpha$-methyl-D-glucoside; citrate, fumarate, and malate are used as a sole source of carbon. *C. michiganensis* subsp. *michiganensis* is distinguished from other subspecies of *C. michiganensis* by the following characteristics: NaCl tolerance is 6%; gelatin is weakly liquefied; $H_2S$ is produced; growth occurs with mannose, cellobiose, and succinate as the sole source of carbon but not mannitol and propionate.

Beside tomato, natural infection has been observed on *Solanum douglassii, S. nigram,* and *S. triflorum*. Ten other species mainly of *Solanum* can be infected by artificial inoculation.

### 13.6.3 DISEASE CYCLE

Bacterial canker of tomato is a typical seed-borne disease. The seeds are internally infected and/or externally infested in fruits bearing on the systemically infected tomato plants. The bacterium starts to multiply when contaminated seeds germinate and resides on the surface of growing plants as an epiphyte. These bacteria may infect plants through wounds as well as natural openings such as stomata and trichomes. Diseased plant residues also become the primary inoculum source. Although the pathogen in the residues dies when the plant tissue is completely decomposed, woody stems may be left intact, at least in part, until the next crop season when they are buried deep in soil. Tomato roots may be infected by the bacterium surviving in such a way in soil. The bacterium surviving on the surface of agricultural equipment and implements may also become the inoculum source.

Secondary cycles of the disease usually occur by dispersal of the pathogen from diseased plants through rain splash. Careless plant management such as pruning undesirable lateral shoots with an undisinfected knife or fingernails may also result in a wide spread of the disease.

### 13.6.4 CONTROL

1. Healthy seeds are planted or a bulked seed is treated in hot water at 55°C for 25 min to eradicate the contaminated pathogen.
2. Nursery and field soil is disinfected by chloropicrin, steam treatment, or solarization.
3. Agricultural equipment and implements are disinfected with sodium hypochlorite.
4. Pruning is performed during fine weather conditions with disinfected knives and fingernails.
5. Crop rotation of at least 4-year intervals is practiced.
6. Foliage infection is prevented by spraying copper or kasugamycin compounds.
7. Tomato seedlings to be planted for seed production may be treated with novobiocin by immersing either only the root system or the entire plant in the solution.

### 13.6.5 SIMILAR DISEASES

Wilting symptoms of tomato plants may be induced by the following diseases: bacterial wilt (*P. solanacearum*), soft rot (*E. carotovora* subsp. *carotovora*), Fusarium wilt (*F. oxysporum* f. sp. *lycopersici*), Verticillium wilt (*V. dahliae*), and heavy infection of root knot nematodes. Bacterial canker can be differentiated from other bacterial diseases by the absence of macroscopic exudation from vascular tissue or soft rot appearance on affected foliage and stems and from fungal diseases by conspicuous bacterial streaming from discolored tissue under a microscope.

## 13.7 Angular Leaf Spot of Cucumber

Angular leaf spot of cucumber was first discovered in the United States in 1913, and the pathogenic organism was identified in 1915 by Smith and Bryan. The first occurrence in Japan was recorded in 1957 in Kochi prefecture in Shikoku island and has subsequently spread across the country as greenhouse cultivation of cucumber becomes popular.

## 13.7.1 SYMPTOMS

Symptoms appear on cotyledons, leaves, leaf petioles, stems, and fruits. Lesions formed on cotyledons are soft, transparent, round- to irregular-shaped spots. On leaves the disease appears as light yellow, water-soaked angular spots. Under favorable humid environments, conspicuous droplets of bacterial ooze appear, which in dry weather become a white encrustation. As the disease advances, lesions turn grayish and fall out, leaving the plant with a ragged appearance (Fig. 13.9).

Light yellow to light brown lesions also form on petioles and stems with abundant exudation of bacterial ooze. Affected leaf petioles sometimes become softened and droop. When the stem is severely infected, a whole plant is blighted and killed. On fruits, small, round and water-soaked spots are formed with abundant secretion of milky exudate which later turns brown and encrusts. Such fruits become extensively soft rotted later. Mild infection on young fruits may result in malformed or twisted fruits.

**Fig. 13.9** Bacterial leaf spot of cucumber caused by *Pseudomonas syringae* pv. *lachrymans.*

## 13.7.2 CAUSAL ORGANISM: *PSEUDOMONAS SYRINGAE* pv. *LACHRYMANS* (Smith and Bryan 1915) Young, Dye, and Wilkie 1978

In addition to the general characteristics of genus *Pseudomonas,* (Chapter 3, Section 3.6.1), *P. syringae* has the following common properties: cells are motile by one to several polar flagella; pyoverdins are produced in most pathovars but not all; poly-β-hydroxybutyrate granules are not accumulated; nitrates are reduced; protocatechuate is cleaved; levan is formed on nutrient agar containing 5% sucrose; reactions are negative in arginine dihydrolase, oxidase, and potato rot; gamma-aminobutyrate, L-aspartate, caprate, fumarate, glucose, L-glutamate, L-glutamine, glycerol, mucate, pyruvate, and succinate are used as a sole source of carbon; tobacco hypersensitivity reaction is positive; this organism is a member of rRNA I group.

*P. syringae* pv. *lachrymans* is differentiated from other pathovars of the species by the following characteristics: hydrolysis of arbutin and aesculin and gelatin liquefaction are positive; the bacterium uses erythritol, inositol, mannitol, D-sorbitol, quinate, L-tartrate, and trigonelline as a sole source of carbon, but not anthranilate, betaine, DL-homoserine, DL-lactate, and D-tartrate.

Besides cucumber (*Cucumis sativus*), this bacterium attacks various species of cucurbitaceae such as *Benincasa hispida, C. melo, Citrullus lanatus, Cucurbita* spp., *Lagenaria siceraria, Luffa cylindrica,* and *Momordica charantia.*

Host specificity of bacteriophages is relatively narrow. A study in Japan indicates that around 70% of strains collected across the country could be phage-typed by three phage strains.

## 13.7.3 DISEASE CYCLE

Severe infection occurs at a temperature range between 20°C and 25°C and in humid conditions. The primary inoculum source is the infected and/or infested seeds. In internally affected seeds, the bacterium can survive more than 2 years in seed coat. The infection ratio of seedlings derived from such seeds sometimes ranges from 19 to 25%. The pathogen can also survive for long periods in plant residues buried in soil and on the surface of agricultural equipment and implements such as supporting poles, wires, and vinyl-film. All of these infested materials become the primary inoculum source.

Secondary cycles occur on leaves, stems, and fruits by rain splash, drops of moisture, irrigation water, or farming practices. High humidity is the most important factor for inducing an epidemic of the disease. Severe damage occurs in relative humidity above 90%. Lesions formed in relative humidity below 85% remain as minute dots, without enlarging to typical large angular spots.

### 13.7.4 CONTROL

1. Use of healthy seeds or disinfection of a bulked seed by sodium hypochlorite or calcium hypochlorite is the primary step of control.
2. Diseased plant residues are burned or taken away from cucumber fields and buried in soil.
3. Nursery or field soils are disinfected by chloropicrin, steam heat, or solarization.
4. Greenhouses are sufficiently ventilated to keep the leaf surface dry.
5. Good results may be obtained by spraying kasugamycin and copper compounds in the early stage of disease development.

### 13.7.5 SIMILAR DISEASES

Marginal leaf blight caused by *P. viridiflava* and *P. fluorescens (P. marginalis* pv. *marginalis*) may often occur in mixture with angular leaf spot under greenhouse conditions. The disease is readily distinguishable from angular leaf spot by the V-shaped large lesions formed along leaf edges around hydathodes and the absence of exudation of bacterial ooze.

## 13.8 Bacterial Ring Rot of Potato

Bacterial ring rot of potato was first discovered in Germany in 1913 and subsequently spread to Europe and the United States by the 1940s. In Japan the disease was first detected in 1947 in Hokkaido, but its occurrence has successfully been suppressed by domestic quarantine systems to produce and distribute disease-free tubers.

## 13.8.1 SYMPTOMS

The disease usually appears in the field at a later stage of growth after flowering. Severely infected tubers do not sprout, or sprouted shoots are usually killed before appearing above the ground. Even when infected plants could emerge from soil, they show stunting with a small number of leaves and finally die. When the disease develops in the later stage of growth, only one or a few stems but not all develop symptoms. Yellowing and rolling develop from lower leaves, which finally wilt and die (Fig. 13.10). The process of wilting is slow, particularly in humid conditions.

Brown stripes extending to leaf petioles may appear on infected

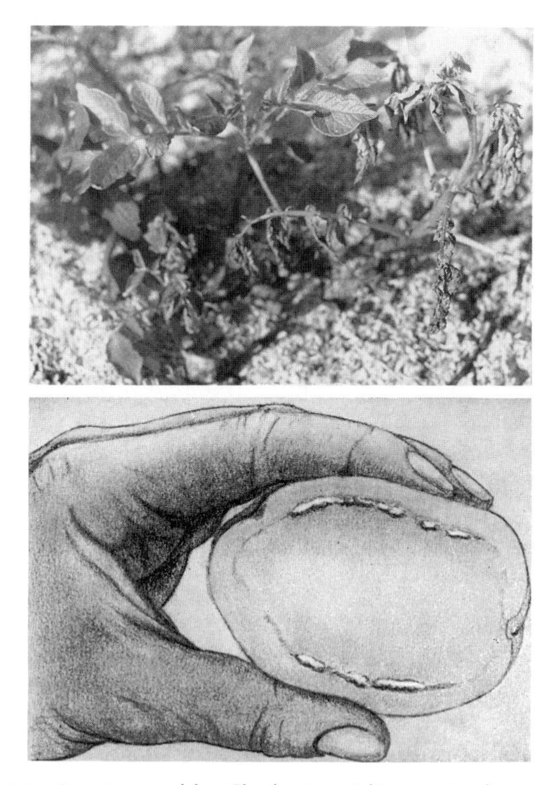

**Fig. 13.10** Potato ringrot caused by *Clavibacter michiganensis* subsp. *sepedonicus.* (top) Aboveground symptoms; (bottom) bacterial exudation from diseased ring zone by squeeze. (Courtesy of Verlag Paul Parey.)

stems. On a transverse section of the diseased stem at several centimeters above ground level, milky bacterial ooze may be secreted from vascular tissue when the stem is squeezed. Tubers are infected through stolons. Diseased tubers externally appear the same as healthy ones, but section of tubers reveals a pale yellow ring a few millimeters wide along the vascular tissue. When the tuber is squeezed, a creamy pulp of tissue comes out of the vascular area (Fig. 13.10). In tubers the disease continuously advances in storage, sometimes forming cracks on the surface and finally resulting in extensive rot.

### 13.8.2 CAUSAL ORGANISM: *CLAVIBACTER MICHIGANENSIS* subsp. *SEPEDONICUS* (Spieckermann and Kotthoff 1914) Davis, Gillaspie, Vidaver, and Harris 1984

In addition to the properties of *Clavibacter* and *C. michiganensis* (Chapter 3, Section 3.6.4 and Chapter 13, Section 13.6.2), the bacterium is characterized by the following features: growth on agar plate is slow, taking several days for forming colonies 1 mm in diameter; better growth is available between 20°C and 25°C; starch is hydrolyzed but not gelatin; maximal NaCl concentration for growth is 3–4%; maximal growth temperature is 34°C; acetate and succinate but not propionate are used as a sole source of carbon; acid is produced from mannitol but not from lactose, trehalose, inositol, and salicin.

Natural infection occurs only on potato. By artificial inoculation, 25 species, mainly of *Solanum,* are infected. Eggplant is also infected by inoculation, and this technique may be used for diagnosis.

### 13.8.3 DISEASE CYCLE

The most important source of primary inoculum is infected seed tubers. It is said that a knife that cut a diseased tuber can serially infect 20 to 24 tubers. The bacterium can survive until the next crop season in the diseased plant residues that remain in or on the ground and cause the primary infection. Besides tubers, the bacterium may also survive for 6 to 7 months on infested agricultural equipment and implements and cause primary infection.

Secondary infection occurs by the pathogen dispersed from the primary foci through wounds either of foliage or root systems produced by insects, plant growth, or farming practices. However, the significance of secondary cycles in an epidemic of potato ringrot is not substantial compared to the primary cycles derived from infected tubers. The most favorable soil temperature for infection ranges between 19° and 28°C. At temperatures below 16°C and above 31°C, multiplication and spread of the bacterium in plant tissues are reduced, and development of symptoms is significantly delayed.

### 13.8.4 CONTROL

1. Use of disease-free seed tubers is essential for successful cultivation of potato.
2. The knife used to cut tubers should be disinfected by sodium hypochlorite solution after cutting each tuber.
3. Crop rotation with at least 3-year intervals is practiced in fields where the disease occurred.
4. Agricultural equipment and implements are disinfected by sodium hypochlorite solution.
5. Field sanitation is performed by removal of the diseased plants or plant residues.

### 13.8.5 SIMILAR DISEASES

Wilting symptom is also produced by bacterial wilt caused by *P. solanacearum*. Bacterial ringrot is readily distinguishable because bacterial wilt is much faster in the process of disease development than ringrot and affects all stems sprouting from a tuber. A convenient diagnosis of both diseases can be made by examining gram-reaction of bacteria which exude from vascular tissue of a cut surface.

## 13.9 POTATO SCAB

Potato common scab was discovered in 1825 in the United States, but the cause was not determined until 1891, when Thaxter published *Oospora scabies*. This organism was transferred by Gussow (1914) to *Actinomycetes* and then by Waksman and Henrici (1943) to *Streptomyces* as *S. scabies*. The disease is one of the few plant diseases caused by actinomycetes but its distribution is worldwide. Another type of potato scab was found in Maine in 1953 in a soil as acid as pH 4.3. This disease is different from common scab in the susceptibility to acid, i.e., common scab rarely occurs in soil with pH lower than 5.2. These two scab diseases are caused by different pathogens.

## 13.9.1 SYMPTOMS

In general, scabs are characterized by brown to dark brown, irregularly concentric, wrinkled cork layers around some depression in the center. From detailed characteristics, however, the symptoms are divided into the three distinct types ordinary scab, pitted scab, and russet scab, depending on the species or strain of the pathogen, potato cultivar, and soil environment like humidity and pH (Fig. 13.11).

The typical corky symptoms of common scab are formed by repetition of the development of a defense cork layer against infection, hyphal penetration of the defense layer, and development of another new cork layer. Pitting or depression of the scabs results from necrotic collapse of host cells by phytotoxins produced by the causal organism. The toxins are assumed to be the metabolites of phenylalanine such as phenylacetate and oxyphenylacetate.

**Fig. 13.11** Common scab of potato caused by *Streptomyces scabies.* [Courtesy of Dr. N. Tashiro.]

## 13.9.2 CAUSAL ORGANISM: *STREPTOMYCES SCABIES* (Thaxter 1891) Lambert and Loria 1989; *STREPTOMYCES ACIDISCABIES* Lambert and Loria 1989

Besides these two pathogens, several other species of *Streptomyces* have also been listed as the cause of potato scab. They are *S. griseus, S. olivaceus, S. aureofaciens,* and *S. flaveolus,* although the virulence of these organisms is significantly lower than that of *S. scabies.*

*S. scabies,* the cause of common scab, had been treated as *species incertae sedis* or uncertain species since 1974 because there was no extant of type strain. It was renamed in 1989 with the designation of neotype strain. *S. acidiscabies* was also named in 1989 to the pathogen of potato scab occurring in acid soil.

In addition to the general properties of genus *Streptomyces* (Chapter 3, Section 3.6.6), both actinomycetes have the characteristics described below.

### *Streptomyces scabies*

Spores are formed as a spiral chain consisting of 20 or more spores, $0.5 \times 0.9–1.0$ μm in size, smooth and gray; brown melanoid pigment is formed on tyrosine nutrient agar and peptone iron agar plates; the lowest pH for growth is 5.0; the organism uses L-arabinose, D-fructose, D-glucose, D-mannitol, rhamnose, sucrose, D-xylose, and raffinose but not xanthine as a sole source of carbon; the organism is susceptible to 20 μg/ml streptomycin and 0.5 μg/ml crystal violet; mol% G + C of the DNA is 70. The formation of gray and spiral spore-chains is a distinctive character of the pathogen.

### *Streptomyces acidiscabies*

Spores are formed as flexuous chains consisting of 20 or more spores, $0.4 \times 0.5–0.6$ or $0.9–1.1$ μm in size, smooth and white (reddish on certain high-pH media); soluble pigment is red above pH 8.3 but golden yellow below pH 8.3; this organism uses L-arabinose, D-fructose, D-glucose, D-mannose, rhamnose, sucrose, D-xylose but not raffinose as a sole source of carbon; lowest pH for growth is 4.0; the organism is tolerant to 20 μg/ml streptomycin and 0.5 μg/ml crystal violet; mol% G + C of the DNA is 71. The pathogen is distinguished from other pathogenic *Streptomyces* in the production of red pigment, white to reddish spore mass, and resistance to acid.

In the literature, the host range of *S. scabies* includes the following plants besides potato: beet, cabbage, carrot, eggplant, mangel, onion, parsnip, radish, rutabaga, salsify, spinach, and turnip. Because of the complexity of classification of the genus *Streptomyces,* however, the interrelationships of the strains isolated from these plants should be reexamined from the viewpoint of molecular taxonomy and/or chemotaxonomy.

### 13.9.3 DISEASE CYCLE

Field trials in Europe have shown that planting of scabbed seed tubers does not necessarily result in the infection of their progeny tubers. On the basis of these data, the idea has been proposed that common scab is soil-borne rather than seed-borne, and scabbed tubers have a minor importance as the inoculum source. However, recent studies in Japan have pointed out the importance of scabbed seed tubers based on the following data: high correlations are observed between the rate of scabbed areas on seed tubers and disease incidence on new progeny; and the disease severity can be reduced by disinfection of seed tubers (Table 9.4).

The suppressive effect of soil microorganisms on common scab has been well documented; hence, chemical disinfection of soil often induces significant increase of the disease. The possibility of latent infection of seed tubers has been suggested from the data that the disease occurred in severe form even when apparently scab-free seed tubers were planted in fields where soil was chemically disinfected (Table 9.4).

Secondary cycles of the disease have no significance from a practical point of view because infection takes place only for several days at the initial stage of tuber formation.

### 13.9.4 CONTROL

1. Seed tubers are disinfected by thiophanate methyl-streptomycin compounds, copper compounds, or PCNB.
2. Resistant potato cultivars are planted.
3. Field soil is disinfected by chloropicrin, PCNB, or solarization.
4. Fields are irrigated for 4–6 weeks at the time of tuber initiation.
5. Long rotation with cereal crops or grasses is practiced.
6. Soil is acidified against common scab by sulfur to pH 5.2. It is said that slow acidification by the application of acid-producing fertilizers like ammonium sulfate is more desirable than rapid acidification by sulfur.
7. Compost is applied after being completely decomposed.

# 13.10 Halo Blight of Kidney Beans

Halo blight of kidney beans was first discovered in the United States in 1926, and shortly afterward its distribution became worldwide. In Japan, incidence of the disease was first recognized in Shizuoka prefecture in 1958, and it occurred in severe form in Hokkaido in the early 1970s.

## 13.10.1 SYMPTOMS

The disease can be seen throughout the growing stages of beans. Infection occurs on leaves, stems, pods, and seeds. Internally and/or externally contaminated seeds germinate, with development of lesions on cotyledons and primary leaves. Round to irregular and water-soaked lesions form on cotyledons, and yellowish brown to reddish brown and water-soaked lesions with a yellow halo form on primary leaves. Plants severely infected in this stage are stunted and die.

In trifoliate leaves, dark green to reddish brown and irregularly shaped lesions are formed on leaflets. These lesions accompany a wide yellow halo in most cases. The chlorotic halo is particularly distinct when infection took place on young expanding leaves. When young plants are severely infected, leaflets often show a mosaic patterns of dark and pale green areas and become malformed. The symptom is readily distinguishable from that of virus disease by the presence of small, water-soaked lesions along leaf veins (Fig. 13.12).

**Fig. 13.12** Halo blight of kidney bean caused by *Pseudomonas syringae* pv. *phaseolicola*.

On stems, lesions appear as reddish brown elongated streaks. Symptoms on pods are dark green to reddish brown, round, water-soaked or greasy spots with a somewhat depressed center (Fig. 13.12). Infected seeds are usually symptomless or a slight fading may be observed on the seed coat.

## 13.10.2 CAUSAL ORGANISM: *Pseudomonas syringae* pv. *phaseolicola* (Burkholder 1926) Young, Dye, and Wilkie 1978

Besides the general features of genus *Pseudomonas* and *P. syringae*, this bacterium has the following characteristics: hydrolysis of arbutin is negative; the ability to use mannitol, raffinose, and rhamnose as a sole source of carbon and sodium nitrate as a sole source of nitrogen is different depending on the strain; it belongs to rRNA I group.

Besides kidney bean, natural infection occurs on soybean, Adzuki bean, hyacinth bean, pea, and kudzu. The presence of two different types is known in terms of pathogenicity to *Vigna radiata* (or *Phaseolus aureus*) and the ability to grow on mannitol and sodium nitrate as a sole source of carbon and nitrogen. The type that infects only *V. radiata* can utilize these substances, whereas the ordinary type that infects many legumes cannot. The strain that naturally attacks kudzu (*Pueraria lobata*) is distinct in that it has the most potent capacity to produce ethylene among microorganisms ever known. The interrelationship has been suggested among three bacteria, i.e., this bacterium, *P. syringae* pv. *glycinea*, and *P. syringae* pv. *mori*. In the pathogen of halo blight of kidney bean, two races have been differentiated by pathogenic responses to Red Mexican UI-25.

## 13.10.3 DISEASE CYCLE

The disease cycle of halo blight of kidney bean is shown in Fig. 13.13. The disease is a typical seed-borne disease, and the primary inoculum source is infected and/or infested seeds. The bacterium infects bean seeds through lesions formed on pods and invades tissues of the seed coat without producing any visible symptoms. External infestation of seeds occurs during threshing by adherent of dusts of the diseased tissues either on the surface or in wounds. There is no difference between externally contaminated seeds and infected ones with respect to the importance as the inoculum source.

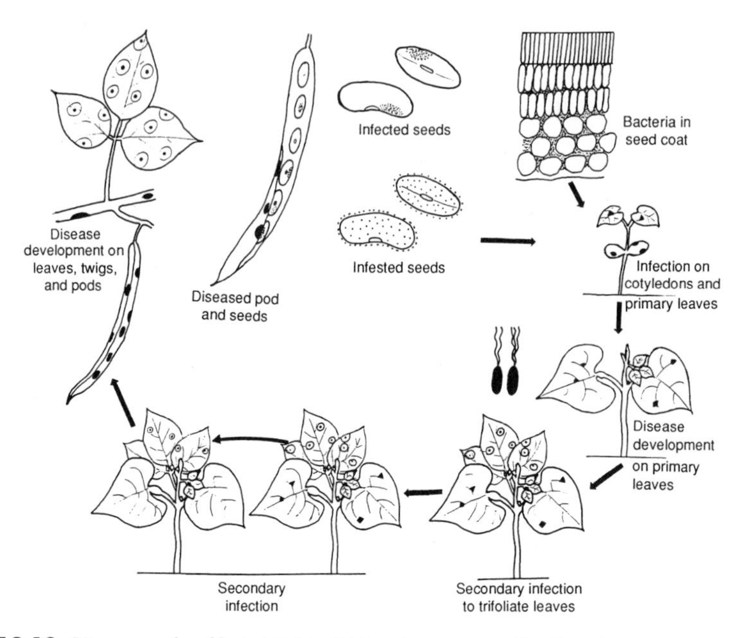


**Fig. 13.13** Disease cycle of halo blight of kidney bean caused by *Pseudomonas syringae* pv. *phaseolicola.*

The secondary cycle takes place by dispersal of the bacterium from the diseased plants that are derived from the primary infection to neighboring plants by rain splash.

## 13.10.4 CONTROL

1. Planting of disease-free seeds is the primary method of control. When cleanness is in doubt, bulked seeds may be dipped into sodium hypochlroite solution for disinfection. The treatment is, however, totally ineffective for internally infected seeds.
2. Field hygiene is maintained by removing diseased plants.
3. Plants are protected from infection by foliar spray of kasugamycin or copper compounds.

## 13.10.5 SIMILAR DISEASES

Bacterial diseases of kidney beans include bacterial blight (*X. campestris* pv. *phaseoli*) and bacterial brown spot (*P. syringae* pv. *syringae*). Halo blight is differentiated from these diseases by development of water-soaked, relatively small lesions with a characteristic halo that usually occurs at a relatively early stage of growth.

## 13.11 Fire Blight

The occurrence of fire blight was first observed in North America in the late 18th century, but the cause was suspected to be insect damage or freezing damage. Thomas J. Burrill revealed in 1880, on the basis of his study of more than 10 years, that the disease is caused by a bacterium, and in 1882 he named it *Micrococcus amylovorus*. Incidence of the disease was recorded in England in 1958 and in continental Europe in 1966. It is said that the disease also distributes in Central America and South America, but its damage has not been serious. The occurrence of fire blight has not been confirmed in Asia.

### 13.11.1 SYMPTOMS

The disease may be called by different names depending on the affected plant parts.

*Blossom blight.* Blossoms just after flowering are affected, developing a water-soaked appearance, and finally die turning dark brown to black.

*Twig and leaf blight.* The disease rapidly progresses on succulent twigs or new shoots, moving 15–30 cm in a few days. Blighted twigs turn brown to dark brown in apple and dark brown to black in pear. Leaves are infected either through wounds or through natural openings, like stomata and hydathodes, and exhibit necrotic lesions which spread to midrib and petiole, turning them black. The lesion further spreads over entire leaf, resulting in its death (Fig. 13.14).

*Fruit blight.* Usually unripened fruits are affected. Infection occurs directly either through wounds or through lenticels of the fruits, or through infected spur. Diseased fruits may appear water-soaked red in apple but water-soaked dark green in pear. Milky to amber-colored bacterial exudate may be observed on the surface of the infected fruits (Fig. 13.14).

*Limb and trunk blight.* On susceptible cultivars, the disease advances from shoots downward to twigs, branches, scaffold limbs, and trunk, developing cankers. Cankers are slightly sunken and surrounded by irregular cracks in bark from where bacterial ooze appears.

**Fig. 13.14** Fire blight of apple caused by *Erwinia amylovora.* (Courtesy of Dr. R. N. Goodman.)

*Collar and root blight.* These symptoms are caused by infection at the base of the trunk or roots by the pathogen running down the surface of trunk or by spreading of the trunk cankers. Development of canker at collar or ground level causes immediate death of entire plant, and infection of the root system induces slowdown of the aboveground growth.

### 13.11.2 CAUSAL ORGANISM: *ERWINIA AMYLOVORA* (Burrill 1882) Winslow, Broadhurst, Buchanan, Krumwiede, Rogers, and Smith 1920

In addition to the general properties of genus *Erwinia* (Chapter 3, Section 3.6.3), this bacterium has the following characteristics: nicotinic acid is required as a growth factor; levan is produced from sucrose; gelatin is liquefied; growth does not occur at 36°C; production of $H_2S$ and nitrate reduction is negative; trehalose, citrate, formate, and lactate but not mannose, melibiose, mannitol, glycerol, aesculin, salicin, tartrate, galacturonate, and malonate are used as a sole source of carbon. The mol% G + C of the DNA is 53.6–54.1. This bacterium is serologically homologous; hence, serological technique is most efficient for diagnosis of the pathogen.

*E. amylovora* infects a number of plants of the family Rosaceae. Host plants include 180 or more species, including apple, pear, quince, pyracantha, hawthorn, and cotoneaster. The presence of a strain that attacks only raspberry has been reported.

### 13.11.3 DISEASE CYCLE

The disease cycle of fire blight is shown in Fig. 13.15. The source of inoculum for primary infection may be bacteria residing epiphytically on or in gemmisphere or those exuding from holdover cankers. The pathogen in the epiphytic phase is considered to be an important cause of blossom blight in orchards where oozing cankers are scarcely observed or in new orchards where no disease has been recorded previously. Primary infection may also take place by grafting of infected scions or pruning by contaminated scissors.

Secondary cycles of fire blight may result from dispersal of the pathogen from infected blossoms and cankers by rain splash, wind, insects, birds, or humans. Therefore, secondary infection is far more important than the primary infection because of its serious damage. The major routes of secondary infection vary, depending on countries or regions because meteorological factors, the time and amount of rainfall in particular, are different in each geographic area.

### 13.11.4 CONTROL

1. Introduction of the pathogen into a disease-free area and subsequent establishment are prevented by combined control measures of quarantine procedures, distribution of disease-free nursery plants, and field sanitation.
2. Pruning is performed with knives disinfected with sodium hypochlorite solution, and pruned twigs are burned.
3. The bark of cankers on limbs and trunks that form bacterial ooze may be removed by surgery followed by disinfection with chemical compounds containing zinc chloride.
4. Copper compounds or streptomycin compounds are sprayed according to the data of disease forecasting systems.

### 13.11.5 SIMILAR DISEASES

In Hokkaido, a bacterial disease named bacterial shoot blight occurs on pear twig of a cultivar "Mishirazu," causing blight of blossoms, young

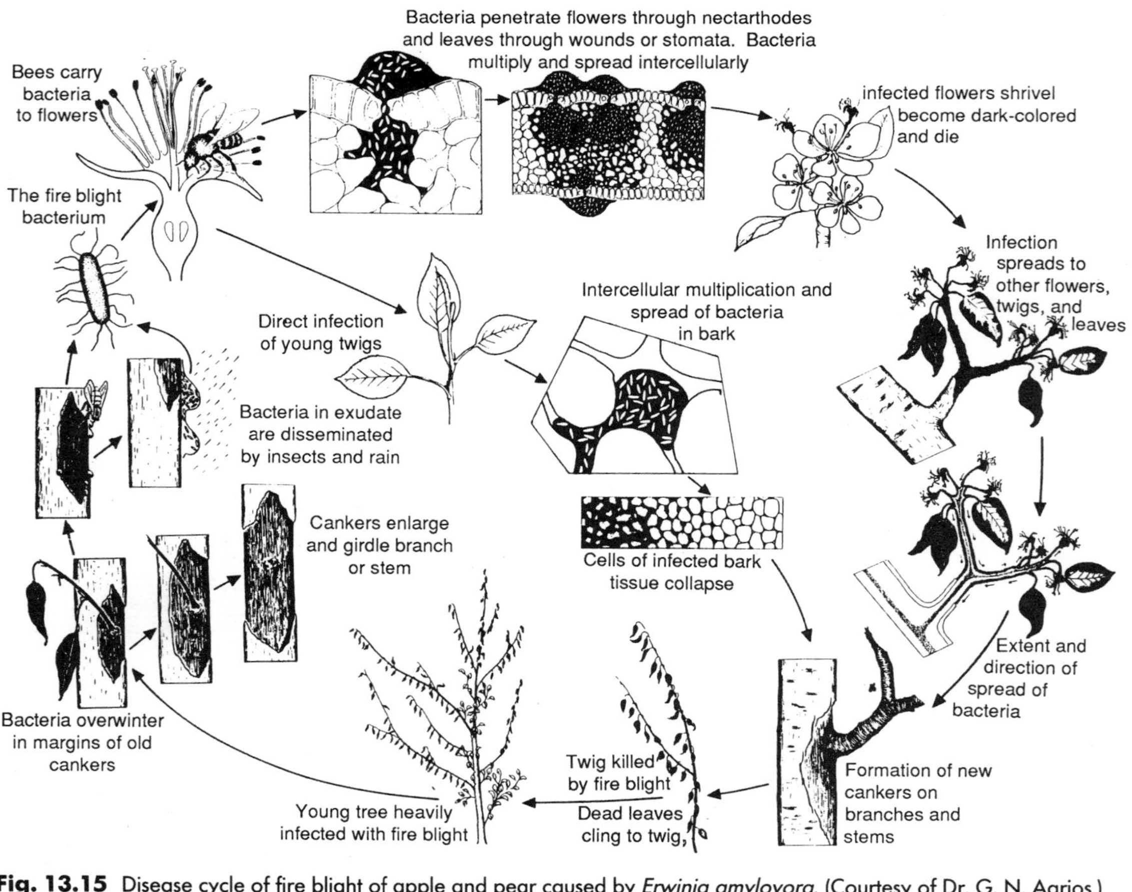


**Fig. 13.15** Disease cycle of fire blight of apple and pear caused by *Erwinia amylovora.* (Courtesy of Dr. G. N. Agrios.)

fruits, leaves, and shoots. The symptoms of the disease are identical to those of fire blight. The pathogenic bacterium is also identical with *E. amylovora* except for some bacteriological as well as serological properties. This bacterium infects certain cultivars of Japanese pear and Chinese pear but not cultivars of apple. Therefore, it is considered to be a distinct pathovar of *E. amylovora*.

## 13.12 Crown Gall

Crown gall has been known in the United States and Europe since the 1880s. Smith and Townsend proved it to be a bacterial disease and named the causal bacterium *Bacterium tumefaciens*. Soon after that, the disease was recognized to be worldwide in its distribution. It is generally understood that the disease was introduced into Japan in 1890 by young cherry trees imported from the United States.

### 13.12.1 SYMPTOMS

Unlike other hyperplastic diseases of plants, galls of this disease are referred to as *tumors* because of the specific mechanisms of disease development or transformation of host cells by pathogenic genes of the causal organisms.

In general, tumors of varying sizes appear at the soil level, or the crown. In some plants such as grape, raspberry, and blackberry, tumors are formed on trunks or twigs 1 m or more above the ground. The texture of tumors depends on the plant or plant parts; tumors formed on annual herbaceous dicots are usually soft and readily rotted when host plants age, but those on perennial woody plants are woody and continue to enlarge unless insects bore into galls destroying its tissues. Tumors are round, smooth, and pale green to light brown at the early stage, but become semiround to irregular, rough, and dark brown later (Fig. 13.16).

Tumor formation may sometimes be fatal on small young trees but usually does not inflict serious damage on large aged trees. Tumor tissues readily girdle small and thin stems and substantially block the vascular tissue. In large trees, however, even a large tumor rarely girdles a thick trunk, and the vascular system remains intact in part so that water can be conducted.

**Fig. 13.16** Crown gall of rose caused by *Agrobacterium tumefaciens* (top) and hairy root of rose caused by *A. rhizogenes* (bottom). (Courtesy of Dr. K. Ohta.)

## 13.12.2 CAUSAL ORGANISM: *AGROBACTERIUM TUMEFACIENS* (Smith and Townsend 1907) Conn 1942

The general characteristics of genus *Agrobacterium* are given in Chapter 3, Section 3.6.2. *A. tumefaciens* has been classified into two biovars by the bacteriological properties shown in Table 13.3. The former biovar 3 was recently classified as an independent species, *A. vitis*.

The bacterium attacks a great number of plants, including 643 species of 93 families. The host range of individual strains, however, varies from wide to narrow, depending on Ti plasmids. *A. vitis* is different from *A. tumefaciens* in its host range restricted to grape and other phenotypic characters (Table 13.3).

Pathogenicity of agrobacteria is determined by plasmids: the strains carrying Ti plasmid induce crown gall, those carrying Ri plasmid induce hairy root, and those cured of these plasmids become avirulent or saprophytes. Development of tumors orginates in the autonomous prolifera-

**Table 13.3**
Comparison of Bacteriological Properties of Biovars of *Agrobacterium tumefaciens* and *A. vitis*

| Characteristics | Biovar 1 | Biovar 2 | *A. vitis* |
|---|---|---|---|
| 3-ketolactose | + | − | − |
| Litmus milk | alkali | acid | alkali |
| Growth on selective medium of New and Kerr[a] | − | + | − |
| Growth on selective medium of Schroth *et al.*[b] | + | − | − |
| Growth at 35 C | + | − | +(d)[c] |
| Activities of: Oxidase | + | −(d) | + |
| Arginine dihydrolase | + | − | + |
| Phosphatase | +(d) | −(d) | − |
| Tolerance to 2% NaCl | + | − | + |
| Production of $H_2S$ | + | − | − |
| Use of: ethanol | + | − | −(+) |
| melezitose | + | − | − |
| erythritol | − | + | − |
| tartrate | − | + | + |
| malonate | − | + | + |
| tyrosine | − | + | − |

[a] P. B. New and A. Kerr, 1971, *J. Appl. Bact.* **34,** 233.
[b] M. N. Schroth, J. P. Thompson, and D. C. Hildebrand, 1965, *Phytopathology* **55,** 645.
[c] Values in parentheses from personal communication of Dr. H. Sawada.

tion of tumor cells that are transformed by T-DNA located on Ti plasmid (see Chapter 8). After transformation takes place, therefore, the tumor grows by itself regardless of the presence of the pathogen. This is the essential difference in crown gall from other bacterial gall diseases such as olive knot disease. In this respect, crown gall and hairy root (*Agrobacterium rhizogenes*) are identical in the mechanisms of disease development, although symptoms are distinct in that the latter forms a number of hairlike thin roots.

### 13.12.3 DISEASE CYCLE

Crown gall is a typical soil-borne disease. The pathogen can survive in soil for long periods in the absence of host plants. The bacterium is a wound parasite, and the presence of new injures at infection sites is essential for the transformation of the host cells. Wounds may be created by nursery practices like transporting, grafting, or pruning or by insect damage (Fig. 8.4).

The incubation period is usually 5 days to several weeks, depending on the plant, plant part, growth stage, and nutrition. Tumors rapidly enlarge on plant or plant parts that are actively growing but remain inactive on plants in dormancy. Therefore, infection late in autumn usually results in latent infection.

### 13.12.4 CONTROL

1. The majority of infections occur at nurseries. Therefore, it is important to produce healthy young plants by careful nursery management, including grafting and transplanting.
2. Biological control measures may be applied before transplanting. Root systems of young plants are dipped into a suspension of *A. radiobacter* strain K84 at a concentration higher than $10^8$ cells/ml. Because the ratio of strain K84 and the causal organism should be $>1:1$ for successful control, the concentration of the antagonist may be altered, depending on the degree of contamination in the field soil or expected density of the pathogen.
3. Nursery and field soil may be disinfected by chloropicrin. Because the pathogen has the ability to grow in soil, however, the effect of soil disinfection may not continue for long time where fields are recontaminated with conducive soil through wind or cultural practices.

4. For chemotherapy of diseased plants, Bacticin (see Chapter 12) may be smeared on small galls or after removal of large tumors.

### 13.12.5 SIMILAR DISEASES

Besides crown gall, many bacterial plant diseases produce galls on twigs, limbs, and trunks. In galls of these hyperplastic diseases, bacterial pockets are found in light green to light yellow soft parenchyma tissues surrounded by white woody tissues (Fig. 10.9). Thus, crown gall is readily distinguished by examining the section of tumor.

## 13.13 Bacterial Canker of Stone Fruit

Bacterial canker of stone fruit is caused either by *P. syringae* pv. *syringae* or *P. syringae* pv. *morsprunorum*. Disease caused by the former bacterium is equivalent to bacterial gummosis of cherry (*P. ceraci*) discovered by Griffin (1911) in the United States and the latter to bacterial canker of stone fruit trees (*P. prunicola*), discovered by Wormald (1930) in England. In Japan, the pathogen of bacterial disease of cherry and apricot discovered in 1930 in Kyoto was suggested to be *P. ceraci*. Another bacterial disease was reported on Mume or Japanese apricot (*Prunus mume*) in 1937 without identification of the causal bacterium. In 1969, bacterial canker caused by *P. syringae* pv. *morsprunorum* was reported on Mume.

### 13.13.1 SYMPTOMS

Symptoms and disease severity vary depending on the plant species, cultivar plant part, and weather conditions, particularly rain.

*Twigs.* In the early spring, slightly sunken and dark brown lesions appear on twigs beneath infected spurs. In dry weather, the surface of lesions becomes rough and cracked. Severe infection on twigs results in shoot blight and death of the infected branches. When trees break dormancy in spring, gums often appear from cankered regions on the limbs (Fig. 13.17).

*Buds.* Bud blast usually occurs in severe form on cherry and apricot in dormancy. Both flower and leaf buds are attacked. Infected buds fail to

**Fig. 13.17** Bacterial canker of cherry caused by *Pseudomonas syringae* pv. *morsprunorum*. (Courtesy of Horticultural Research International, East Malling, United Kingdom.)

develop new shoots and die. The damage to buds may often be noted in the blooming stage by a small number of scattered flowers (Fig. 13.17).

*Flowers.* Infected blossoms appear water soaked and wilt, turn dark brown, and droop from twigs. The disease spreads from blossoms to spurs and further to twigs, causing canker symptoms.

*Leaves.* Angular or round, water-soaked and dark green spots 1–3 mm in diameter develop on leaves. These lesions later turn reddish brown at the edge and brown to dark brown in the center and finally fall out, producing shot-hole symptom.

*Fruits.* Spots on fruits are round, dark brown to black, slightly depressed, and 2–3 mm in diameter. In severely infected fruits, spots often coalesce to form irregularly shaped, large lesions with cracks on the surface.

### 13.13.2 CAUSAL ORGANISM: *PSEUDOMONAS SYRINGAE* pv. *SYRINGAE* Van Hall 1902; *PSEUDOMONAS SYRINGAE* pv. *MORSPRUNORUM* (Wormald 1931) Young, Dye, and Wilkie 1978

These two bacteria have the general characteristics of genus *Pseudomonas* and *P. syringae* and are characterized by the properties listed in Table 13.4.

### 13.13.3 DISEASE CYCLE

The bacterium survives as an epiphyte on the plant surface throughout the growing season and enters through cracks at the base of leaf petioles, leaf scars, and wounds and produces cankers on twigs and spurs. Latent infection often takes place in late autumn and disease develops in the next spring. Primary infection mainly occurs by bacteria which exude from canker lesions or reside in healthy buds as an epiphyte. Secondary cycles occur by dissemination of the pathogens in rain from lesions formed on twigs and leaves. Therefore, primary and secondary cycles may occur at the same time on new shoots developing early in the season. Blossom blight usually originates in the infected buds but may also result from flower infection through nectaries.

**Table 13.4**
Comparison of Bacteriological Properties of *Pseudomonas syringae* pv. *syringae* and *P. syringae* pv. *morsprunorum*

| Characteristics | pv. *syringae* | pv. *morsprunorum* |
|---|---|---|
| Longevity on nutrient-sucrose agar | 8–14 days | 2–6 days |
| Growth in sucrose broth | Yellow | White |
| Activities of: protease | + | − |
| β-glucosidase | + | − |
| tyrosinase | − | + |
| Utilization of: L-leucine | + | − |
| L-tartrate | + | − |
| antranilate | + | − |
| Pathogenicity: pear | Depressed, black | No symptom |
| cherry and lemon | Depressed, black | Brown, shallow, small |

### 13.13.4 CONTROL

1. Disease-free young plants are planted.
2. Infected twigs are pruned and pruned twigs are burned. Scissors are disinfected with sodium hypochlorite or invert soap.
3. When the disease starts to appear, streptomycin compounds, copper compounds, or dithianon can periodically be sprayed at about 10-day intervals.

### 13.13.5 SIMILAR DISEASES

A bacterial disease of peach caused by *P. syringae* pv. *persicae* has been reported in France. It produces symptoms that are similar to those of bacterial canker. The relationship between these two diseases is a matter for elucidation in the future. Another bacterium that is reported to infect peach fruits by artificial inoculation is *P. syringae* pv. *papulans,* the causal bacterium of blister spot of apple fruits. There is no report of natural infection by this bacterium, however. Bacterial shot hole caused by *X. campestris* pv. *pruni* is usually distinguished from bacterial canker because it does not show bud blast and blossom blight.

## 13.14 Bacterial Canker of Kiwi

Bacterial canker of kiwi was first discovered in California in 1980, and the pathogenic bacteria was identified as *P. syringae* in 1983. The disease distributes in other parts of the world. In Japan, a similar disease was discovered in Shizuoka and Kanagawa prefectures in 1985 and soon found in kiwi fields across the country. The disease is similar in symptoms to bacterial canker reported from California but is more destructive and has not yet been reported in other regions of the world.

### 13.14.1 SYMPTOMS

*Trunks and limbs.* The symptoms become apparent from late winter to early spring when vines break dormancy. Red-rusty droplets appear on buds, joints of branches, forks of boughs, leaf scars, and pruning scars. Affected bark becomes deeply shriveled and dried, and cracks are often formed on the affected branches.

*Leaves.* Infection is likely limited on young leaves expanding in spring. Small water-soaked light green spots scattered on leaf lamina become subsequently brown to dark brown and angular in shape, 2–3 mm in diameter, and delineated by small veins. Bright yellow halos are formed around the spots. Bacterial exudation may be observed on the lower surface of the leaf. When severely infected, the whole leaf may become blighted and shriveled (Fig. 13.18).

*Canes.* The infected part of the cane becomes dark green and water soaked. Longitudinal cracks are often formed. The lesions elongate, and whole shoots become wilted and blighted, producing exudation of bacterial ooze (Fig. 13.18).

*Flowers.* Infected flowers turn brown and wither without opening. Infected flowers that open may have petals not fully developed. Necrotic lesions may be formed on sepals also.

### 13.14.2 CAUSAL ORGANISM: *PSEUDOMONAS SYRINGAE* pv. *ACTINIDIAE* Takikawa, Serizawa, Ichikawa, Tsuyumu, and Goto 1989; *PSEUDOMONAS SYRINGAE* pv. *SYRINGAE* Van Hall 1902

*P. syringae* pv. *actinidiae* has the following characteristics in addition to the general properties of genus *Pseudomonas* and *P. syringae:* fluorescent pigment is not produced on King B medium; activities of β-galactosidase, urease, and gelatin liquefaction are negative; growth in sucrose broth is white; utilization of xylose, erythritol, DL-lactate, caprate, trigonelline, and L-histidine is negative. This bacterium naturally infects only kiwi plant. Peach and Mume can be infected by artificial inoculation.

The California strain (*P. syringae* pv. *syringae*) differs from *P. syringae* pv. *actinidiae* in positive reactions in production of fluorescent pigment, liquefaction of gelatin, and pathogenicity to cowpea, sliced onion, and carrots. The detailed properties of this bacterium are shown in Table 13.4. The relationships of these two bacteria in bacterial canker of kiwi are equivalent to those between *P. syringae* pv. *syringae* and *P. syringae* pv. *morsprunorum* in bacterial canker of stone fruits.

### 13.14.3 DISEASE CYCLES

The disease appears in the two forms winter canker and spring canker. Winter canker involves damage to the main vine structure and

**Fig. 13.18** Bacterial canker of kiwi caused by *Pseudomonas syringae* pv. *actinidiae*. Top, infected leaves; bottom, bacterial ooze exuded from infected cane. (Courtesy of Dr. S. Serizawa.)

overwintered canes and has direct effects on yield by reducing the size of productive vines.

Spring canker has less direct effects on the year's yield but is potentially important as the cause of winter canker the following year. The inoculum source of primary cycles is the bacteria exuding from canker lesions on canes, limbs, and trunks. Incidence of the disease is intensive where the mean temperature in early spring is below 15°C accompanied by frequent rains. Primary infection takes place through stomata and wounds and continues until late spring when secondary infection begins. Mature leaves and canes become resistant in summer and scarcely develop symptoms even through wounds.

### 13.14.4 CONTROL

1. Disease-free young plants are planted.
2. Windbreaks are set around orchards to protect leaves and canes from injuries caused by strong wind.
3. Unnecessary cane elongation in late autumn is avoided.
4. When pruned canes show exudation of bacterial ooze from the cut surface, they are cut back to the healthy portion.
5. Scissors for pruning are disinfected by sodium hypochlorite solution.
6. Spring canker may be controlled by spraying streptomycin, kasugamycin, or copper compounds at 10-day intervals. Streptomycin or kasugamycin solutions may be injected into trunks late in autumn to control winter canker.

### 13.14.5 SIMILAR DISEASES

There is a bacterial blossom blight of kiwi that occurs only on young flowers and is characterized by browning of petals and dropping of premature flowers. The pathogens are considered to be a complex of *P. syringae*, *P. viridiflava*, and *P. syringae* pv. *actinidiae*.

## 13.15 Citrus Canker

From the existence of citrus canker lesions on the oldest citrus herbaria collected in 1827-1831, the origin of the disease is considered to be Asia, probably South China, Indonesia, and India, where *Citrus* species are presumed to have originated. In Japan, the disease was first discovered in Fukuoka prefecture in 1913. Citrus canker was introduced into the United States on infected nursery plants and produced severe damage in the Mexican Gulf states. A citrus canker eradication campaign was conducted for more than 20 years, beginning in 1915 in the southern states of the United States (see Chapter 12). Since then, the disease has been designated as one of the most important diseases of the international quarantine. Citrus canker is widely distributed in Asia and South America. The disease has also been observed in the 1980s in Australia and in the United States where eradication was once declared.

### 13.15.1 SYMPTOMS

Citrus canker occurs on leaves, twigs, and fruits throughout the growing season. Severe infection often occurs after rainstorms in spring and typhoons in late summer. The damage is particularly severe in young trees that flush many angular shoots, resulting in significantly retarded growth or sometimes death of the affected plants.

Canker lesions appear on leaves, twigs, and fruits as light yellow, raised, and spongy eruptions on the leaf surface. As the lesions further enlarge, the spongy eruptions begin to collapse, and brown depressions appear in their central portion, forming a craterlike shape. The edges of lesions remain raised above the surface of host tissue and are characterized by a greasy appearance. As the disease advances, the central portions become grayish-white, hard, and appear as corky, dead tissues with a rough surface, surrounded by yellow halos. Canker lesions vary in size depending on the susceptibility of the host plants. Lesions as large as 5–10 mm in diameter are not uncommon on susceptible cultivars (Fig. 13.19). The difference in resistance among cultivars appears more distinctly on twigs than on leaves. Lesions on moderately resistant cultivars cease to enlarge at 1 to 2 mm in diameter and finally drop off in the form of dry brown scales. When small young fruits are infected, canker spots coalesce to form large lesions with deep cracks and cause premature drop.

### 13.15.2 CAUSAL ORGANISM: *XANTHOMONAS CAMPESTRIS* pv. *CITRI* (Hasse, 1915) Dye 1978

In addition to the general characteristics of genus *Xanthomonas* and *X. campestris* (Chapter 3, Section 3.6.1 and Chapter 13, Section 13.1.2), this bacterium has the following properties: growth on nutrient agar is good;

**Fig. 13.19** Citrus canker caused by *Xanthomonas campestris* pv. *citri.*

gelatin liquefaction and starch hydrolysis are positive. The bacterium produces indoleacetic acid that induces hypertrophy of affected plant cells. The ability to use mannitol as a sole source of carbon and phage susceptibility is different depending on the strain and correlates with the origin of host plants in Japan; the mannitol-negative and Cp2 phage-sensitive strain mainly from Unshu orange (*Citrus unshu*) and mannitol-positive and Cp1 phage-sensitive strain from other citrus plants. The phage-bacteria relationship of this bacterium is simple, and 97–98% of the strains distributing across Japan can be identified with two phages of Cp1 and Cp2.

### 13.15.3 DISEASE CYCLE

The disease cycle of citrus canker is described as it occurs in the climate of the northern hemisphere (Fig. 13.20). The disease used to occur at any season on susceptible young trees. Its epidemic on mature

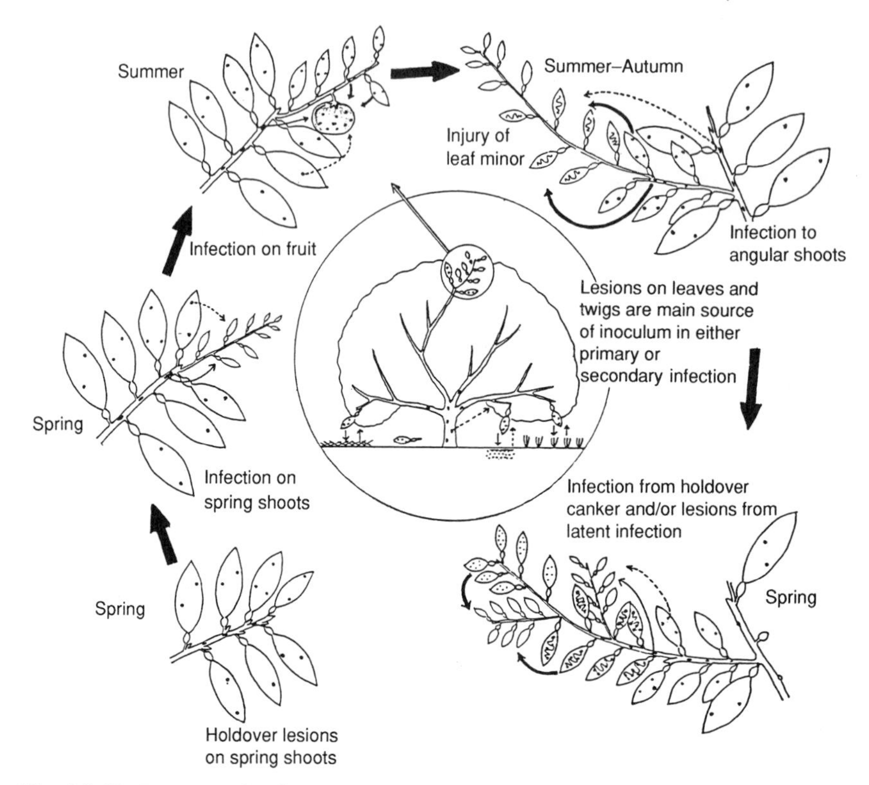


**Fig. 13.20** Disease cycle of citrus canker caused by *Xanthomonas campestris* pv. *citri.* Wide arrows, major infection route; narrow arrows, minor infection route; dotted arrows, negligible infection route.

plants, however, is rather sporadic, i.e., the disease occurs in severe form every decade.

Primary infection occurs on the first flush in spring by the pathogenic bacterium dispersed from the holdover cankers on angular shoots. Therefore, the incidence of citrus canker is not severe on mature trees, which develop fewer angular shoots in late summer to autumn. Spring canker may also develop on the margins of holdover lesions and forms the inoculum source of primary infection.

Secondary infection of the disease occurs in rain by the pathogen which is splashed from new lesions formed on young shoots. Rainstorms or typhoons bring about epidemic in the presence of diseased leaves derived from primary infection. Secondary infection to angular shoots is greatly facilitated by leaf miner (*Phyllocnistis citrella*) (see Chapter 10, Section 10.1.1).

Many species and cultivars of *Citrus* show high susceptibility to citrus canker by artifical inoculation but varying degree of resistance in natural infection. The resistance of Unshu orange is derived from the physiological resistance of twigs: poor formation of lesions on spring twigs results in low density of the carryover of the pathogen and decline of disease incidence as well.

### 13.15.4 CONTROL

1. Windbreaks are set around citrus groves.
2. Angular shoots holding canker lesions are pruned.
3. Periodic spraying of insecticides is performed to control leaf miner.
4. Copper compounds or streptomycin compounds may be sprayed every week during the active growth of new shoots. These control measures are focused on preventing primary infection on spring shoots.

### 13.15.5 SIMILAR DISEASES

The nursery disease found in Florida in 1984 was referred to as citrus bacterial spot or Xanthomonas leaf spot, and the pathogen was named as *X. campestris* pv. *citrumelo*. The disease can be differentiated from citrus canker by a distinctive water-soaked appearance and the absence of hypertrophic eruption of the lesion at the early stage of disease development. The pathogen of canker B in Brazil or canker C in Mexico was designated as *X. campestris* pv. *aurantifolia* and is distinguishable from *X. campestris* pv. *citri* by a narrow host range mainly limited to Mexican lime and lemon under natural conditions. Another bacterial brown spot caused by *P. syringae* was found in western Japan in 1989. This disease forms brown spots 5–10 mm in diameter in the early spring where mean temperature is between 15°C and 20°C. The disease occurs only on Hassaku (*Citrus grandis*).

## 13.16 Citrus Stubborn Disease

Citrus stubborn disease was first discovered in California in 1915–1917 during a yield survey by orange growers. The little-leaf disease severely occurring in Palestine in 1928 was later identified to be stubborn disease as well. Etiological studies started in the 1930s revealed that the causal agent is transmitted by grafting. This finding lead to the conclusion that stubborn disease is a virus disease. In 1969–1970, *Spiroplasma citri* was discovered in phloem of diseased sweet orange and succeeded in its tansmission by leafhoppers and cultivation in the United States and France. The pathogenic organism was taxonomically characterized in 1973.

### 13.16.1 SYMPTOMS

Stubborn disease remarkably reduces growth vigor of infected citrus plants but usually is not fatal. Severe symptoms may become mild when infected plants are transferred to a temperature lower than 28°C. Symptoms vary depending on plant species, environmental conditions (particularly temperature), and strain of the pathogen. In general, affected plants show stunting and short internodes. Leaves show yellowing and blotchy mottling and become abnormally small. Flowers and fruits also become very small but show no virescence or phyllody. Proliferation appears in the growing point, resulting in distortion of some plant parts (Fig. 13.21). In severely infected plants, the roots die, leading to lethal wilting. Coinfection with mycoplasmlike organism may induce different symptoms.

### 13.16.2 CAUSAL ORGANISM: *SPIROPLASMA CITRI* Saglio, L'Hospital, Lafleche, Dupont, Bove, Tully, and Freundt 1973

In addition to the general properties of genus *Spiroplasma* (Chapter 3, Section 3.6.7), *S. citri* has the following characteristics: acid production is positive from glucose and fructose but variable from mannose by strains; arginine dihydrolase and phosphatase reactions are positive; tetrazolium chloride is anaerobically reduced; maximum growth temperature is 32°C; colonies formed on solid media containing 20% horse serum and 0.8% agar show a typical umbonate appearance of 60–150 μm in diameter; mol% G + C of the DNA is 25–27%.

*S. citri* is an omniparasite. At least 35 species of 12 families have been reported as natural hosts. They include Madagascar periwinkle, lettuce, black mustard, horseradish, cabbage, watermelon, cherry, peach, and pear in addition to *Citrus* spp. Another 38 species of 19

**Fig. 13.21** Citrus stubborn disease caused by *Spiroplasma citri* (navel orange). (Courtesy of Dr. D. J. Gumpf.)

families, including onion and leek, have been added as host plants by leafhopper transmission.

### 13.16.3 DISEASE CYCLE

Citrus stubborn disease appears in severe form in the hot and relatively arid regions of the world such as the southwestern United States, Middle East, and Mediterranean countries. In these regions, the mean maximum summer temperature is above 35°–40°C and rainfall is infrequent. In contrast, only mild symptoms or no symptoms develop in regions where the mean maximum temperature is lower than 28°C.

Many spiroplasmas usually have the epiphytic phase on flowers or other plant organs, but *S. citri* has been detected only from phloem of

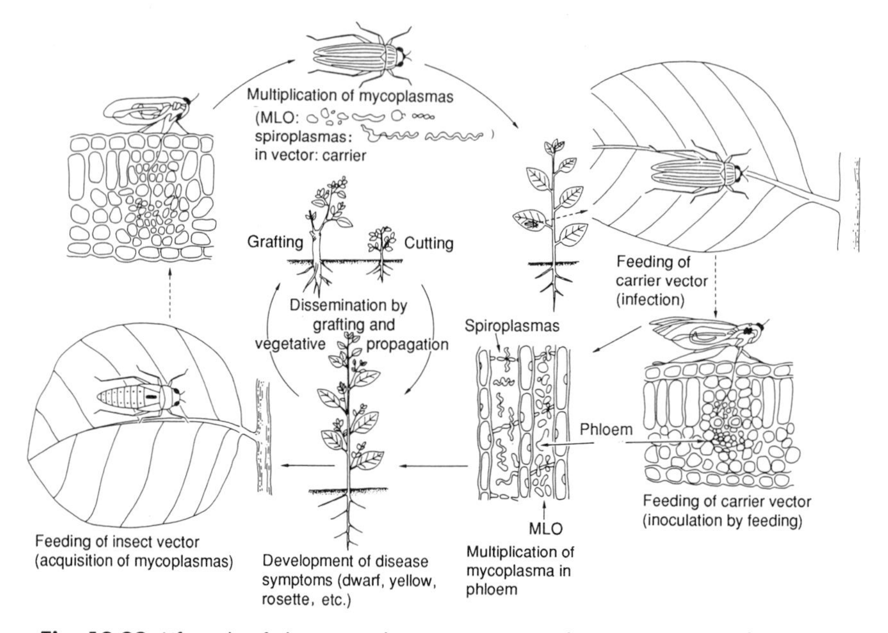


**Fig. 13.22** Life cycle of plant mycoplasmas (MLO, spiroplasmas). (Courtesy of Dr. Y. Doi.)

infected plants. Infected wild plants become the primary inoculum source from which insect vectors can acquire and disseminate *S. citri*. Biennial and perennial plants have a more significant role as the reservoir of *S. citri* than annual plants because the latter dies relatively soon after infection.

*S. citri* can be transmitted to susceptible plants by leafhoppers belonging to at least six species in three genera (*Circulifer* sp., *Macrosteles* sp., and *Scaphytopius* spp.). *S. citri* is characteristic in the feature that viability is strengthened by passage through insect vectors in its life cycle. The virulence of *S. citri* may be lost when the parasitic phase in plants is maintained for a long time (Fig. 13.22).

## 13.16.4 CONTROL

1. Application of insecticides is effective for reducing the vector populations in the field but not for reducing the incidence of stubborn disease because vectors rapidly transmit *S. citri*.

2. *S. citri* is susceptible to antibiotics such as tetracycline, but its application is not effective to control the disease in field conditions.
3. No resistant *Citrus* spp. or cultivars are available.
4. Field hygiene, including removal of infected host plants and weed control, is the only practical control measure against citrus stubborn disease at the moment.

## 13.17 Pierce's Disease of Grapes

Pierce's disease of grapes was first discovered in vineyards in southern California in 1884. The disease is distributed today in the southeastern United States, Mexico, Costa Rica, and Venezuela. It is also considered to occur in most areas of Central America as well as northern parts of South America. This disease is a major factor limiting cultivation of *Vitis vinifera* and *V. labrusca* grapes in the Gulf states of the United States. The incidence of Pierce's disease of grapes has not been detected in countries outside the continental Americas.

### 13.17.1 SYMPTOMS

Symptoms and disease severity vary with the age, species and cultivars of plants, environmental conditions, and it may take 3 months to 4–5 years for a vine to die.

On leaves, at first chlorotic spots appear in marginal areas of the leaf blade and gradually enlarge, causing withering and drying of surrounding tissue. The brown discoloration and wither further spread along leaf veins toward leaf petiole, covering the entire leaf blade, which eventually dries and dies (leaf scorch). Affected leaf blades defoliate, leaving the petiole still attached to vines. The same symptoms spread to the adjacent leaves on the same vines. The foliation of a part or on entire vine is delayed in spring, and several leaves on top of spring shoots develop interveinal chlorotic mottling or sometimes deformation. The new shoots grow slowly, causing dwarf and death of the tips. Affected vines fail to mature evenly, leaving green premature canes or green island on a dark brown mature cane. Premature coloring occurs on berries on affected vine. Affected plants may be stunted or wilted as the root system progressively dies (Fig. 13.23).

**Fig. 13.23** Pierce's disease of grapevine caused by *Xylella fastidiosa.* (Courtesy of Dr. A. H. Purcell.)

## 13.17.2 CAUSAL ORGANISM: *XYLELLA FASTIDIOSA* Wells, Raju, Hung, Weisburg, Mandelco-Paul, and Brenner 1987

Pierce's disease had been believed to be a virus disease since the late 1930s. In 1978, however, *X. fastidiosa* was isolated on culture media, and Koch's postulates with this bacterium were fulfilled by inoculation tests.

*X. fastidiosa* has the following characteristics besides the general properties of genus *Xylella* (Chapter 3, Section 3.6.8): positive reactions are found in gelatin liquefaction, $\beta$-lactamase activity and use of hippurate as a sole source of carbon; negative reactions are found in the production of indole and $H_2S$, activities of $\beta$-galactosidase, lipase, amylases, coagulase, and phosphatase; mol% G + C of the DNA is 51.0–52.4.

The host range that was investigated by transmission of "Pierce's disease virus" includes 75 species of 23 families, of which 36 species of 18 families were demonstrated to be natural hosts. The omniparasitic nature of *X. fastidiosa* has also been revealed by subsequent investigations. The bacterium induces the symptoms of leaf scorch, wilt, and/or dwarf on a number of plants such as alfalfa, apricot, blackberry, citrus, elm, peach, periwinkle, platanus, plum, macadamia, maple, mulberry, oak, raspberry, and sycamore.

*X. fastidiosa* colonizes in xylem vessels of affected plants and blocks translocation of water and minerals by the aggregates of bacterial cells and tylosis. The water deficiency thus induced may be the major mechanism of leaf scorch symptoms. The role of plant toxic metabolites produced by the bacterium is also suggested.

The latent period is 8 to 12 weeks in young plants. When infection occurs in September or later, symptoms do not appear until May or June of the following year. On 5-year-old vines, the disease develops in 3 to 15 months.

## 13.17.3 DISEASE CYCLE

*X. fastidiosa* disseminates by grafting as well as by insect vectors such as sharpshooter leafhoppers (Cicadellidae) and spittlebugs (Cercopidae). In Florida, vectors usually transmit the bacterium vine-to-vine in vineyard; hence, the role of wild plants as the source of primary inoculum is not significant. In California, however, the pathogen is transmitted by vectors from infected wild plants around vineyards. The danger of spread by cuttings or scions is not likely because those taken from infected vines do not survive long. Pierce's disease may not become a serious problem unless infected wild plants are established around vineyards. Therefore, the incidence of disease may occur in severe form in such a case that the susceptible cultivars are introduced in areas where the bacterium has already been established in grapes or wild plants.

## 13.17.4 CONTROL

1. The use of resistant cultivars is the most promising and only way to control Pierce's disease of grapes at the moment.
2. Heat treatment may be applied to cuttings and propagation buds by immersing in water at 45°C for 3 hr.

3. Application of tetracycline is empirically effective but not from a practical point of view.
4. Application of insecticides to control insect vectors has not been effective enough to control the disease.

## 13.18 Bacterial Blight of Mulberry

Bacterial blight of mulberry was first reported in the United States in 1893 and is the earliest record of the disease caused by *Pseudomonas.* In Japan, this disease was confirmed as early as 1901. It appears in April to May and rapidly spreads in the rainy season of June to July. Incidence of the disease ceases in summer but may occur again in autumn in mild form.

### 13.18.1 SYMPTOMS

The symptoms appear on leaves and twigs. Lesions formed on leaves are small, water-soaked, irregularly shaped, and brown to dark brown spots. These lesions may or may not be surrounded by a yellow halo and enlarge to angular lesions which often fall out. When dark brown and more-or-less sunken spots are formed on midribs or petioles, the affected leaves may be curled or distorted on the basis of unbalanced growth of leaf lamina (Fig. 13.24). On twigs, dark brown to black, elongated, and somewhat sunken streaks appear, exuding bacterial ooze from the surface in humid weather. These streaks sometimes coalesce to girdle the shoot and induce die-back and stunting of young trees.

### 13.18.2 CAUSAL ORGANISM: *PSEUDOMONAS SYRINGAE* pv. *MORI* (Boyer and Lambert 1893) Young, Dye, and Wilkie, 1978

In addition to the general properties of genus *Pseudomonas* and *Pseudomonas syringae,* the bacterium has the following characteristics. Positive reactions are found in hydrolysis of arbutin and use of D-gluconate, 2-ketoglutarate, inositol, mannitol, sorbitol, and trigonelline as a sole source of carbon. Negative reactions are obtained with gelatin liquefaction and use of erythritol, DL-homoserine, DL-lactate, quinate, and L-

**Fig. 13.24** Bacterial blight of mulberry caused by *Pseudomonas syringae* pv. *mori*. (Courtesy of Dr. M. Sato.)

tartrate as a sole source of carbon. This organism is a member of rRNA group I.

This bacterium infects only mulberry by natural infection, but kidney beans and *Vanieria cochinchinensis* (*Cudrania javanensis*) are infected by artifical inoculation.

### 13.18.3 DISEASE CYCLE

The disease cycle in Japan is shown in Fig. 13.25. The most important source of primary inoculum is the lesion on twigs. Bacteria exuding from

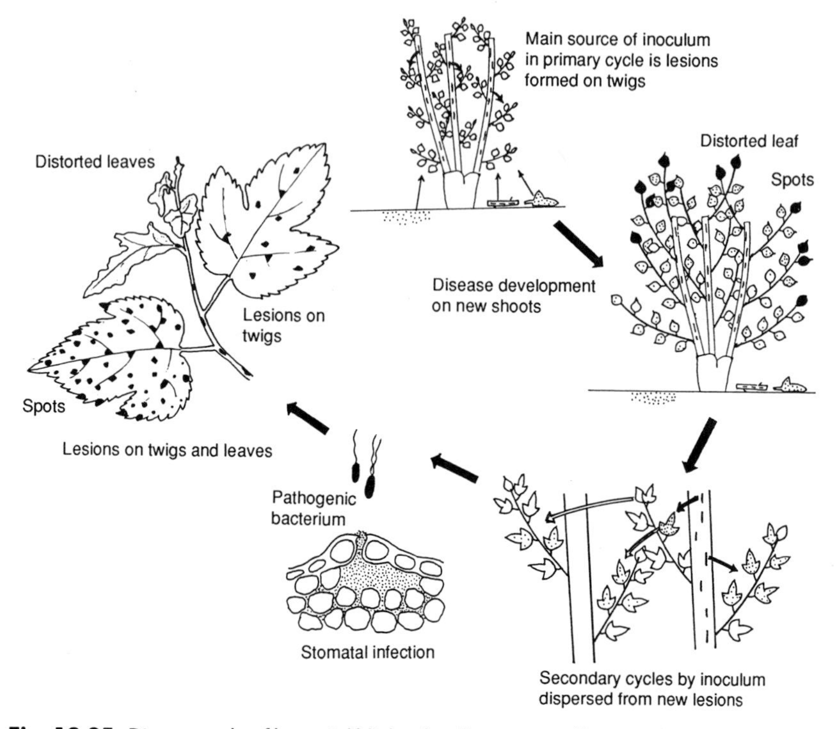


**Fig. 13.25** Disease cycle of bacterial blight of mulberry caused by *Pseudomonas syringae* pv. *mori.*

the shoot lesions are disseminated by rain splash. The inoculum source may also be found on pruned shoots and fallen leaves remaining on the ground. Secondary infection occurs in rain by the bacterium exuding from new lesions formed on developing shoots. Dense planting, application of excess nitrogen, careless field hygiene, and proliferation of lower shoots increase the incidence of the disease.

## 13.18.4 CONTROL

1. The use of resistant cultivars is the most promising measure for control of the disease.
2. Cultural control measures including proper fertilization, plant density, field sanitation, and method of pruning are performed.
3. When the disease starts to appear, copper, oxytetracycline, streptomycin, or copper hydroxy nonylbenzenesulfonate compounds are sprayed every 10 days.

# Further Reading

Agrios, G. N. (1988). "Plant Pathology." Third Edition. Academic Press, Inc., San Diego.

Chupp, C. and Sherf, A. F. (1960). "Vegetable Diseases and Their Control." The Ronald Press Company, New York.

Fahy, P. C. and Persley, G. J. (1983). "Plant Bacterial Diseases. A Diagnostic Guide." Academic Press, Sydney.

Fletcher, J. T. (1984). "Diseases of Greenhouse Plants." Longman Group Limited. Longman House, Burnt Mill, Harlow, England.

Hayward, A. C. (1991). Biology and epidemiology of bacterial wilt caused by *Pseudomonas solanacearum*. *Ann. Rev. Phytopath.* **29,** 65–87.

Hiruki, C. ed. (1988). "Tree Mycoplasmas and Mycoplasma Diseases." The University of Alberta Press, Edmonton, Alberta.

Hopkins, D. L. (1989). *Xylella fastidiosa:* Xylem-limited bacterial pathogen of plants. *Ann. Rev. Phytopath.* **27,** 271–290.

International Rice Research Institute. (1988). "Rice Seed Health." IRRI, Manila.

Mew, T. W. (1987). Current status and future prospects of research on bacterial blight of rice. *Ann. Rev. Phytopath.* **25,** 359–382.

Mukhopadhyay, A. N., Singh, U. S., Kumar, J. and Chaube, H. S. ed. (1992). Volume I Diseases of Cereals and Pulses; Volume II Diseases of Vegetables and Oil Seed Crops; Volume III Diseases of Fruit Crops; and Volume IV Diseases of Sugar, Forest, and Plantation Crops. Prentice Hall, Englewood Cliffs, New Jersey.

Ou, S. H. (1972). Rice Diseases. Commonwealth Mycological Institute, Kew, Surrey, England.

Perombelon, M. C. M. and Kelman, A. (1980). Ecology of the soft rot erwinias. *Ann. Rev. Phytopath.* **18,** 361–387.

Stall, R. E. and E. L. Civerolo. (1991). Research relating to the recent outbreak of citrus canker in Florida. *Ann. Rev. Phytopath.* **29,** 399–420.

Stapp, C. (1961). "Bacterial Plant Pathogens. Oxford University Press, London.

Van Der Zwet, T. and Keil, H. L. (1979). "Fire Blight. A Bacterial Disease of Rosaceous Plants." Agriculture Handbook No. 510 of United States Department of Agriculture.

Whitcomb, R. F. and Tully, J. G. (1989). "The Mycoplasmas." *In Spiroplasmas, Acholeplasmas, and Mycoplasmas of Plants and Arthropods.* Vol. 5 Academic Press, Inc. San Diego.

# *Subject Index*

# *Systematic Index*